WITHDRAWN FROM
KENT STATE UNIVERSITY LIBRARIES

LAND RECLAMATION

An End to Dereliction?

Papers presented at the Third International Conference on Land Reclamation: An End to Dereliction?, held at the University of Wales College of Cardiff, Cardiff, UK, 2–5 July 1991.

LAND RECLAMATION

An End to Dereliction?

Edited by

M. C. R. DAVIES

*Division of Civil Engineering,
School of Engineering,
University of Wales College of Cardiff,
Cardiff, UK*

ELSEVIER APPLIED SCIENCE
LONDON and NEW YORK

ELSEVIER SCIENCE PUBLISHERS LTD
Crown House, Linton Road, Barking, Essex IG11 8JU, England

Sole Distributor in the USA and Canada
ELSEVIER SCIENCE PUBLISHING CO., INC.
655 Avenue of the Americas, New York, NY 10010, USA

WITH 36 TABLES AND 105 ILLUSTRATIONS

© 1991 ELSEVIER SCIENCE PUBLISHERS LTD
© 1991 CROWN COPYRIGHT—pp. 3–39

British Library Cataloguing in Publication Data

International Conference on Land Reclamation
(3rd: 1991: University of Wales)
Land reclamation: an end to dereliction?
I. Title II. Davies, M. C. R. (Michael C. R.)
333.73

ISBN 1-85166-658-3

Library of Congress CIP data applied for

No responsibility is assumed by the Publisher for any injury and/or damage to persons or property as a matter of products liability, negligence or otherwise, or from any use or operation of any methods, products, instructions or ideas contained in the material herein.

Special regulations for readers in the USA

This publication has been registered with the Copyright Clearance Center Inc. (CCC), Salem, Massachusetts. Information can be obtained from the CCC about conditions under which photocopies of parts of this publication may be made in the USA. All other copyright questions, including photocopying outside the USA, should be referred to the publisher.

All rights reserved. No part of this publication may be reproduced, stored in a retrieval system, or transmitted in any form or by any means, electronic, mechanical, photocopying, recording, or otherwise, without the prior written permission of the publisher.

Printed in Great Britain at The Alden Press Ltd, Oxford

PREFACE

The desire to improve the environment in areas where the land has been reduced to dereliction through industrial and other uses has lead to policies – at local, national and international levels – aimed at promoting land reclamation. Dereliction results from many causes. Similarly a breadth of disciplines and skills are employed in the planning and implementation of land reclamation schemes. The necessity for a multi-disciplinary approach to reclaiming land is reflected in the contents of this book, which considers all stages of land reclamation projects from the initial policy decisions to the management of completed schemes.

The papers published in this book were presented at the 3rd Land Reclamation Conference to be organised under the auspices of the Standing Local Authority Officers' Panel on Land Reclamation, held at the University of Wales College of Cardiff in July 1991. The major theme of the conference was "An end to dereliction?" – are the necessary planning, engineering and scientific skills available to regenerate areas and improve the environment? The purpose of this conference was to establish the current state-of-the-art by bringing together those associated with all aspects of the reclamation of land.

The wide range of disciplines covered by the contents of this book necessitated the skills of specialist reviewers. For this assistance I am most grateful to Basil Evans, Martin Hooker, Pat Sheppard Williams and Paul Wright.

MICHAEL C. R. DAVIES
University of Wales College of Cardiff

CONTENTS

Preface v

SECTION 1 : PLANNING AND LAND USE

Derelict Land - Recent Developments and Current Issues
(Keynote Address)
    R. MABEY   3

An Evaluation of Surveys of Soil Contamination in the
City of Swansea, South Wales
    E.M. BRIDGES   40

A View of Two New Government Initiatives, Scottish
Enterprise and Central Scotland Woodland Company
    A.S. COUPER   50

The Use of Derelict Land - Thinking the Unthinkable
    D. HARTLEY   65

Planning for Reclamation
    D.M. HOBSON   75

Technical Notes

Treating Dereliction, The Lessons Learned
    G. DOUBLEDAY   82

The Reclamation of Hard Rock Quarries: Policy and
Planning Implications
    B. SNAITH and P. GAGEN   84

Considerations in the Redevelopment of Derelict Land:
Experiences of an Urban Development Corporation
    J.S. GAHIR, C. DUTTON and H.L.M. JONES   86

Roumanian Concepts and Achievements
    R. SOFRONIE and L. OTLACAN   88

SECTION 2 : ENGINEERING TECHNIQUES

Keynote Address
    D.G. GRIFFITHS   93

Landform Replication as an Approach to the Reclamation
of Limestone Quarries.
    D. BAILEY and J. GUNN   96

Garden Festival Wales - Ebbw Vale 1992
    C.D. GRAY   106

Albion Lower Tip Landslide - Control of Groundwater
Levels Using Well Points and Bored Drains.
    J.D. MADDISON   117

Gas Control Measures for Development of Land Affected
by Landfill Gas
    D.P. ROCHE and G.B. CARD   125

Redevelopment of Cymmer Colliery, Porth
    N.J. TAYLOR   135

The Application of Strata 3 to the Assessment
of Contaminated Land - A Case Study
    B.R. THOMAS, L.M. GREENSHAW, S.P. BENTLEY and
    A.S. STENNING   146

Windsor Park Housing Scheme : The Reclamation of an
Ex-Gasworks Site in Beckton
    P. TREADGOLD and S. TILLOTSON   158

ADLIS - An Air Photo Land Information System for
Derelict Land Studies
    C.T. YONG and J. UREN   174

Reclamation of a Backfilled Sand Pit
    P.J. WITHERINGTON   183

Technical Notes

Clean Cover Reclamation: The Need for a Rational
Design Methodology
    I.J. ANDERS   193

Evaluating the Potential for Subterranean Smouldering
    T. CAIRNEY   196

The Use of Dynamic Compaction in Land Reclamation
    T. GORDON   201

The Subsurface Drainage of Disturbed Soils
    R.A. HODGKINSON   202

Engineering Techniques for Environmental Control
of the Coastal Zone
    R.A. PEREVERSEV and J.F. FARBEROV   204

Subsidence and Drainage on Abandoned South Wales
Mines Sites - Hazards and Engineering Solutions
    M.J. SCOTT and I. STATHAM   206

Reclaiming the Fertile Lands of the Lake Faguibine System
    M.M. VIERHOUT   208

A Case History - Brett Gravel's Faverstam Quarry
    L.M.S. WILLIAMS   210

SECTION 3 : CONTAMINATED LAND

Behaviour of Pollutants in Soils
    R.A. FAILEY and R.M. BELL   215

Redevelopment of a Former Steelworks
    D.L. JONES   228

Environmental Assessments of Reclaimed Land in the
United States of America
    M. LEONARD and K. PRIVETT   235

Risk Analysis as a Guide to Optimal Reclamation in
Strategies for Contaminated Land
    M. LOXHAM   241

An Assessment of the Efficiency of Remedial Treatment
for Metal Polluted Soil
    S. MUSGROVE   248

Land Contamination by Hazardous Gas - The Need for a
Coordinated Approach
    S.J. EDWARDS and C.F.C. PEARSON   256

A Review of Methods to Control Acid Generation in
Pyritic Coal Mine Waste
    I.D. PULFORD   269

Ranking of "Problem" Sites and Criteria for Clean-up of
Contaminated Sites
    M.A. SMITH   279

Reclamation of the Former Laporte Chamical Works,
Ilford, Essex
    M. SURY and A. SLINGSBY   289

Topsoil from Dredgings: A Solution for Land Reclamation
in the Coastal Zone
    B.R. THOMAS and M.S. de SILVA   299

Slurry Trench Cut-Offs for Gas and Leachate Control
    S. JEFFERIS   310

Technical Notes

Burning Chemical Waste Site: Investigation, Assessment
and Reclamation
    D.L. BARRY   319

Redevelopment of a Gas Works Contaminated Site
    J. BROMHEAD and D. ROCHE   321

Landfill Gas Problems
    D. HOBSON   323

Reclamation of Gas Works Site for Housing Development
in Salisbury : A Case Study
    E.J. WILSON   325

SECTION 4 : BIOTECHNOLOGY

Evaluation of Composted Sewage Sludge/Straw for the
Revegetation of Derelict Land
    S.L. ATKINSON, G.P. BUCKLEY and J.M. LOPEZ-REAL   329

In Situ Biological Treatment of Contaminated Land -
Feasibility Studies and Treatment of a Creosote
Contaminated Site
    P. BARRATT and P. HAROLD   336

Site Factors Affecting Tree Response on Restored
Opencast Ground in the South Wales Coalfield
    N.A.D. BENDING, A.J. MOFFAT and C.J. ROBERTS   347

Techniques for Reclaiming Metalliferous Tailings
    J.P. PALMER   357

Revegetation of Reclaimed Land
    P. SAMUEL   366

Re-establishing Earthworm Populations on Former
Opencast Coal Mining Land
    J. SCULLION   377

Technical Notes

Reconstruction of Lime Quarries in the Mediterranean
Region of Spain
    M.P. ARAMBURU MAQUA and R. ASCRIBANO BOMBIN   387

Tree Establishment on Derelict Land: Getting it Right
    A.S. COUPER   389

The Use of Sewage Sludge as a Fertiliser in the
Afforestation of Opencast Coal Spoils in South Wales
    A.J. MOFFAT, N.A.D. BENDING and C.J. ROBERTS   391

Minimal Soil Amelioration as a Trigger Factor in the
Revegetation of a Derelict Landscape near Sudbury,
Canada
    K. WINTERHALDER   393

SECTION 5 : LAND USE AND MANAGEMENT

Ravenhead Renaissance - A Case Study
    G. ASHWORTH CBE   397

Leisure Based After-Use of Restored Land
    D.K. HEMSTOCK   402

Tinsley Park Lives Again
    D. HUNTER and M. STOTT   411

Index of Contributors   421

SECTION 1 : PLANNING AND LAND USE

DERELICT LAND - RECENT DEVELOPMENTS AND CURRENT ISSUES

ROBIN MABEY

ABSTRACT

This paper reports the main changes that have taken place in the operation of the DLG scheme since 1988 in England. It reports on the 1988 Survey of Derelict Land, the main studies and reports and sets out proposed changes.

INTRODUCTION

Since the last Conference in 1988, there have been a number of significant events affecting the Derelict Land Grant (DLG) Programme. The first part of the paper reports on these in chronological order. A note on the Derelict Land Act 1982 is at Annex I.

1988 SURVEY OF DERELICT LAND

This survey was the third national survey of derelict land. The earlier surveys were carried out in 1974 and 1982.

The main conclusions of the Survey[1] were as follows. Annex II provides relevant tables.

Total Derelict Land

The Survey recorded 40,500 ha of derelict land as at 1 April 1988, 31,6000 ha (78%) of which were considered to justify reclamation. The amount of derelict land decreased by about 11% between 1982 and 1988, and the area justifying reclamation fell by 8%. This compares to increases of 6% and 4% respectively between 1974 and 1982.

Types of Derelict Land

Derelict spoil heaps were the most extensive type of derelict land accounting for 29% of the total followed by general industrial dereliction which accounted for 21%.

A relatively high proportion of both general industrial dereliction and derelict colliery spoil heaps were considered to justify reclamation - 94% and 93% respectively. In contrast, the proportion of derelict metalliferous spoil heaps considered to justify reclamation was only 27%, with 63% of the total spoil heaps category justifying reclamation. Between 73% and 92% of the other categories of dereliction were considered worth reclaiming.

There was a decrease in each broad type of dereliction between 1974 and 1982 except for derelict spoil heaps and other forms of dereliction. Similarly there was a decrease in each broad type of dereliction, except other forms of dereliction, between 1982 and 1988. General industrial dereliction made up about two-thirds of other forms of dereliction in 1988.

Location of Derelict Land

Over half (52%) of derelict land was in the three most northerly regions - 22% in the North West and 15% in both Yorkshire and Humberside and the North. The South East (8%) and East Anglia (1%) had the smallest amounts.

Derelict land was found almost equally in urban and rural area but urban areas had 60% of the derelict land justifying reclamation. Inner city areas contained 17% of the derelict land and 22% of the derelict and justifying reclamation.

Most of the military dereliction (90%), derelict metalliferous spoil heaps (86%) and excavations and pits (68%) were found in rural areas. In contrast, much derelict land related to mining subsidence (37%), general industrial dereliction (37%) and other forms of dereliction (25%) was found in inner city areas.

The amount of derelict land decreased in all regions between 1982 and 1988 except in Yorkshire & Humberside where it increased by 13%. This compares with the period between 1974 and 1982 when dereliction increased in all regions except the North, Yorkshire & Humberside and East Anglia. Between 1982 and 1988, the decrease in the amount of derelict land was much greater in rural areas than in urban areas - 17% compared to 4%.

Ownership of Derelict Land

The private sector owned 43% of the derelict land, local authorities 17% and other public sector bodies 25%. Sixty nine percent of the private sector land was considered worth reclaiming compared with 96% for the local authority land and 86% for the other public sector land.

Fourteen per cent of the derelict land and 13% of all the derelict land justifying reclamation was recorded on land registers.

There was no significant change in the pattern of ownership of derelict land between 1982 and 1988.

Derelict Land Reclaimed

Between 1982 and 1988, 14,000 ha of derelict land was reclaimed, 61% by local authorities. Ninety per cent of the local authority reclamation was carried out with the aid of DLG.

Derelict spoil heaps made up 23% of the total amount of land reclaimed and general industrial dereliction 22%. Over a quarter of the land reclaimed was in the North West region, and the three northern regions together accounted for 59%.

Uses of Reclaimed Derelict Land

By 1 April 1988, 12,600 ha of the land reclaimed had been brought back into beneficial use. Twenty seven per cent was for hard end use (i.e. mainly for industry, commerce and residential use). Public open space was the most common end use accounting for another one third of the total in use.

The South East was the only region for which hard end uses exceeded soft end uses (i.e. mainly for sport and recreation, public open space and agriculture and forestry. In other regions, the highest proportion of hard end use for reclaimed land was 34%.

Hard end uses accounted for a much lower proportion of the land reclaimed by local authorities than by other agencies, including the private sector - 20% compared to 40%.

1988 REPORT OF THE NATIONAL AUDIT OFFICE ON DERELICT LAND

This report was the first external examination of the DLG system. It was wide-ranging in scope and the National Audit Office (NAO) consulted a range of interests concerned with DLG, as well as considering the Department's own performance. The main conclusions were as follows[2]:

> "Since 1981, the Department has given greater priority to reclaiming land for industrial, commercial or housing development, particularly in the inner cities. Consultants

and local authorities have reservations about the emphasis placed on such 'hard' end uses; while not wholly sharing those reservations, the Department will be reviewing priorities in 1989.

NAO examination suggested that rolling programmes, first established in 1986-87 to provide continuity of funding in areas of extensive dereliction, have generally worked well although the Department have not yet reviewed the effectiveness and it is probably too early to come to any firm conclusions.

An Internal Audit examination in 1985 found serious weaknesses in appraisal, control and monitoring procedures. The Department have since taken steps to improve these procedures, issued new guidance to staff, and commissioned research into the new appraisal methods.

Improvements are needed in monitoring at regional offices and at headquarters; and defined targets and established performance indicators are need to assist management of individual projects and the grant scheme as a whole. The Department recognised the need to improve their management information, but there have been difficulties and long delays in developing a satisfactory computerised system.

The derelict land grant programme has been operating for more than 20 years and currently costs about £80 million a year. The Department are not fully in a position to assess its effectiveness, but have some measures in hand which are intended to help them do so within 2 years.

Dereliction is a continuing process and there is some evidence to suggest that, despite the efforts of both public and private sectors, the growth in dereliction is outstripping reclamation programmes.

Available surveys of derelict land are out of date although a new survey is being conducted this year. The NAO welcome the intention to shorten the interval between national surveys and to consider some sample updating between surveys.

While recognising the programme's successes, a 1987 consultants' report noted that, for a significant proportion of projects, intended hard end uses had not been achieved at the time of their survey, often due to lack of demand for land. The provisional results of a further study tended to confirm the earlier findings, but the Department pointed out that there was inevitably some delay between completion of reclamation and commencement of development and that the speed of development was closely influenced by the policy priority of applying grant in run-down areas.

There is little reliable information, but some doubts as to the extent to which the availability of grant significantly influences reclamation and investment decisions. The Department intend to examine this issue with the research commissioned into appraisal methods.

Local authorities believe that substantial benefits have been achieved from their reclamation programmes. They have criticised the programme's procedures and priorities as between hard and soft end uses, but the Department have emphasised the need for proper appraisal of projects and consider that their priorities are sufficiently flexible.

The Department need to do more to follow up completed projects and secure recoveries of grant which may be due on disposal of reclaimed sites.

There is no obligation on owners to reclaim land rendered derelict by factory or works closure, and substantial amounts of derelict land grant are paid for this purpose; the Department intend to review preventive measures for industrial dereliction, although to extend restoration conditions to all industrial activities would require a change to the operation of the planning system.

Controls over mineral workings were reviewed in detail in the mid 1970s and further measures flowing from this review (and reflected in the Town and Country Planning (Minerals) Act 1981) are being implemented; these and other measures to be introduced in 1988 will continue the process of tightening controls which dates back to 1948.

Local authorities and representatives of industry have expressed conflicting views about the adequacy of powers available to local authorities to compel derelictors to reclaim land.

The Department have assured the NAO that they will be looking at issues arising from the new measures to control mineral operations within 5 years of full implementation.

The NAO concluded that although the planning regime for mineral operations has been progressively tightened over the last 40 years, heavy calls on derelict land grant resources will continue. These arise especially from historic dereliction and on cessation of activities not covered by satisfactory restoration conditions or orders".

The report was subsequently considered by the Public Accounts Committee of the House of Commons. They endorsed the report of the C & AG.

1989 DOE RESPONSE

In July 1989 the Department responded to the Committee. The relevant extract from the Treasury Minute[3] are at Annex III.

1989 REVIEW OF DERELICT LAND POLICY

Derelict Land Policy was reviewed in the early 1980s and this resulted in Circular 28/85. It was decided in 1988 that this policy should be reviewed and this review was carried out by officials during 1989. The report of the review group was published for consultation in September 1989[4].

In the light of its consideration of the review, the Government proposed a number of changes to the DLG programme. These were as follows:

- to re-define the National policy objectives of the programme so as to provide for both reclamation of sites for development and restoration for environmental improvement: priorities would be determined in the context of local derelict land programmes and projects selected for grant assistance on the basis of these programmes

- to encourage local authorities to draw up strategic programmes for derelict land reclamation

- to give further incentives to voluntary organisations such as Groundwork Trusts to undertake environmental improvement including the re-introduction of non local authority derelict land grant in the urban programme

- areas and aimed particularly at small scale environmental schemes

- to allow local authorities to carry out reclamation for development in two stages, the first to remove dereliction and improve the site environmentally and the second to provide for the specific needs of as assured developer

- to recognise nature conservation as a legitimate end use for a derelict site where appropriate

- to encourage the development of new technologies for the treatment of contaminated or otherwise derelict sites

The review suggested that new measures were required to prevent owners from letting land fall into dereliction and that these needed to be developed in more detail. It also raised the question of simplifying the various derelict land grant rates now available in different areas and suggested a number of options. The Government indicated that further consultation papers would be prepared on these matters.

The public consultation that took place on the review demonstrated extensive interest in derelict land issues. Key features from the response were as follows:

Programme Objectives and Priorities

There was a general welcome for the key recommendation that the current priority for hard end use reclamation in inner city areas should be dropped and replaced by a system in which priorities should be determined locally in the context of a reclamation strategy drawn up by the local authority. Local authorities also welcomed the proposal that the number of authorities granted rolling programme status should be increased.

There was less support for the proposed change to the definition of derelict land. Most local authorities said they encountered no difficulty with the current definition and argued that there was merit in maintaining the same definition in order to be able to compared progress between Surveys.

Role of the Voluntary Sector

There was a general welcome for the recognition that voluntary groups could play a greater role in reclamation. This could be done directly if the level of assistance available to groups such as Groundwork Trusts was enhanced. Also voluntary organisations could advise local authorities and private landowners about the suitability of particular schemes and for example the effect on nature conservation. Many called for guidance to be issued by the Department requiring local authorities to consult the voluntary sector before grant was approved on a reclamation scheme.

Role of the Private Sector

The private sector organisations who responded were generally receptive to playing an increased role in reclamation, but felt that there was a need to improve the way in which the various grant systems operated. There was particular criticism of the form of DLG available to them which often left the developer with a net loss after the land had been reclaimed. There was also support for the re-introduction of private sector DLG into the 57 Urban Priority Areas in cases where schemes were not eligible for City Grant.

Grant Rates

The local authorities were unanimous in opposing any change to grant rates which would reduce the current 100% rate available to most of them and were also opposed to the alternative proposal not to grant aid land acquisition. They said that if the grant rate was reduced, this would put in jeopardy the major reclamation schemes which often small local authorities could not contribute towards from their own resources.

Preventive Measures

Despite a recognition of the difficulties involved both the local authorities and the voluntary sector responses welcomed the suggested measures including the use of restoration conditions in a planning consent and a possible charge on the owners who held land derelict for long periods. The private sector responses thought that some of these measures could be punitive and unfair and required further thought. They favoured tax incentives to restore derelict land.

1990 ENVIRONMENT COMMITTEE ON CONTAMINATED LAND

The Environment Committee of the House of Commons considered the problems of contaminated land during 1990. Although contaminated land is not necessarily derelict, much derelict land is contaminated. The Committee considered the review of derelict land policy. They made the following observations[5]:

> "The review of derelict land policy is very much in line with our thinking and we support its general direction. Nevertheless, we are not sure how far the DOE has taken on board the pollution implications of contaminated land. As currently drafted, the policy statement takes account of public health and landscape implications of contaminated land, but makes no explicit mention of water pollution and broader environmental considerations. **We recommend** that derelict land policy should be reformulated to take greater account of the environment broadly in line with the DOE view. But we go further, in seeking that greater priority be given to environmental concerns. **We recommend** that for policy purposes, the second component of the DOE objective on dereliction should be re-worded to include an explicit reference to threats to the natural environment, as well as to public health and safety. **We further recommend** that the DOE should, in allocating grants, give priority to those sites which are significantly contaminated an polluting or potentially so".

1990 GOVERNMENT RESPONSE

The Government response was published in July 1990 and referred to the Committee's recommendations as follows[6]:

> "We welcome the Committee's suggestion that derelict land policy should be reformulated to take greater account of the environment. We also accept the Committee's recommendation that DLG objectives should be widened to include a reference to the need to deal with contamination which is a threat to the natural environment as well as to public health and safety. Derelict Land Grant already makes a significant contribution towards the reclamation of contaminated site, but the Department of the Environment will now have regard to policy objectives which will lead to greater priority being given to sites which are significantly contaminated and a source of pollution provided that they are in a derelict state".

FUTURE RESTORATION OF COLLIERIES

In 1990, the Government came to an agreement with the British Coal Corporation that the Corporation would reclaim collieries which closed in the four years from 1 April 1990. In a PQ dated 5 April 1990, the Parliamentary Under Secretary of State Colin Moynihan said:

> "I regard this as an important step forward in placing environmental responsibilities where they belong. I believe that as a result of this agreement, land will be reclaimed to an acceptable standard more quickly than would otherwise be the case. We shall keep the arrangement under review and will be considering how the problems of coalfield dereliction may best be tackled in the longer term".

The text of the agreement is at Annex IV.

EXPENDITURE ON DERELICT LAND

Expenditure on derelict land grant in England is currently (1991/92) £87.9 million. However, this figure excludes reclamation work undertaken by UDCs and under the City Grant and Urban Programme. When this is taken into account, current total expenditure is of the order of £150m-£200m.

The DLG programme has had two special features. The first dealing with coalfield reclamation derives from the Government's response to the Commission on Energy and the Natural Environment (CENE)[7]. The second deals with the problem of old limestone mines in the West Midlands. Expenditure on these is currently (1991/92) £12.5 million and £6.3 million respectively.

ACHIEVEMENTS OF THE DLG PROGRAMME

In 1989/90, 1183 ha of derelict land were reclaimed - some 618 ha for hard end use development and some 565 ha for soft end use.

The outputs of the programme however go wider than these basic statistics suggest. There are though, no comprehensive data available. What is known however is that hard end uses comprise a variety of developments. The survey undertaken by Roger Tym & Partners[8] suggests that a considerable total of investment, economic activity and housing provision has been attracted to reclaimed land and that this has occurred largely in urban areas and to a significant extent in inner cities. Leverage ratios i.e. the ratio of private sector investment attracted by DLG ranges from 1:5.6 residential, 1:7 mixed/other, 1:10 industry/commerce.

After value also provides a measure of the success of this aspect of the programme. DLG is paid to local authorities on a gross basis and the increased value of the reclaimed land (the after value) is clawed back when the land is disposed of. In 1989/90 some £11 million was repaid to the Department as after value.

Annex V provides some basic statistics about the programme.

CURRENT ISSUES

The reviews and reports described above have provided a stimulus for a number of current developments in the DLG programme. These are as follows.

SURVEYS OF DERELICTION

In the light of the comments of the NAO, the Department has recently appointed consultants to examine a number of matters about future surveys. The research will assess the current arrangements for conducting Derelict Land Surveys, and the methods by which local authorities collect and collate information to update the Survey data for their own purposes.

The main aims of the research are:

> To review the objectives of the Department's Derelict Land Surveys and the use to which they are put. In the light of these objectives consider what information is required to meet these needs.

> To review the arrangements for conducting the 1988 Derelict Land Survey, and assess the extent to which these methods provided the information necessary to fulfil the objectives and needs identified above.

> In the light of this analysis, to consider alternative approaches for the future conduct of Surveys and make recommendations having regard to the financial and operational constraints faced by local authorities.

To assess the scope for local authorities in receipt of DLG to maintain an up to date register of derelict land. Also to assess the potential for linking relevant data from other sources such as the contaminated land registers, waste disposal survey, and the vacant land survey.

The research project will lead to the establishment of an improved information base which will assist the Department to determine priorities for the Derelict Land Programme in relation to other policy areas. Such information will also provide an improved basis for allocating funds between the regions.

RESEARCH

Over the period 1985-1991 there have been a number of ad hoc studies into derelict land matters.

During the mid eighties when derelict land policy emphasised reclamation for developed or hard end uses, research projects tended to concentrate on assessing the effectiveness of reclamation for such use. The objective of one such study, Evaluation of Derelict Land Grant Scheme[8], was "to evaluate and assess the benefits and disadvantages resulting from reclamation schemes; the distribution of benefits according to type of scheme, end use of the reclaimed site and location. In addition the study was to describe examples of successful reclamation so that good practice might be published".

The report found that environmental and safety objectives of DLG scheme were met in virtually every case, but that objectives relating to the provision of development land take longer to achieve. This research suggested that the emphasis on reclaiming sites for hard end use only was somewhat misplaced and recommended that policy should be more flexible and that each scheme should be viewed in the context of the authorities overall strategy.

The consultants recommended that sponsors of proposed reclamation schemes should be encouraged to carry out feasibility studies and that in this context the Department might consider extending private sector DLG to site surveys. It was also felt that there was room for substantial improvement in the design and management of land reclaimed to soft end use. The report advocated that the Department should monitor systematically the post-reclamation outcomes of all schemes.

In other studies the emphasis has been on examining the cost effectiveness of reclamation for soft end use. In one study, "Cost Effective Management of Reclaimed Derelict Sites"[9], the main objectives of the research were "to identify cost effective strategies for the management of reclaimed derelict sites; and assess the extent to which those management strategies are effected by the methods and costs of the initial reclamation".

The report concluded that restoration to nature conservation was a highly cost effective option. The traditional public open spaces of mown grass with standard trees, particularly in small urban sites were costly to reclaim and maintain and performance was poor. Larger sites were more cost effective as a result of economies of scale. Recreation sites had high capital costs and maintenance requirements. Often the latter were not fulfilled resulting in low performance.

Among the recommendations made by the report was the suggestion that the opportunities which derelict land provides for nature conservation should be recognised. It also proposed that the possibility of deriving income from land reclamation sites should be investigated and that management plans for all sites should be prepared.

Recent research has focused on the operation of the derelict land programme. Consultants were commissioned to review "Project Appraisal Systems and Practices"[10]. The research concentrated on four urban policy regimes, one of which was DLG.

The research found current guidance on appraisal procedure was appropriate to the scale and nature of projects. The consultants felt, however, that there was a tendency for the appraisal criteria to address detailed technical issues and costs with a general lack of middle ground appraisal. This was considered to reflect a general absence of clear strategies and objectives within the local authorities.

Recommendations were made on the importance of developing a strategic approach, streamlining the annual bid, improving monitoring and evaluation, dealing with the issue of additionality and enhancing skills and training.

Following on from this research the Department commissioned a further study on the "Strategic Approach to Derelict Land Reclamation"[12]. The central aims of the research were to "identify the extent to which a strategic approach to derelict land is being followed by local authorities, to identify the advantages of such an approach, and to consider the potential for, and constraints on the further development and wider adoption of the strategic approach".

The researchers found that the extent to which local authorities have adopted a structured, systematic, strategic approach to programme preparation varied considerably. The study established that the development of a strategic approach to land reclamation while not a panacea can produce benefits. Such an approach enables more and better reclamation, better both in the sense of effectively serving policy and in the sense of more efficient utilisation of resources.

The report recommends that the Department should encourage local authorities to adopt such an approach and guidance should be prepared indicating how local authorities should proceed.

Research Strategy

The Department has decided to develop a strategic approach to research needs, and in the Spring of 1991 a paper was published for consultation. This set out the following approach.

That in future the value of the research commissioned by the Department could be enhanced by ensuring that the programme is continuous and well structured and that benefits could be gained by commissioning research which collects data over time. For example it would be relevant to know what is happening on reclaimed sites five, ten, and fifteen years after reclamation. It is also considered that a wider remit in terms of the issues which are addressed in research could provide a more comprehensive setting for appraising the effectiveness of policy and enabling systematic consideration of alternative means of dealing with the problem of dereliction. It is anticipated that in the future emphasis should be on integrated data collection and analysis. This will include consideration of the use of Geographic Information Systems.

On reviewing previous research and identifying the information requirements of the derelict land programme it is anticipated that future research will focus on the following areas: how land becomes derelict and how this can be prevented; the social, economic and environmental impacts of financial aid on derelict land; completions and post completions monitoring and the use, or possible use, of marketing techniques in the disposal and development of reclaimed sites; monitoring recent policy changes; review of the practice and research into methods of treatment of selected types of dereliction; cost benefit analysis of the 80kN sq m standard; potential extent of future dereliction and the level of resources required; potential sources of income from reclaimed derelict sites and how this could be used to supplement DLG funds should be examined.

POLICY DEVELOPMENT

The Department has decided to publish a series of Derelict Land Grant Advice Notes to publicise policy and good practice. Consultation took place on the first of these dealing with general policy matters in the period Nov 1990-Jan1991. Once published, the Advice Note will replace Circular 28/85.

Key features of the Advice Note include first a revised definition of objectives. Secondly support for a more strategic approach with up to 75% of a regions' DLG allocation being available for rolling programmes of reclamation. Thirdly the re-introduction of non-local authority DLG into the 57 areas. Fourthly an extension of the cost limit for small clearance schemes from £15000 to £30000; and fifthly, greater support for the voluntary and private sectors. These changes reflect the Government's wish to broaden the scope of DLG and make it a more attractive and flexible grant regime.

Emphasis is also placed on supply side matters. Local authorities are asked to formulate land use policies to encourage the development of derelict land by market forces.

Little attention to date has been placed on measures to prevent land becoming derelict or to require the derelictor to pay for the dereliction that has been caused. Over the period 1982-88 while some 14,000 ha were reclaimed, a further 8,800 ha became derelict. This suggests action is needed to prevent land becoming derelict and this matter is under consideration, and the Department hopes to publish a consultation paper.

It is also intended to consult on arrangements for financing the programme and on further Advice Notes. These will include guidance on the detailed operation of the DLG scheme and good practice notes.

© CROWN COPYRIGHT

REFERENCES

1. <u>Survey of Derelict Land in England 1988</u> Vols I and II 1991 HMSO ISBN 0 11752305 4, ISBN 0 11 752306 2.

2. <u>Report of the Controller & Auditor General</u>. Department of the Environment: Derelict Land Grant. 28 October 1988 HMSO ISBN 0 10 268988 1.

3. <u>Treasury Minute of the 13th-19th Reports from the Committee of Public Accounts 1988-89</u> CM 747 HMSO ISBN 0 10 107472 7.

4. <u>A Review of Derelict Land Policy</u>. Department of the Environment September 1989.

5. House of commons Environment committee. <u>first Report contaminated Land Volume 1.</u> 24 January 1990 HMSO ISBN 0 10 274590 0.

6. Department of the Environment. The Government's response to the First Report from the House of commons Select Committee on the environment. <u>Contaminated Land</u>. July 1990. HMSO ISBN 0 10 111612 8.

7. <u>Coal and the Environment</u>. The Government's Response to the Commission on Energy and the Environment's Report on "Coal and the Environment". May 1983. Cmnd 8877. HMSO ISBN 0 10 188770 1.

8. Department of the Environment. <u>Evaluation of Derelict Land Grant Schemes</u>. 1987. HMSO ISBN 0 11 752005.

9. Department of the Environment. <u>Cost Effective Management of Reclaimed Derelict Sites</u>. 1990. HMSO ISBN 011 752258 9.

10. Department of the Environment. <u>Project Appraisal Systems and Practices</u>. Aston University.

11. Department of the Environment. <u>Strategic Approach to Derelict Land Reclamation</u>. Aston University.

ANNEX I

DERELICT LAND GRANT SCHEME

ENABLING POWERS

1. Section 89 (2) of the National Parks and Access to the Countryside Act 1949, as substituted by section 3 of the Derelict Land Act 1982, confers specific powers on local authorities to carry out works to enable derelict, neglected or unsightly land to be reclaimed, improved or brought into use. The term "local authority" embraces not only county and district councils but also other local planning authorities as defined by the Town and Country Planning Act 1990. The 1949 Act also makes provision for local authorities to acquire land for such purposes either by agreement, or compulsorily. The various other powers under which local authorities may acquire and develop land for particular purposes may of course be exercised in relation to derelict land.

AVAILABILITY OF GRANT

2. In England central government grants are available under section 1 of the Derelict Land Act 1982 to local authorities and to other public bodies, voluntary organisation, private firms and individuals for the reclamation of derelict land. Grant is paid at the appropriate rate on any net loss incurred in carrying out an approved reclamation scheme.

RATES OF GRANT

3. In the Assisted Areas and Derelict Land Clearance Areas grant is paid to local authorities and the English Industrial Estates Corporation at a rate of 100%, and to bodies other than local authorities at 80%. Outside these areas the grant rate is 50% for both local authorities and others, except that in national parks and areas of outstanding natural beauty local authorities may receive 75% grants.

CRITERIA FOR THE PAYMENT OF DERELICT LAND GRANT

4. As a matter of policy the Secretary of State will not normally be prepared to approve grant unless the land to be reclaimed is derelict (defined administratively as so damaged by industrial or other development that it is incapable of beneficial use without treatment). However grant may be paid to local authorities in respect of the costs of acquisition of land which is not derelict but is required for purposes connected with reclamation works, and in relation to land which, though not derelict, is likely to become so as a result of actual or threatened collapse of disused underground mining operations other than coal.

RECLAMATION WORKS

5. The works which are eligible for grant are those which the Department is satisfied are required for the purpose of reclamation, depending on the particular circumstances and nature of the scheme. As a general guide this would include the works necessary to bring the site to the equivalent of a green field state. For grant purposes the definition of green field state for schemes with a "hard" after use is "a site which either has the bearing capacity of surrounding non-derelict land or a bearing capacity of 80 kN per sq m whichever is the lesser". 80 kN is equivalent to 3/4 ton per sq ft and is sufficient for warehousing and light industrial development. For environmental schemes a lower load bearing capacity may well be acceptable.

ANNEX II

DERELICT LAND SURVEY 1988

THE AMOUNT OF DERELICT LAND AND THE AREA JUSTIFYING RECLAMATION AT 1 APRIL 1988: BY INNER CITY, OTHER URBAN AND RURAL DISTRIBUTION AND TYPE OF DERELICTION.

Hectares

Type of Derelict Land	Inner City		Other Urban		Rural		Total		Area Justifying Reclamation as Percentage of Derelict Land
	Derelict Land	Area Justifying Reclamation	Derelict Land	Area Justifying Reclamation	Derelict Land	Area Justifying Reclamation	Derelict Land	Area Justifying Reclamation	
Colliery Spoil Heaps	498	482	2,075	2,027	2,122	1,863	4,695	4,372	93%
Metalliferous Spoil Heaps	199	199	451	366	4,119	708	4,769	1,273	27%
Other Spoil Heaps	265	263	925	877	1,240	681	2,430	1,821	75%
Excavations and Pits	415	413	1,510	1,375	4,039	2,581	5,964	4,369	73%
Military etc Dereliction	67	66	183	160	2,326	1,835	2,576	2,061	80%
Derelict Railway Land	1,047	1,008	2,127	1,985	3,239	2,053	6,413	5,046	79%
Mining Subsidence and Land Affected by Underground Mining Operations	378	378	269	245	388	306	1,035	929	90%
General Industrial Dereliction	3,135	3,022	3,702	3,527	1,645	1,425	8,482	7,974	94%
Other Forms of Dereliction	1,028	1,013	1,798	1,618	1,305	1,155	4,131	3,786	92%
Totals	7,032	6,844	13,040	12,180	20,423	12,607	40,495	31,631	78%

CHANGES IN THE AMOUNT OF DERELICT LAND BY TYPE OF DERELICTION: 1974, 1982 AND 1988.

Hectares

	1974	1982	1988	% Change 1974-82	% Change 1982-88
Spoil Heaps	13,100	13,300	11,900	+2%	—11%
Excavations and Pits	8,700	8,600	6,000	—2%	—31%
Military etc Dereliction	3,800	3,000	2,600	—20%	—15%
Derelict Railway Land	9,100	8,200	6,400	—10%	—22%
Other Forms of Dereliction	8,600	12,500	13,600	+47%	+9%
Total	43,300	45,700	40,500	+6%	—11%

Note: Percentages are rounded to the nearest integer and areas to the nearest 100 hectares. Component figures may not sum to the independently rounded totals.

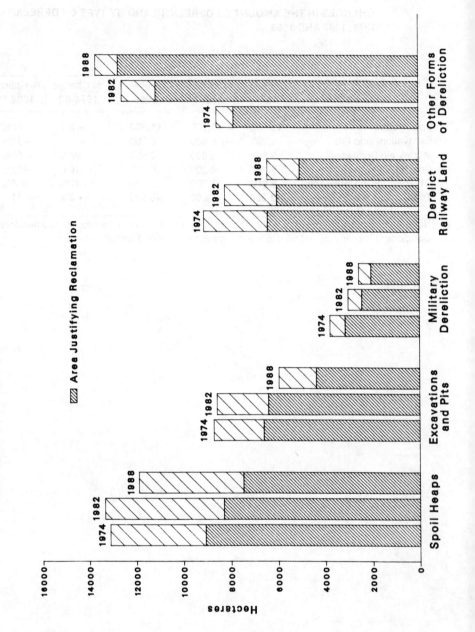

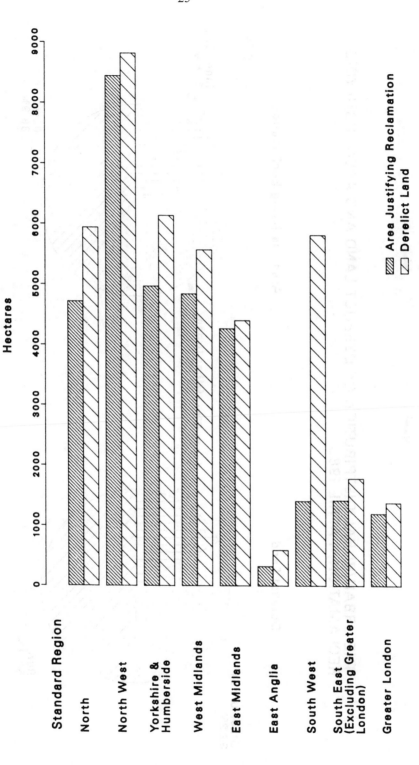

THE URBAN/RURAL DISTRIBUTION OF DERELICT LAND AND AREA JUSTIFYING RECLAMATION, 1 APRIL 1988

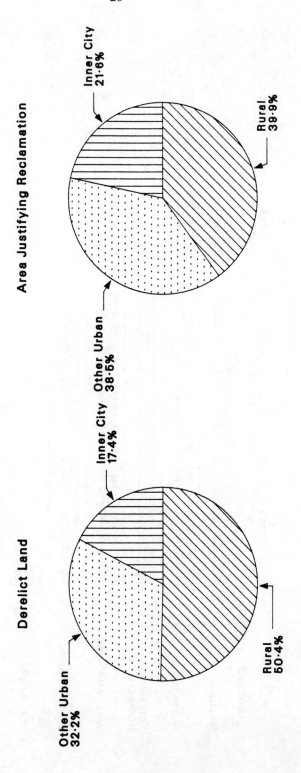

CHANGES IN THE AMOUNT OF DERELICT LAND BY REGION: 1974, 1982 AND 1988.

Hectares

	1974	1982	1988	% Change 1974-82	% Change 1982-88
North	9,400	7,300	5,900	—22%	—19%
North West	8,000	10,000	8,800	+25%	—12%
Yorkshire and Humberside	5,500	5,400	6,100	—2%	+13%
West Midlands	4,700	5,800	5,600	+24%	—4%
East Midlands	5,200	5,200	4,400	+1%	—15%
East Anglia	1,800	800	600	—55%	—26%
South West	6,400	6,600	5,800	+3%	—12%
South East (excluding Greater London)	2,000	2,500	1,800	+24%	—29%
Greater London	300	2,000	1,400	+567%	—30%
England	43,300	45,700	40,500	+6%	—11%

Note: Percentages are rounded to the nearest integer and areas to the nearest 100 hectares. Component figures may not sum to the independently rounded totals.

CHANGES IN THE AREA JUSTIFYING RECLAMATION BY REGION: 1974, 1982 AND 1988.

Hectares

	1974	1982	1988	% Change 1974-82	% Change 1982-88
North	7,800	5,600	4,700	—28%	—16%
North West	7,200	9,000	8,500	+25%	—6%
Yorkshire and Humberside	4,600	4,100	5,000	—12%	+22%
West Midlands	4,200	5,200	4,900	+24%	—6%
East Midlands	4,700	4,800	4,300	+2%	—10%
East Anglia	1,300	600	300	—54%	—45%
South West	1,800	1,400	1,400	—22%	0%
South East (excluding Greater London)	1,200	2,000	1,400	+67%	—30%
Greater London	300	1,600	1,200	+433%	—27%
England	33,100	34,300	31,600	+4%	—8%

Note: Percentages are rounded to the nearest integer and areas to the nearest 100 hectares. Component figures may not sum to the independently rounded totals.

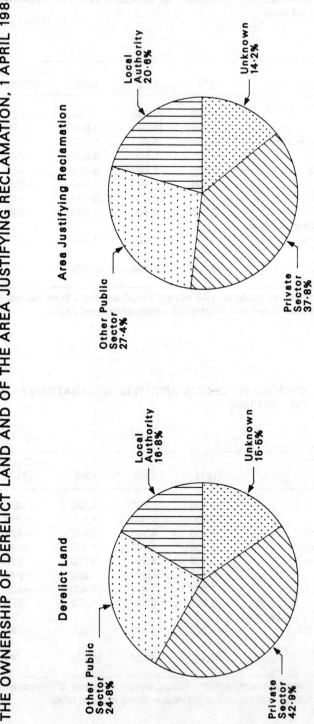

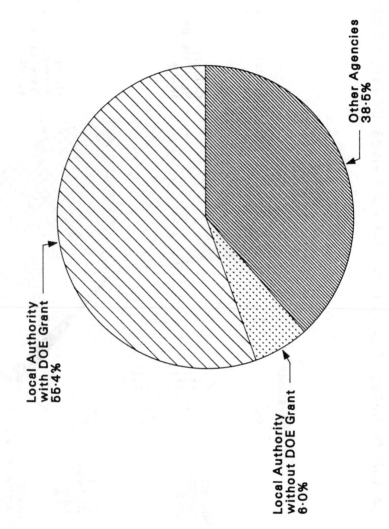

THE AMOUNT OF DERELICT LAND RECLAIMED 1982-88, BY AGENCY

THE AMOUNT OF DERELICT LAND RECLAIMED 1982-1988, BY TYPE OF DERELICTION

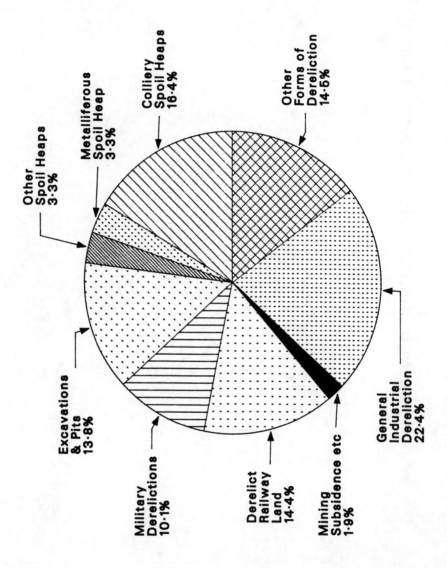

DERELICT LAND RECLAIMED, 1982-88 BY STANDARD REGION

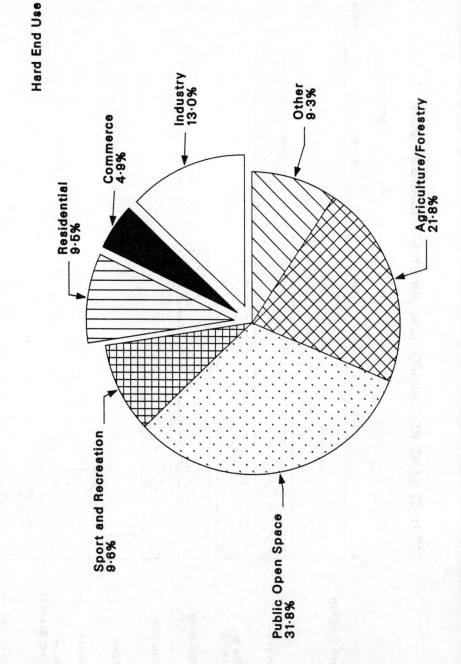

Seventeenth Report
Department of the Environment

ANNEX III

Derelict Land Grant

57. The Treasury and the Department of the Environment (DOE) note the conclusions and recommendations of the Committee.

Effectiveness of the Derelict Land Grant programme

58. DOE agrees with the Committee that it is important to assess the effectiveness of the Derelict Land Grant (DLG) programme and welcomes its conclusion that the Department's arrangements for the derelict land grant scheme have improved since 1985. However, DOE does not accept the Committee's unqualified claim that the Government is unable to make such an assessment at present. Much information is already available and additional measures designed to monitor the programme's outputs in more detail are already in place. DOE accepts the Committee's view that continuing efforts will be needed to maintain progress on developing further quantitative and qualitative measures of performance and already has steps in hand to achieve this.

59. Amounts of land reclaimed each year by the programme and targets for future years are published in the annual report on DOE Inner City Programmes and in the Public Expenditure White Paper. Thus, since 1982, the grant has assisted in the reclamation of over 8,000 hectares of derelict land and this has undoubtedly contributed to the reduction in the total level of dereliction which is indicated by the provisional results of the 1988 Derelict Land Survey. Those results suggest that the total amount of derelict land in England in 1988 was 40,500 hectares, a reduction of about 11 per cent since 1982.

60. For 1989–90 regional targets have been set for the amount of land to be reclaimed for both soft and hard end use. These targets are related to the amounts of DLG allocated to each region for the financial year. In addition, DOE regional offices have begun to monitor projects completed since 1983 and intended for hard end use, in order to assess the extent to which such development has taken place.

61. The Department has also set in hand, with the local authorities concerned, a review of six rolling programmes which have been in operation for at least three years. This review is intended to draw general conclusions on the effectiveness of the rolling programme approach and also to reassess the individual programme strategies. The outcome will be known in time to influence the programme from 1990–91 onwards.

62. The Department intends to analyse and publish the information on targets and outputs in its report on the programme for 1989–90 and further refine these for publication in the 1990–91 report, as suggested by the Committee.

Computerisation programme

63. DOE accepts that monitoring and management information systems need to be strengthened, particularly at DOE headquarters. The design of the computer system is well advanced and steps are in hand to speed up its implementation, although this will involve some additional costs. Trials of an initial phase of the system are now expected to begin in one region in January 1990, with full implementation of the whole system in all ten regional offices and at headquarters being completed by the autumn of 1990, ahead of the timetable reported by the Comptroller and Auditor General (C&AG). Benefits will be apparent from the first stage of implementation onwards and the full benefits of computerisation will begin to flow as soon as implementation is complete.

Final claims and grant recovery

64. DOE agrees that the submission of final claims and the recovery of grant, following a local authority disposal of a reclaimed site, should receive a high priority. DOE is continuing to press local authorities to submit final claims and the historical backlog of outstanding claims is now much reduced to a total of 79, worth £17 million, at the end of May 1989, compared to 185 claims, worth £36 million, in October 1988. The Department has written to those authorities with the worst record of outstanding final claims, informing them that, if these are not submitted by a set date, the Department proposes to suspend the payment of grant on interim claims for all derelict land schemes carried out by those authorities.

65. Grant recovery has also been given greater priority, both in DOE regional offices and at headquarters, resulting in a record recovery of over £11 million in 1988–89. Targets for grant recovery, as suggested by the Committee, are now set and published in the annual Public Expenditure White Paper.

Surveys of derelict land

66. The results of the 1988 Derelict Land Survey are being analysed and a full report will be published later in the year. The Department will consult with the Local Authority Associations about future survey arrangements including the C&AG's suggestion for sample surveys in the years between national surveys.

Appraisal procedures

67. A number of steps have already been taken recently to improve procedures. Since April 1988 private sector reclamation within Urban Programme areas has been supported by City Grant, which allows a more detailed appraisal of the developers' costs and returns than is currently possible under present DLG appraisal procedures. Secondly, since April 1989, private sector applicants for DLG are asked to sign a formal declaration that the grant is needed in order to make the project viable.

68. The Department will be examining further the appraisal procedures for private sector DLG applicants in order to ensure that grant is only paid where it is most needed. The recommendations of a consultant's report on appraisal procedures employed in respect of various inner city regeneration schemes, including DLG, will be taken into account. Their final report will be submitted to the Department shortly.

Review of programme priorities

69. The DOE notes the Committee's recommendation that the objectives of the DLG programme should be reviewed. In fact, as the Committee were told, this is already being done in the review which the Parliamentary Under Secretary of State at the Department of the Environment announced in the House of Commons on 20 July 1988. The terms of reference were:

"to review the causes and incidence of derelict land and the measures needed to deal with it and to prevent dereliction in the future having regard to value for money".

70. This review is also addressing the issue of prevention and control of new dereliction, as the Committee recommends, including the consideration of new safeguards to minimise the cost to the Exchequer of dealing with new industrial dereliction. DOE also accepts the Committee's view that the Department, in its review of the working of the 1981 Minerals Planning Act over the next two years, should pay particular attention to any further changes which might minimise reclamation costs to the Exchequer.

71. The review of the derelict land programme will also examine the need to achieve the right balance between hard and soft end uses and between projects in rural and urban areas, taking into account the cost implications and the need for appropriate appraisal techniques.

72. DOE Ministers will be reporting to Parliament on the outcome of this review.

ANNEX IV

MR MOYNIHAN'S STATEMENT

My Department and the Department of Energy have held a number of discussions with British Coal about the problems of coalfield dereliction. I am glad to say that as part of the outcome of those discussions the British Coal Corporation have accepted responsibility for the restoration of collieries which close in the 4 years following 31 March 1990.

I regard this as an important step forward in placing environmental responsibilities where they belong. I believe that as a result of this agreement land will be reclaimed to an acceptable standard more quickly than would otherwise be the case. We shall keep the arrangement under review and will be considering how the problems of coalfield dereliction may best be tackled in the longer term.

In large part, these additional liabilities will be eligible for Deficiency Grant under the Coal Industry Act 1990, so there will be no adverse impact on the Corporation's finances.

The undertaking does not apply in Scotland and Wales where the derelict land grant systems operate on a different basis.

The full terms of the agreement are as follows:

> 1. British Coal will continue their current practice of consulting with the relevant local planning authority(ies) about the appropriate planning future for the site of any colliery which closes.

2. In respect of any deep mine started before 1 July 1948 which closes within 4 years commencing 1 April 1990, British Coal will agree to an acceptance of responsibility for the restoration of such sites to a soft end use. This acceptance of responsibility will not affect in any way the Corporation's present practice of seeking appropriate planning consents for redevelopment of the site as soon as appropriate following any announcement of closure.

3. When the planning consent provides only for restoration to a soft end use, British Coal will demolish all buildings, plant and machinery structures or erections and will restore the colliery surface area, together with adjoining or associated tip areas used solely for the deposit of mineral waste for that colliery, in accordance with a scheme agreed with the mineral planning authority, within a period of 24 months from the date when mining operations have permanently ceased, or such longer period as may be provided for in the agreed scheme.

4. Where the planning consent provides that the whole site or part of the site is appropriate for redevelopment, the land will be valued as if no responsibility for restoration to a soft end use existed, but to discharge their voluntary undertaking, British Coal will contribute to any subsequent derelict land grant aided scheme the notional cost of restoration of the whole site to a soft end use to a standard acceptable to the mineral planning authority. In a situation where the cost of such restoration is unable to be agreed between British Coal and the mineral planning authority, then such costs will be determined on the basis of an independent valuation.

DERELICT LAND GRANT: STATISTICS ANNEX V

TABLE 1. EXPENDITURE: £ million

	1979/ 1980	1980/ 1981	1981/ 1982	1982/ 1983	1983/ 1984	1984/ 1985	1985/ 1986	1986/ 1987	1987/ 1988	1988/ 1989	1989/ 1990	1990/ 1991 est.
Local authority	22.37	30.02	29.68	60.34	66.44	66.56	71.19	76.75	72.01	66.53	61.12	66.95
Non local authority	-	-	0.02	0.76	1.75	3.30	4.02	5.37	8.76	11.58	4.32	2.92
Total	22.37	30.02	29.7	61.10	68.19	69.86	75.21	82.12	80.77	78.11	65.44	69.87

TABLE 2. LAND RECLAIMED: hectares

	1979/ 1980	1980/ 1981	1981/ 1982	1982/ 1983	1983/ 1984	1984/ 1985	1985/ 1986	1986/ 1987	1987/ 1988	1988/ 1989	1989/ 1990	1990/ 1991 est.
For development:	N/A	N/A	N/A	N/A	N/A	594	559	381	588	912	618	555
Other (open space etc):	N/A	N/A	N/A	N/A	N/A	749	461	522	697	575	565	710
Total	1236	1403	1742	1444	1364	1343	1060	903	1285	1487	1183	1265

TABLE 3. REGIONAL DISTRIBUTION OF DLG IN RECENT YEARS (INITIAL ALLOCATIONS):
£ million

Region	1988/89	1989/90	1990/91
Northern	12.6	8.2	9.0
North West	19.3	15.8	16.0
Merseyside	7.5	6.0	6.5
Yorks/Humberside	9.7	12.0	12.6
West Midlands	15.0	13.0	14.3
East Midlands	8.3	8.2	8.6
London	1.5	0.8	1.0
South West	2.2	2.0	2.2
South East	0.7	0.7	0.6
Eastern	0.5	0.5	0.7
Total	77.1	67.0	71.5

TABLE 4. COALFIELD (CENE) RECLAMATION: DLG EXPENDITURE: £ million

1983/84	1984/85	1985/86	1986/87	1987/88	1988/89	1989/90	1990/91 estimated
5.77	5.05	5.09	6.87	7.57	8.70	10.68	12.54

TABLE 5. WEST MIDLANDS LIMESTONE SCHEMES: DLG EXPENDITURE: £ million

1983/84	1984/85	1985/86	1986/87	1987/88	1988/89	1989/90	1990/91 estimated
0.62	2.08	2.95	5.13	4.94	5.63	5.67	6.25

TABLE 6.   AFTERVALUE REPAID TO THE DEPARTMENT OF THE ENVIRONMENT: £ million

1986/87	1987/88	1988/89	1989/90	1990/91 forecast
2.00	2.30	10.35	11.44	11.00

AN EVALUATION OF SURVEYS OF SOIL CONTAMINATION IN THE CITY OF SWANSEA, SOUTH WALES

E. M. BRIDGES
University of Wales, Swansea SA2 8PP

ABSTRACT

This contribution reviews surveys of contamination of the soil by toxic metals in Swansea over the past thirty years and presents new information on the background concentrations of these metals throughout the city. A knowledge of these background values is helpful for any interpretation of individual samples during preparatory work for restoration projects and for monitoring the success of restoration after completion of the civil engineering works. Results of several surveys are compared and the problems of monitoring environmental contamination by non-ferrous metals are raised.

INTRODUCTION

The City of Swansea has had a long and interesting industrial history which reached its peak of international importance in the middle years of the nineteenth century. Industrialisation, based on metal smelting, came early to the Swansea region and the development from 1717 onwards of the Lower Swansea Valley to become the world centre of non-ferrous metal smelting, brought undoubted benefits to the town. Unfortunately, when the industry declined, a legacy of dereliction and pollution remained from this phase of the City's history (1,2,3,4). For almost sixty years the area was left derelict as there was no provision for clearance of the ruins of the former works or their massive waste heaps. The soils of the surrounding area had been eroded and contaminated with metals, so few plants grew on the valley sides. The Lower Swansea Valley became known as one of the most concentrated areas of industrial dereliction in Britain. In 1961, the valley became the focus of an exciting initiative to remove the dereliction and to revitalise the area. Thirty years later the task has been completed so successfully that little now remains of the former eyesores left by previous industries. However, experience has shown that metallurgical industries without fail leave contamination in and around those areas where they were situated. The extent of this contamination and to what degree of severity the soil of the city had been polluted were unknown. This contribution shows that evidence for soil contamination was slow to emerge and the standards against which the degree of pollution could be judged were unavailable until the last decade.

The soils

Many soils of Swansea have a slowly permeable subsoil and suffer from prolonged waterlogging during the winter months, and in general the soils of the City of are naturally lacking in lime. Extremely acid soils (pH < 4.0) occur on Kilvey Hill and Townhill where a strong natural tendency to acidification has been intensified by industrial pollution. Throughout most of the city, soils would be naturally acid (pH 4.1-5.9) but the pattern is made complex by liming amenity areas and gardens where soils have been made neutral or only slightly acid (pH 6.0-7.0). Where over-enthusiastic liming has occurred, and in places where lime has been released from old buildings, soils may become alkaline (pH > 7.0). The significance of acidity is that at lower pH most toxic metals become more soluble and available for plants. Similarly, with poor drainage, soils may become anaerobic and in reducing conditions toxic metals become more soluble and can be taken-up by plants more easily.

PREVIOUS SURVEYS

In 1961 the Lower Swansea Valley Project was set up "to investigate the physical, social and economic situation in the Lower Swansea Valley, to understand the reasons which had inhibited its development in the past and to provide the information necessary for its future development". During the life of the project, only a brief report was submitted about soil conditions and this dwelt mainly upon the erosion which had occurred and continued to affect the area until the early 1960s (5,6). Although numerous samples of the various metalliferous dumps were examined, only two soils were analysed for toxic metals and this was in connection with two grass trials carried out by the (then) National Agricultural Advisory Service. These soils were found to be strongly acidic (pH 4.5) and contained total amounts of copper 900 mg/kg, lead 2700 mg/kg, zinc 2700-9000 mg/kg, nickel 27 mg/kg, chromium 27-90 mg/kg and tin 90 mg/kg. Plant-available copper amounted to 70 mg/kg and zinc 280 mg/kg. The grass trials were laid down to ascertain the most suitable fertilizer rates to encourage plant growth on the eroded, polluted soils. It was noted at the time that during the time of the experiment no change took place in the distribution of copper in the soil, but the available zinc appeared to become concentrated in the upper 2.5 cm of the soil, presumably by the action of the plants.

In 1975, following a survey for the Welsh Office, it was reported that average available concentrations in the soil of copper 12 mg/kg, zinc 101 mg/kg, cadmium 2.6 mg/kg, nickel 3.9 mg/kg and lead 4.2 mg/kg on seven test farms in the Swansea Valley were all in excess of background values, but still within the accepted safe limits. (The safe limits quoted were copper 20 mg/kg, lead 20-50 mg/kg, zinc 300-400 mg/kg, nickel 10 mg/kg and cadmium 10 mg/kg.) (7).

Redistribution of the metals in the soil was discovered during a study of the cycling of elements by the trees planted to improve the visual amenity of the Lower Swansea Valley (8,9). Fifteen years after planting the trees, some metals had preferentially accumulated near the soil surface, others had been leached lower in the soil profile. Lead (maximum concentration 1985 mg/kg) accumulated at the soil surface, and the supposition is that the trees had removed lead from the soil and redistributed it as the leaves fell to the ground each autumn. Maximum concentrations of copper 443 mg/kg, zinc 1107 mg/kg, nickel 265 mg/kg and cadmium 21.9 mg/kg all occurred deeper in the soil, being precipitated by the slightly high pH of the lower soil horizons.

Beneath the planted coniferous trees a deep layer of partially decomposed organic matter had accumulated within which could be distinguished a fresh litter (L) layer and an older fermentation (F) layer. The F layer contained higher concentrations of copper 62.4 mg/kg, lead 156.7 mg/kg, zinc 1107 mg/kg, nickel 57.7 mg/kg and cadmium 3.6 mg/kg than occurred in the litter layer. Evidence from studies elsewhere have indicated that lower rates of organic decomposition occurred where soils were influenced by a neighbouring smelting plant as numbers of soil fauna are reduced and mychorrhizal interactions are inhibited by the presence of the metals (10).

In preparation for the excavation of the lake in marshy ground alongside the Nant-y-Fendrod stream in the Lower Swansea Valley, the city council commissioned a survey of the metal content of marsh soils. This marsh had been subject to the leachates from surrounding tips of zinc, copper and steel wastes. Samples taken from 12 boreholes revealed that 15 per cent of the material was grossly contaminated: up to 168 mg/kg of cadmium, 3,086 mg/kg of copper, 8,000 mg/kg of lead and 89,000 mg/kg of zinc demonstrated that this was some of the most highly contaminated soil in the entire project area. With this information it was possible to plan the safest and most economical means of dealing with it without removing from the Lower Swansea Valley. Compared with mean values elsewhere, and the available guidelines for soil contamination, this material was hazardous and required careful handling to minimise its environmental impact (11).

Since the publication of a paper comparing the metallic content of soils in rural districts and urban areas in Scotland, it has been appreciated that urban areas generally have enhanced concentrations of toxic metals (12). To test the wider validity of this observation, research scientists at Imperial College, London examined the content of garden soils in 53 locations in Britain, including Swansea (13,14). Samples of household dusts were also examined in an attempt to relate the health of the population to metal concentrations in the immediate environment of the inhabitants.

The results of this survey, the most extensive investigation of its kind to be carried out on garden soils and household dusts confirmed the elevated concentrations of copper, lead, zinc and cadmium in the soils of urban areas throughout the United Kingdom. It also found that lead concentrations in household dusts are significantly and positively related to lead concentrations in garden soils, to the area of exposed soil around the house, the presence of lead-based paints in the house and recent redecoration. The concentration of lead was greatest in the soils and dusts of older properties.

Davies (15) has sampled soils around the nickel refinery at Clydach, just beyond the northern boundary of the city. His results revealed an approximately eliptical pattern of enhanced nickel concentrations aligned along the valley. Visual interpretation of a three-dimensional diagram appears to suggest a peak of more than 160 mg/kg occurs at the refinery, rising from background values. Goodman (16) from a transect across Swansea indicated 3 mg/kg acetic acid extractable nickel as the background concentration for Gower and notes that the nickel concentrations in the Swansea Valley were low despite the high content of nickel in the nineteenth century smelter wastes.

THE PRESENT SURVEY

The survey reported here followed the protocol evolved for a survey of background levels of soil contamination in Walsall by JURUE (17). Using the 1 km squares of the National Grid as a framework, soil samples were collected from 171 sites within the City of Swansea. Up to five samples per grid square were taken from city centre areas, but smaller numbers from each grid square on the urban-rural fringe. Within each grid square the sampling was influenced by dominant land use, samples being taken from private gardens, industrial areas, public open space, woodland and commonland areas. At each site 10 subsamples of the surface soil to a depth of 5 cm were taken with a stainless steel trowel and bulked to form the sample for analysis. After removal of any surface plant debris, samples were placed in labelled plastic bags and taken to the laboratory where they were allowed to air-dry. In preparation for analysis, the samples were ground using a "Tema" mill and passed through a 2 mm sieve. Total metal concentrations were determined on solutions extracted from the soil samples by hot concentrated Analar nitric acid. Analysis took place using a Phillips SP9 atomic absorption spectrophotometer with standard laboratory methods for calibration and quality control. Values for pH were obtained on 2.5 : 1 distilled water : soil suspension and texture of soils was determined manually.

As might be expected from its history, considerable residual contamination remains in the soils of the Lower Swansea Valley, but surface soil contamination in the valley is now only slightly worse than in many of the surrounding urban areas. The mean values for Swansea (copper167 mg/kg, lead 535 mg/kg, zinc 574 mg/kg and cadmium 3.35 mg/kg) indicate that soil metal contamination in the city, despite its history of metal smelting, is not much worse than that in many other British towns and cities (18). However, certain aspects of the results give cause for concern.

Average values can be misleading in a survey of contaminated land (19) as occasional hotspots can distort the values, but they do provide a general indication of the overall problem. The mean values for total copper obtained from Swansea soils occurred in the upper part of the natural range (2-250 mg/kg) with average figures for several grid squares exceeding these values. For lead, the natural range lies between 2 and 300 mg/kg so that a considerable part of the city's soils are significantly contaminated with this metal There is a concentration of the higher values in the former smelting districts of the Swansea Valley, but hotspots occur in other places (Fig. 1). A similar pattern emerges for zinc, natural range 1-900 mg/kg, where the former industrial area has higher values. The most disturbing feature is the distribution of cadmium, natural range 0.01-2 mg/kg, where many sites were found to be well above safe values (Fig. 1).

Compared with the tentative guidelines issued by the DoE Interdepartmental Committee for the Redevelopment of Contaminated Land (20), samples from many of the grid squares exceed the "stringent" trigger concentration recommended if redevelopment as housing with private gardens were to be contemplated, but only a few grid squares contain sites exceed the "lenient" threshold recommended for parks and public open space. Of the individual samples analysed, 19 per cent of lead and 28 per cent of cadmium values exceeded the lenient guideline. Copper and zinc are phytotoxic and are normally estimated on an extractable basis, figures for which are usually between 25 and 33 per cent of total values (21). Guidelines in use in the Netherlands are also given on the maps for comparative purposes.

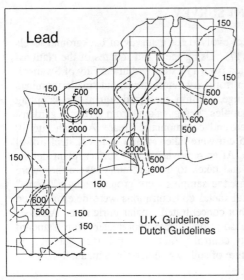
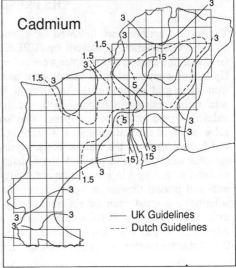
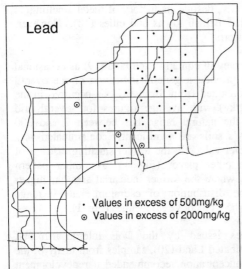
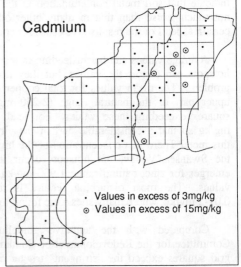

Figure 1. Upper maps show isopleths of average concentrations (mg/kg) of lead and cadmium in Swansea soils. Lower maps show distribution of samples containing lead and cadmium in excess of the trigger concentrations proposed by DoE.

Worst-case concentrations

The concept of "worst-case concentrations" has been proposed by JURUE to indicate whether any significant proportion of an area under investigation exceeds a concentration which is equivalent to the worst 2.5 per cent. The statistical approach taken is that if it had been possible to take 40 samples from each 1 km grid square, one of them might be expected to contain a soil metal concentration as high as the worst-case concentration. This figure has been calculated using the population mean plus two standard deviations. Calculations were based on the sum of soil metal concentrations in four adjacent 1 km grid squares. The measurements were subjected to a logarithmic transformation as the data possessed a log-normal form with a long tail leading to higher values, rather than taking the linear-normal Gaussian distribution. When the calculation was completed for the four squares, the value was assigned to their central point. The figures for the next intersection were then obtained by omitting the two lower or left hand squares and adding the two on the other side.

Distribution of the potential worst-case concentrations of copper contamination occupy a broad band following the Tawe valley from the mouth of the river to the northern boundary of the city. The highest values reach a peak over the area formerly occupied by the copper smelters, northeast of the city centre. Values then decline steadily northwards. A subsidiary area of potential high values is associated with the Swansea Trading Estate, but values are lower than for the tract following the Tawe valley. Outside these two areas the worst case values are no more than expected for soils uncontaminated by industrial activity (Fig. 2).

The potential worst-case concentration of lead contamination also reflect the past industrial activity in the Lower Swansea Valley but the effect is more widespread throughout eastern Swansea. Virtually the whole area of eastern Swansea has potential worst-case contamination values which fail to meet the criteria adopted in the Netherlands. The less stringent British guidelines are broken around Llansamlet, the town centre and extending towards Port Tennant where made ground around the docks, composed of metalliferous slags, could influences values. Smaller areas where breaches of guidelines occur are at the Swansea Trading Estate and an anomalous area centred upon West Cross but the latter case is no worse than occurs in many other cities, and well below areas where lead is mined (Fig. 2),

The worst-case concentration of zinc occurs around the dockland area where the statistical approach suggests high values of contamination by this metal. The residual contamination of the Llansamlet area, observed in the average values, is less strongly shown in the calculated worst cases which are spread over a wider area. Secondary areas of high values occur associated with the Swansea Trading Estate and southern Sketty and the Mayals districts (Fig. 2).

In the case of cadmium, highest worst-case concentrations occur in the Llansamlet district, extending southeastwards to Trallwn; a secondary area occurs associated with the docks where the upper guidelines of the DoE again are breached. Worst case concentrations which breach the stringent DoE recommendations for redevelopment occur throughout the city but of more concern are the very high values, between two and three times higher than the lenient DoE limit, which are liable to be encountered (Fig. 2).

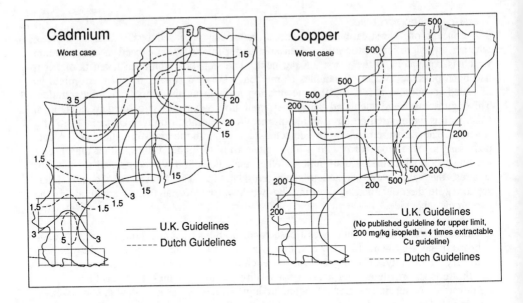
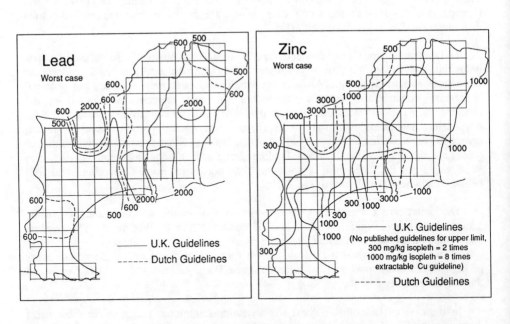

Figure 2. Isopleths for calculated "worst-case concentrations" of copper, lead, zinc and cadmium in Swansea soils.

DISCUSSION

This paper has reviewed the several surveys of toxic metal contamination of soils in the City of Swansea and has presented figures and maps which demonstrate the distribution of copper, lead, zinc and cadmium concentrations throughout the city. As was expected residual contamination still occurs in and around the Lower Swansea Valley where the eighteenth- and nineteenth-century metalliferous works were situated. However, compared with surveys of the whole of Wales (22,23), the current survey shows that contamination of soils in the city is much wider than might be expected from a consideration of industrial activity alone. The values obtained are comparable with those of the Imperial College survey and can be shown to come from a statistically similar population, indicating successive surveys can achieve reproducible results.

The urban area of Swansea was greatly extended during the 1960s and 1970s as many new houses were built. The metalliferous wastes of the Lower Swansea Valley were a cheap and readily available source of hardcore for the building industry. Between January 1962 and December 1965 it was estimated that a volume of 231,869 cubic yards weighing some 278,243 tons was removed from the tips in the central part of the valley (24). Some of this material is now beneath the M4 motorway, but considerable quantities were used for foundations of houses and driveways throughout the city. As a result, the incidence of toxic metal contamination is much wider than anticipated from a consideration of industrial activity alone.

Large sums of public money have been spent on land reclamation in the Lower Swansea Valley and a new landscape with thriving commercial and industrial premises has replaced the former dereliction. Beneath this restoration lie the toxic wastes interred in a protective sandwich of calcareous steel wastes and covered by local till deposits. The continuing satisfactory performance of the restoration should be assessed. The problem is what to monitor. Repeated soil surveys of the type described do not provide a satisfactory answer as much of the covering material is fresh uncontaminated material, some of which has been brought in from outside the area. Analysis would indicate whether underlying toxicity was migrating upwards into the soil. It might be possible to utilise remote sensing apparatus to indicate this form of recontamination as described by Coulson and Bridges (25).

A second line of investigation would be to monitor the metallic content of drainage waters seeping from the restored land. This should certainly be part of a monitoring scheme, but although difficult to ascertain exactly the route of water through restored ground, it can be modelled. Over a period of years, the content of drainage waters should represent an equilibrium reached within the restored land. However, care must be taken to ensure that the samples measured are not diluted by surface or groundwater. Measurements taken by the National Rivers Authority for the Tawe and the Nant-y-Ffin tributary indicate dramatic reductions in pollutants following diversion of the Nant-y-Ffin around the site of the former Imperial Zinc Works. Until 1974 this stream was sufficiently acid (pH 3.0-3.8) to destroy aquatic life and acidify the Tawe below its confluence. The metallic pollutants were unacceptably high in amount, but copper concentrations fell from 1974 onwards when the works closed. Cadmium was also well in excess of environmentally acceptable concentrations, but this too fell to a mean of 0.062 mg/l by 1982, and remains satisfactorily low.

Thus, it would seem that a combination of surveys of soil metal conditions and the analyses of seepage waters should provide the best means of monitoring restorations of sites contaminated by former metalliferous works. The maps and figures provided by this survey indicate the distribution of contamination and give a background against which the real hotspots may be judged, and their danger to the health of the population assessed.

ACKNOWLEDGEMENTS

The author wishes to record his thanks to all those people who kindly allowed samples to be taken from their gardens. Mr P. Bevan is thanked for undertaking the chemical analyses of the soil samples and the National Rivers Authority for supplying data about the metal content of stream waters. Acknowledgement is also made of the advice provided by the Geochemical Research Group at Imperial College, London. Mr G.B. Lewis drew the illustrations and the text was prepared by "Words" Word Processing Services, Swansea.

REFERENCES

1. Hilton, K.J., *The Lower Swansea Valley Project*, Longmans, London, 1967.

2. Bridges, E.M., Desecration and Restoration of the Lower Swansea Valley. In *Management of Uncontrolled Hazardous Waste Sites*, Hazardous Materials Control Institute, Silver Spring, Maryland, 1984, pp. 553-559.

3. Bridges, E.M. and Morgan, H., Dereliction and Pollution. In *The City of Swansea: Challenges and Change*, ed R.A. Griffiths, Alan Sutton, Stroud, 1990.

4. Bridges, E.M., *Surveying Derelict Land*, Cambridge University Press, Cambridge, 1988.

5. Bridges, E.M., Eroded soils in the Lower Swansea Valley. *Journal of Soil Science*, 1969, **20**, 236-245.

6. Bridges, E.M. and Harding, D.M., Micro-erosion processes and factors affecting slope developement in the Lower Swansea Valley. *Transactions of the Institute of British Geographers Special Publication* No 3, 1971, pp. 149-163.

7. Welsh Office, *Report of a collaborative study on certain elements in air, soil, plants and animals in the Swansea-Port Talbot area together with a report of a moss-bag study of atmospheric pollution across South Wales*, Cardiff, 1975.

8. Chase, D. S., Botanical and economic aspects of revegetating parts of the Lower Swansea Valley. Unpublished Ph.D. thesis, University of Wales, Swansea, 1978.

9. Bridges, E.M., Chase, D.S. and Wainwright, S.J., Soil and plant investigations since 1967. In *Dealing with Dereliction*, eds R.D.F. Bromley and G. Humphrys, University College of Swansea, 1979.

10. Hutton, M., Impact of airborne metal contamination on a deciduous woodland system. In *Effects of Pollutants at the Ecosystem Level*, eds P.J. Sheehan, D.R. Miller, G.C. Butler and Ph. Bourdeau, SCOPE 22, Wiley, Chichester, 1984.

11. Davies, R.L., Environmental monitoring and control. In *Dealing with Dereliction*, eds R.D.F. Bromley and G. Humphrys, University College of Swansea, 1979.

12. Purves, D. and Mackenzie, E.J., Trace element contamination of parklands in urban areas. *Journal of Soil Science*, 1969, **20** 288-290.

13. Culbard, E.B., Thornton, I., Watt, M., Wheatley, S., Moorcroft, S. and Thompson, M., Metal contamination in British urban dusts and soils. *Journal of Environmental Quality*, 1988, **17**, 226-234.

14. Thornton, I., Abrahams, P.W., Culbard, E., Rother, J.A.P. and Olson, B.H., The interaction between geochemical and pollutant metal sources in the environment: implications for the community. In *Applied Geochemistry in the 1980s*, ed I. Thorton and R.J. Howarth, Graham and Trotman, London, 1986.

15. Davies, B.E., Trace element pollution. In *Applied Soil Trace Elements*, ed B.E. Davies, John Wiley, Chichester, 1990, pp 303-305.

16. Goodman, G.T., Airborne heavy metal pollution. In *Dealing with Dereliction*, eds R.D.F. Bromley and G. Humphrys, University College of Swansea, 1979.

17. JURUE, *Background levels of heavy metal soil contamination in Walsall*, Joint Unit for Research on the Urban Environment, University of Aston, Birmingham, 1982.

18. Culbard *et al., op. cit.*

19. Davies, B.E., Data handling and pattern recognition for metal contaminated soils. *Environmental Geochemistry and Health*, 1989, **11**, 137-143.

20. ICRCL, *Guidance on the assessment and redevelopment of contaminated land.* ICRCL Paper 59/83, Department of the Environment, London, Second Edition, 1987.

21. JURUE, *op. cit.*

22. Davies, B.E. and Paveley, C.F., Baseline survey of metals in Welsh Soils. In *Trace Substances in Environmental Health*, Proceedings of the University of Missouri's 19th Annual Conference on Trace Substances in Environmental Health, 1985, pp. 87-91.

23. Davies, B.E. and Paveley, C.F., Baseline trace metal survey of Welsh soils with special reference to lead. In *Geochemistry and Health*, ed. I. Thornton, Science Reviews Ltd, Northwood, 1988.

24. Hilton, *op cit.*

25. Coulson, M.G. and Bridges, E.M., The remote sensing of contaminated land. *International Journal of Remote Sensing*, 1984, **5**, 659-669.

A VIEW OF TWO NEW GOVERNMENT INITIATIVES, SCOTTISH ENTERPRISE AND CENTRAL SCOTLAND WOODLAND COMPANY

ALAN S COUPER
Lothian Regional Council, Department of Planning,
Landscape Development Unit, Peffer Place, Edinburgh EH16 4BB

ABSTRACT

Two new initiatives have been promoted by the Government aimed at involving the private sector in the renewal of the environment. Both bring an entirely different cultural approach to dealing with the rehabilitation of derelict land. Scottish Enterprise, the successor to the Scottish Development Agency, will combine the functions of the Agency with those of the Training Commission, in one organisation, which will operate locally as a series of private sector led local enterprise companies. From April 1991 they will have responsibility for environmental improvement, urban regeneration and land reclamation.

The Central Scotland Woodland Company is another private sector lead company which will have power to develop woodlands by acquiring, rehabilitating and tree planting derelict and unsightly land between Edinburgh and Glasgow.

INTRODUCTION

To understand the extent of the changes both initiatives bring about, and the bearing they have on Lothian Regional Council's derelict land rehabilitation programme, it is necessary to explain something of the background to them and of Lothian Region as an administrative area.

Scottish Development Agency Act 1975

Following the enactment of the Scottish Development Agency Act 1975 in November 1975 [1], the Scottish Development Agency was set up in Glasgow with the statutory function to develop Scotland's economy and to improve its environment. Under Sections 7 and 8 of the Act are powers for development and improvement of the environment and the treatment of derelict, neglected or unsightly land. Encompassed within Sections 7 and 8 is the power to use the local authority as an agent to undertake such works on the ground.

Regionalisation of the Scottish Development Agency: In 1988, in a move to bring the SDA closer to local communities, seven regional offices were established with the same functions and responsibilities as those held centrally, but with a core unit in Glasgow to direct policy, co-ordinate action and to manage the implementation of projects. The limit of responsibility given to the regional offices was set at £100,000, all major projects over this sum requiring central approval.

Lothian and Edinburgh Regional Office: In two years of operation locally, the SDA's annual expenditure has increased from £6.7m to £11.0m currently out of an overall SDA budget of £184m, demonstrating the direct benefit of regionalisation.

Lothian Region's Derelict Land Problem

Lothian Region covers an area of 1,723 sq km and contains a population of 741,179 people, 408,800 of which reside in the City of Edinburgh. Whilst it has the second largest population of the nine regional authorities in Scotland, it is also one of the smallest in land area.

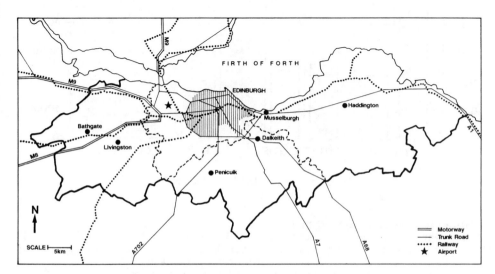

Figure 1.   Plan of Lothian Region.

In 1973 [2] it contained 2450 ha of derelict land, 15.6% of the national problem. From the national survey of 1988, [3], the amount had reduced to 2041 ha, but it represented 16.4% of the national problem.

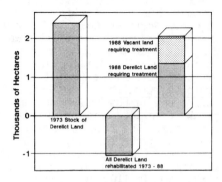

Figure 2.   Comparison of amount of derelict land.

The true impact of this in terms of intensity of derelict land, is that Lothian Region has the highest percentage of any region in Scotland by a considerable margin, a fact few would associate with a region who's capital city is world renowned.

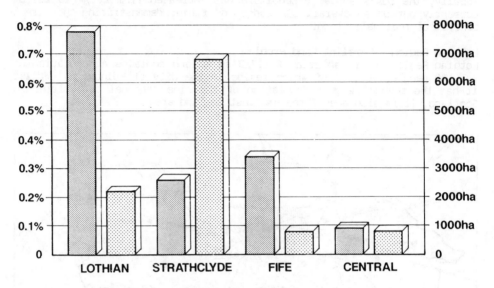

Figure 3.   Intensity of derelict land by region.

Lothian Region Structure Plan 1985
A 10 year programme to rehabilitate 1033 ha was initiated in 1976, [4]. The priorities for implementation, until 1985, were influenced by ease of acquisition, specific purposes or the need to abate a nuisance. It was considered that the programme could make a greater contribution if set within a strategic context. As it was clear that central government funding would not permit the completion of the rehabilitation programme within the lifetime of the Structure Plan, some selectivity was required to direct expenditure to areas where land rehabilitation would assist other social, economic and environmental initiatives.

Consequently, Structure Plan policy identified the Bathgate area, the Edinburgh Green Belt and the Esk Valley as priority areas, [5]. Subsequent decisions by the Regional Council to participate in various joint area initiatives have extended the rehabilitation programme into the remainder of the Region, such as the Leith Project, Better Gorgie/Dalry Campaign and the Granton Industrial Area in the city, and the Central Scotland Countryside Trust and the Dunbar Initiative in the landward area.

Results of a strategic approach: From 1976 to 1991, the Regional Council has operated as agent of the SDA under an agency agreement in rehabilitating 858 ha of derelict land at a cost of £16.98 million, excluding infastructure costs. The pace of activity has fluctuated with resources available, reflecting the priority the SDA gives to employment generating opportunities and to attracting new investors into Scotland. Had there been consistency of funding this effort could by the Lothian Regional Council have resulted in 1100 ha being rehabilitated, increasing the total for Lothian to over 1300 ha..

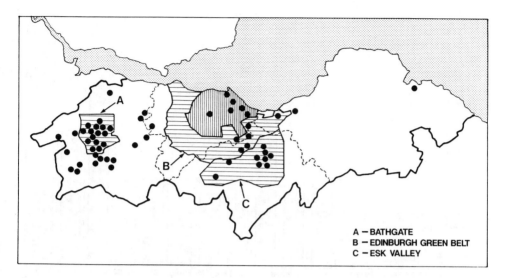

Figure 4.   Lothian Region Structure Plan priority areas.

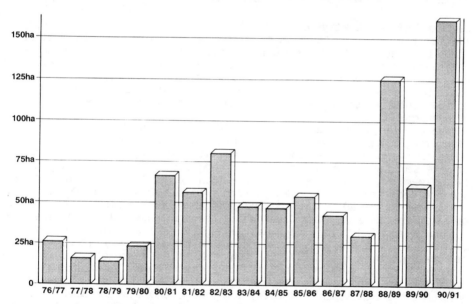

Figure 5.   Pace of activity on the ground.

A rolling programme approach: With the SDA's emphasis on flexibility of resources, there has been no commitment to a rolling programme for rehabilitation of derelict land. The support of joint area initiatives combining a wider spectrum of objectives than environmental improvements is prefered. To achieve a pace of activity on the ground, the Regional Council has had to generate its own rolling programme by obtaining approval for a basket of projects. In adopting this approval the Regional Council can both drive the programme and act as a catalyst.

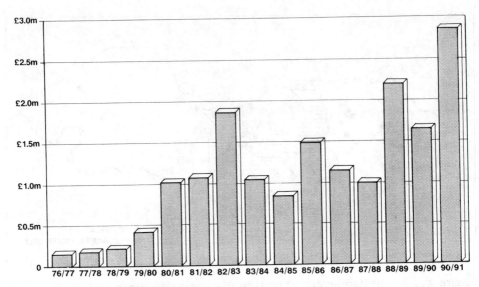

Figure 6.   Annual funding of programme.

Central Scotland Countryside Trust
The Central Scotland Countryside Trust is the result of an initiative first launched in 1978 by Lothian Regional Council to improve the degraded and desolate rural environmental of Central Scotland, [6]. It was born out of a desire to see the rehabilitation of smaller derelict and neglected sites that proliferated in the upland areas of West Lothian. Called Central Scotland Woodlands Project, it eventually led to a Trust being formed.

Since the Trust was formed in 1985 [7], it has been carrying out a wide ranging programme of tree planting, countryside and woodland management and wildlife conservation with the support of public and private sectors, together with local communities. It is run by a board of eleven members, four members from the Countryside Commission for Scotland and one each from Strathclyde, Lothian and Central Regional Councils and Monklands, Motherwell, Falkirk and West Lothian District Councils. It operates as two units, the Trust which is core funded by its members, and a wholly owned trading subsidiary, Central Scotland Woodland Services Limited, which allows commercial opportunities to be exploited, with all profits covenanted to the Trust. Only 6% of its income is provided as core funding, so much reliance is placed upon income from project work. Over 84% of income is derived from the completion of practical environmental improvement projects, a majority funded by the SDA.

Although income from the private sector and sponsorship can be expected to rise above the present 20% level, it is unlikely to exceed that of the public sector. The role of the local authorities and grant aiding bodies is crucial to its success.

Current Success: Over the past five years the average size of planting site has doubled from 0.6ha to 1.2ha with 20 of 1989/90's new woods being over 3 ha in size. The growth of fieldwork has led to a considerable rise in project income to the Trust and its subsidiary trading company. Earnings were £749,000 in 1989/90 compared with £530,000 for 1988/89. It is estimated that the total value of projects the Trust was involved with in 1989/90 is in excess of £1.5 million, [8].

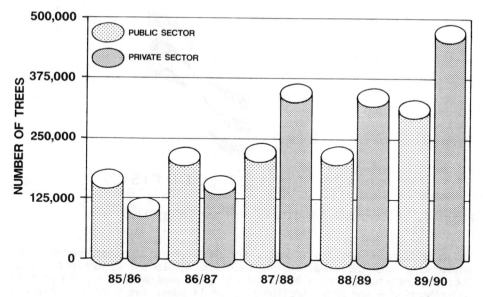

Figure 7.   Comparison of public and private sector tree planting 1985/90

Current Strategy: The Trust has a pivotal role in creating partnerships to provide the necessary resources to bring about the transformation of the environment of Central Scotland. This will possibly take a hundred years or more, so the involvement of local village and business communities is seen as essential to long term success. Its current project strategy, [8], concentrates on 10 key areas;
- planting and management of woodlands;
- landscaping along main transport corridors;
- provision of access to the countryside;
- creation of village nature parks;
- creation and protection of wildlife habitats;
- developing partnership with the local business community;
- employment creation through expansion of activity;
- employment training in countryside skills;
- encouragement of others to learn from the Trust's experience;
- participating and contributing to the wider debate on the Scottish environment.

Future Challenge: The government's launch of the Central Scotland Woodlands Company presents a major challenge to the Trust.

SCOTTISH ENTERPRISE

On 26 July 1989, the Secretary of State for Scotland made an announcement about Scottish Enterprise, following public consultation on a government White Paper, [9], titled "Scottish Enterprise : A New Approach to Training and Enterprise Creation". An invitation to participate, "Towards Scottish Enterprise", [10], was launched by The Industry Department for Scotland in August 1989.

In July 1990, the Enterprise and New Towns (Scotland) Act 1990 was enacted, [11], superceding the Scottish Development Agency Act 1975.

Figure 8.   Corporate identity of Scottish Enterprise.

Under Sections 6 and 7 are powers for development and improvement of the environment and the treatment of derelict, neglected or unsightly land. There is opportunity for a local authority to implement proposals approved under Section 6 and under Section 8 there is power for Scottish Enterprise to act through an agent.

The Role of Scottish Enterprise

Scottish Enterprise will be established on 1 April 1991 as a central body, based in Glasgow, with a network of local enterprise companies with the role of bringing together the training, enterprise and environmental activities currently undertaken by the Scottish Development Agency and the Training Agency.

It is the Government's intention that the Scottish Local Enterprise Companies will have a more comprehensive range of economic and environmental powers than the Training and Enterprise Council in England and Wales, but both networks will operate within a UK training policy framework.

The Local Enterprise Company Network and their Role: Within lowland Scotland, a network of 13 local enterprise companies has been created, that for Lothian covers the full geographical area of Lothian Region and will be called Lothian and Edinburgh Enterprise Limited (LEEL). A separate enterprise organisation will cover the Highlands and Islands area.

Each company will be an independent company, limited by guarantee. It will be run by boards of up to 12 individuals. At least two-thirds of these will be drawn from the private sector, with the remainder being drawn from 'the wider local community', which includes enterprise trusts, new town development corporations, colleges, trade unions, local authorities and voluntary organisations.

Local enterprise companies will be awarded contracts for designing, developing and delivering projects. These contracts will be based on agreed three year business plans and will include specific performance targets. They will be given as much discretion as possible consistent with adequate public accountability. Performance bonuses will also be available as budget enhancements.

Local enterprise companies will be focal points in their area for development of local enterprise and training; the "one door" approach. Each company will be responsible for stimulating the growth of enterprise in its locality. The ultimate objective is the creation of a dynamic

Figure 9.   Enterprise Company areas.

self-sustaining local economy in which investment and training are private sector led and financed.
   Funding: The total budget for Scottish Enterprise's first year of operation is £430m. Of this Scottish Enterprise will retain £69m for its strategic work, the balance of £361m going to the local enterprise companies who will have responsibility for budgets ranging from £3.5-£73.7 million a year, depending on area. They will have authority to approve projects up to £250,000 themselves, over that, Scottish Enterprise will have to approve any project.
   LEEL will have a budget of £50.2m for 1991/92 but at this point the precise figure available for environmental work is not known.
   Strategic Priorities: Scottish Enterprise will in the main undertake initiatives which promote the following strategic priorities, [12]:
1. Improved Human Resources
2. Technology

3. Competitiveness
4. Internationalisation
5. Access to Capital Resources
6. Product Development
7. Environment

Success will often require the integration of a number of these strategic priorities to achieve the required economic or environmental benefit.

Operating Principles and Style: Requests for public funding, whether through Local Enterprise Company business plans or for individual projects, will require to be supported by an appropriate appraisal and justification. The 'additionality' principle of projects that benefit Scotland as a whole is highly important, [12].

Scottish Enterprise seeks to be innovative and to strive at all times for excellence. Wherever possible, it will work under commercial disciplines and use resources as a catalyst to stimulate economic and environmental development.

Environment : Strategic Priority 7

Scotland has significant dereliction as well as some of the finest landscape in Europe. Improvement of the physical environment can be both an end in itself and a contributory factor to economic development. For both reasons it can justify public support. More widely, environmental issues will be increasingly important to people and economies throughout the world. The details of how it's strategy on the environment will be applied and implemented, [12], are;

Objectives:
- To enhance amenity by removing dereliction and enhancing and projecting Scotland's natural amenity.
- To improve the quality of life for the benefit of Scottish residents.
- To assist in the attraction of visitors and new businesses.
- To assist Scottish companies to adapt to new regulations (e.g. through the use of technology) and maintain competitiveness.
- To ensure the sustainable use and re-use of resources to improve Scotland's competitive position in the short, medium and long term.

Methods:
- Land renewal/environmental improvement.
- Integrated environmental and economic projects, e.g. Central Scotland Woodland.
- Enterprise development services, e.g. environmental market opportunity analysis.
- Marketing of Scotland's natural amenity.
- Design and development of specific projects, e.g. energy conservation and recycling by industry.
- Support through environmental area initiatives such as peripheral partnerships and enterprise zones.

Operating Procedures: With the development phase of the companies running in parallel with the dissolution of the SDA, no guidelines are available at present. It is expected that a similar monthly approvals process will be introduced for approving projects and their funding as exists with the SDA. As the majority of projects or the Regional Council's programme are over £250,000 this will involve approval from Scottish Enterprise in Glasgow.

CENTRAL SCOTLAND WOODLANDS COMPANY

In launching this new initiative on 20 January 1989, [13], The Secretary of State for Scotland said;

"Central Scotland Woodlands is an exciting project. Much of Central Scotland is less attractive than it might be because of the effects of urban and industrial development. There is enormous scope for improving the landscape through the creative planting of new mixed woodlands.

We all want to live and work and take our leisure time in an attractive environment. By planting a wide variety of trees in the area between Glasgow and Edinburgh, its amenity will be positively transformed.

We are planning for the future to improve our surroundings and to provide a worthwhile inheritance for future generations. I hope everyone associated with Central Scotland will support this major endeavour."

Purpose of Central Scotland Woodlands Company

The purpose of the Company, [13], which has the full backing of The Scottish Office, is to create an exciting new woodland environment stretching from the foothills of the Campsies to the foothills of the Pentlands, from the eastern fringes of Glasgow to the western fringes of Edinburgh, taking in Falkirk, Cumbernauld and Livingston and sweeping from the valley of the Forth to the valley of the Clyde.

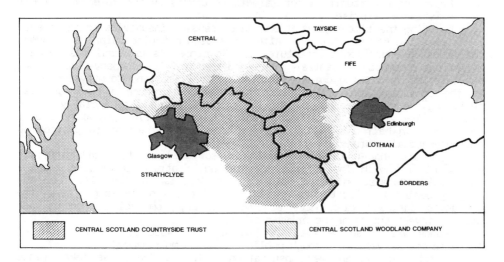

Figure 10. Area of Central Scotland Woodland Company in relation to Central Scotland Countryside Trust.

It aims to change the whole landscape and environment of the Central Belt of Scotland by concentrating of three woodland types:

Community Woodlands: "Community woodlands close to urban areas will be planned from the outset to provide access to an attractive rural haven of peace and quiet on the very doorstep of the local community."

Amenity Woodlands: "Amenity woodlands will be specifically designed to improve the landscape and environment not only near where people live and work, but also throughout the area on better quality land."

Productive Woodlands: "Productive woodlands, mainly coniferous, will be located generally on poorer land in the more remote parts. Careful landscaping and public access is important, but timber production will be the main purpose."

These three main types of woodland planting will be merged with each other in the best way possible to enhance the landscape and compliment other land use interests such as agriculture, nature conservation, recreation, housing and commercial developments.

The Benefits

The Company recognises that the foundations have already been laid by the pioneering work of the Central Scotland Countryside Trust. The new company will seek to make use of the Trust's considerable experience and expertise, especially in creating community woodlands. It sees a wide range of benefits, [13], including;
- Improving the landscape dramatically.
- Creating new wildlife habitats.
- Attracting more visitors to the area.
- Offering new opportunities for tourism, countryside education and recreation.
- Introducing ideal locations for major new leisure facilities close to three-quarters of Scotland's population.
- Opening up attractive locations for other new industrial investments.
- Providing opportunities for farmers to diversify and so enhance their existing enterprises.
- Locating the timber production and processing industries in the area.

Operation of the Company: Although subsequent discussion has changed the proposed operational structure, the Government's proposal was to have a parent company, Central Scotland Woodlands Company with two subsidiaries, a trading company and the Central Scotland Countryside Trust as the community link. Its first task was to produce a business plan outlining the steps needed to make the project become a financially viable reality. The Company has a board representing the main sponsoring bodies but more than half its number are from the private sector to provide an entrepreneurial approach.

Sponsors and Funding: The main sponsor will be The Scottish Office in association with The Forestry Commission, Scottish Enterprise and the Countryside Commission for Scotland. The Scottish Office will fund the Company during the first year of it's operation. In the longer term, the Company's business plan will aim to identify commercial sponsors and attract investment from the private sector.

Funds will also be available through the Woodlands grants and Set Aside scheme as well as through public sector "pump priming".

Timetable: Extensive consultation with the Trust has taken place over the preparation of the Company's business plan, which is currently before the Secretary of State and The Treasury for approval. This is anticipated in April/May, allowing the Company to start operating by Summer 1991.

STRUCTURE PLAN REVIEW 1990

Lothian Region has benefited from its co-ordinating strategic approach to the treatment of derelict land. It improves the appearance of an area, overcomes site development impediments, and contributes to the generation or regeneration of the local economy. The need for the continued effort to secure the rehabilitation of derelict land is not therefore an issue. The

Regional Council has submitted a five year programme to LEEL, [14], based upon the existing priorities for both rural and urban areas. The objectives of the programme are to realise economic development opportunities, remove contamination including treatment of contaminated ground water, and enhance image for attracting investment. However, issues arising from the present review of the Structure Plan, [15], are:

a will development issues change the priorities?
b what priority should be given to the identification and treatment of contaminated land?
c should the Regional Council continue with the methodical treatment of derelict sites to produce land which could be re-used or should it concentrate upon interim measures to speed up the overall improvement of the visual environment?
d should the Structure Plan encourage the export of waste tip materials from Lothian to speed up removal?

The development scenario emerging from the review discussion papers places more pressure on the urban edge and the Green Belt, which suggests the Structure Plan policies should concentrate on treatment priorities to the south and south east of Edinburgh and within the city.

Contamination is an impediment to the re-use of land and should be afforded a high priority for treatment in the derelict land rehabilitation programme. The Structure Plan could acknowledge that there are both known and unrecorded contaminated sites in Lothian; make it clear that these site be identified and registered; and indicate the Regional Council's involvement in the treatment of them.

There is pressure to reduce the impact of tips in the landscape, particularly the retorted oil shale tips of West Lothian, in order to overcome the adverse effect on potential inward investment. One response to the pressure could be "green up" as an interim measure improving appearance yet retaining the longer term resource potential. Nominated tips could be identified for removal while the remainder should be prioritised for interim treatment of "greening up", [16]. In this way the extraction of tips would make a significant contribution to the Regional Council's rehabilitation programme.

The Structure Plan relates to activities with Lothian Region and is not the appropriate document for the promotion of the use of oil shale or colliery tip material outwith Lothian. However, the Structure Plan could include a policy which would indicate support for the export of waste tip materials from Lothian.

THE WAY FORWARD

As soon as LEEL formed a company and was given authority to prepare its business plan by The Scottish Office and the SDA, a liaison group was established with the Regional Council. Through this group the Regional Council was able to make an input into the business plan, [17], by submitting a 5 year rehabilitation programme. Since May 1990 excellent progress has been made on the Regional Council's rolling programme which is a reflection on the successful partnership with the SDA, many senior staff of whom are key players in the embryonic Scottish Enterprise and its local company, LEEL, and the Council's ability to deliver. Of the twenty seven projects on the five year programme, seven have started on site, thus making the broadest possible start to the three key objectives of the programme.

In the enhancement of image objective the benefit of linking Central Scotland Woodlands into the programme was highlighted.

Emerging Factors
The important factors emerging that will help strengthen the means of achieving the strategic objective of reducing Lothian's derelict land problem are:

Partnership is central to success: Scottish Enterprise has made it clear that through partnership it would help build a strong more diverse and sustainable enconomy with a quality of life for all the people of Scotland. It's strategic priority is to enhance amenity by removing dereliction, thus improving the quality of life. That partnership is already establishing well with LEEL.

The Structure Plan must set the strategic approach: Under the planning regulations guiding the preparation of the Structure Plan, the Regional Council has a statutory duty to prepare policies for rehabilitation of derelict land in Lothian. LEEL must recognise the value of the Structure Plan in taking decisions on funding projects.

A more comprehensive policy approach to treatment is needed: Rehabilitation of derelict land is vital to Lothian's economic recovery. In it's review of Structure Plan policy, three issues central to the way derelict land is rehabilitated have been raised:
- contaminated land
- greening versus comprehensive rehabilitation
- export of waste

Widening the policy recognises a more comprehensive and co-ordinated approach to rehabilitation and the contribution this could make to the quality of life. Investment in new techniques to treat ground water contamination due past coal mining activity will also lead to a safer environment.

A clearer idea of the scale of resources is needed: Results from the 1990 national survey will be available soon, which will help determine the resources needed to treat the dereliction problem. Until the register of potentially contaminated sites is complete and some assessment made of treatment costs, a clear idea of the total resources needed in Lothian will not be available until early 1992.

A programme of advance land acquisitions is essential: The excellent progress to date has been due to working on a wide front and having a land bank of acquired sites. LEEL must be persuaded to support advanced funding of land acquisitions to guarantee a sustained pace of activity on the ground.

The integration of the Central Scotlands Woodlands Company's activities into programme: The emergence of the Central Scotland Woodlands Company gives the Regional Council an opportunity to widening the scope of the rehabilitation programme by incorporating their skills or by seeking their own funding in adjacent areas.

Agenda for Action
The emerging factors translate into a seven point plan of action for the Regional Council and LEEL:
1 To update the Regional Council's rolling programme by extending the programme period from 1995 to year 2000.
2 To redefine the precise targets to be achieved in the light of the emerging strategic planning policy.
3 To place a greater emphasis on treatment of contamination and greening of tips.
4 To agree to the continuity of the Regional Council's agency agreement for delivery of projects.
5 To agree a programme of advance land acquisition to guarantee

continuity of pace of activity.
6   To identify the opportunities for integrating the Central Scotland Woodland Company into the rolling programme.
7   To work on as broad and flexible a front as possible.

Conclusion

LEEL is about to embark on implementation of its business plan and aims to be a highly successful enterprise. It has a commitment to quality, impact and efficiency. It is an entirely new kind of economic and environmental development agency and represents a tremendous challenge for Lothian Region.

The acknowledgement by Government of a change for the better to the operational structure of the Central Scotland Woodlands Company, capitalises on the Trust's success and ensures a continuity of effort to deal with Lothian Region's strategic goal, the elimination of derelict land. Having more partners in the process is bound to achieve this objective.

It is a compliment to the Scottish system that the Secretary of State for the Environment is reported, [18], to be considering setting up an English Development Agency to provide momentum to attack urban decay because the SDA is felt to have done a good job in co-ordinating work in Scotland. As the Scottish system moves into a new gear and a new future with Scottish Enterprise, one hopes it will be even better.

REFERENCES

1. The Scottish Office, Scottish Development Agency Act 1975, HMSO, London, 1975, 1975 Chapter 69.

2. The Scottish Office, Derelict Land Survey 1973, HMSO, Edinburgh 1974.

3. The Scottish Office, Scottish Vacant Land Survey, 1988, Commentary, SDD Planning Services, Edinburgh, 1990.

4. Lothian Regional Council Department of Planning, Rehabilitation of Derelict Land 1976-1986, Lothian Regional Council, Edinburgh 1976.

5. Lothian Regional Council Department of Planning, Lothian Region Structure Plan 1985, Lothian Regional Council, Edinburgh 1986, pp 52-53.

6. Lothian Regional Council Department of Planning, Central Scotland Woodlands Project : Report of the Steering Committee, Lothian Regional Council, Edinburgh, 1978.

7. Lothian Regional Council Department of Planning, Central Scotland Woodlands Project, 1979/80 to 1984/85 : A Proposal for the Future - The Establishment of a Trust, Lothian Regional Council, Edinburgh, 1984.

8. Central Scotland Countryside Trust, Annual Report 1989/90, Central Scotland Countryside Trust, Shotts, 1990.

9. The Scottish Office, Scottish Enterprise : A New Approach to Training and Enterprise Creation, HMSO, Edinburgh, 1988.

10. The Industry Department of Scotland, <u>Towards Scottish Enterprise</u>, HMSO, Edinburgh, 1989.

11. The Scottish Office, <u>Enterprise and New Towns (Scotland) Act 1990</u>, HMSO, London, 1990, Chapter 35.

12. Scottish Enterprise, <u>Statement of Policy and Operating Principles</u>, Scottish Enterprise, Glasgow, 1990.

13. The Scottish Office, <u>Central Scotland Woodlands</u>, The Scottish Office, Edinburgh, 1988.

14. Lothian Regional Council Landscape Development Unit, A Five Year Programme to Reduce Lothian's Derelict Land, Lothian Regional Council, Edinburgh, 1990.

15. Lothian Regional Council Department of Planning, <u>Lothian 2005 : The Review of the Lothian Region Structure Plan - Discussion Papers</u>, Lothian Regional Council, Edinburgh, 1991, pp 1-7, 92-100.

16. A.S. Couper, A Planning and Reclamation Policy for the Retorted Oil Shale of West Lothian for the 21 Century, <u>in Reclamation, Treatment and Utilisation of Coal Mining Wastes</u>, ed. A.K.M. Rainbow, Balkema, Rotterdam, 1990, pp 89-101.

17. Lothian Regional Council, <u>Key Issues and Action Required for Lothian and Edinburgh Enterprise Limited</u>, Lothian Regional Council, 1990.

18. Morrison, R., Fresh Attack on Inner City Decay, <u>Construction Weekly</u>, London, 1991, Vol. 3, No. 8, p1.

ACKNOWLEDGEMENTS

This paper represents the views of the author which are not necessarily those of Lothian Regional Council. The author wishes to thank the Director of Planning for allowing publication of this paper and to Departmental colleagues Dr John Sheldon and Christopher Bushe for their helpful comments and assistance.

THE USE OF DERELICT LAND - THINKING THE UNTHINKABLE

DEREK HARTLEY
Teesside Development Corporation
Tees House, Middlesbrough, Cleveland TS2 1RE

ABSTRACT

The Department of the Environment review of Derelict Land Policy (September 1989) comments on local strategic planning for derelict land programmes. The benefits of the Urban Development Corporation approach which is mirrored in joint local authority rolling programmes are recommended.

The Paper looks at how the practical application of the principle in one UDC demonstrates its effectiveness. This is apparent not only in cost effective use of resources and linking infrastructure proposals across a number of derelict sites but by the adoption of end uses which would be unthought-of if considered on one site in isolation. Leisure, housing, nature conservation, commerce and industry are being developed on sites which require an integrated approach in order to function.

The Paper concludes that adequate resourcing and an adoption of the strategic approach in local areas is required if the full benefit of urban reclamation is to be achieved.

INTRODUCTION

An examination of successive derelict land surveys shows that the largest amounts of land reclamation have been achieved in rural areas. The 1988 survey suggested, however, that 60% of the total derelict land justifying treatment is in urban areas and 22% of the total is in inner cities. Present Government policies are biased towards hard end use although unit costs can be ten times as high as soft end use schemes and on average the latter could produce four times the output for the same expenditure. An evaluation is therefore required of the need for reclamation in any particular category.

In the Netherlands, Germany and Denmark for example the greatest priority is given to contaminated land, regardless of whether or not it is in current use or whether it is needed for development. The incidence of contamination is always greater in urban areas. Clearly, therefore, if derelict land funds are to be applied in inner cities on the basis of these policies, it is essential that the work is done in a cost effective manner and should achieve the impact required.

POLICY

In the Department of the Environment booklet "A Review of Derelict Land Policy" (1) the authors conclude in the section on Strategy that "Experience has shown however that the most effective use of resources can be achieved where reclamation schemes are not just considered as individual projects but are seen as part of a local strategy for reclamation and regeneration. This has been the basis of rolling programmes under DLG and the operation of UDC's in areas of concentrated or extensive dereliction."

It must be emphasised that the principle can be applied equally to local authority rolling programmes as well as to those of Urban Development Corporations. However, in this paper the examples given are taken from a UDC and this leads to conclusions specific to UDC's which do not necessarily apply to local authorities.

Practically there can be no question of a radical change in central policy and therefore derelict land reclamation will continue to be within the remit of four distinct sectors ie. UDC's, local authorities, the private sector and voluntary organisations. Accordingly the issue is how should we best utilise the different expertise within these four different regimes so that these agencies complement each other's efforts and add to the impact and effectiveness of the overall programme.

PRACTICE

In strategic planning terms Urban Development Corporations have a number of advantages over other agencies.

TABLE 1
Features of the Regime (2)

	UDCs	Urban Programme	City Grant	DLG
Geographical Coverage				
Targetted exclusively on inner cities	*	*	*	
Priorities for Reclamation				
Able to tackle reclamation/ dev. packages	*		*	
Solely for reclamation				*
Reclamation strategies for areas of concentrated dereliction	*			*
Agents for reclamation				
Provides direct assistance to private sector	*		*	*
Available to support speculative reclamation by public sector	*			*
Types of scheme				
Tends to concentrate on large projects			*	
Tends to concentrate on small projects		*		
Available for projects leading to "soft" end use	*	*		*
Available to support projects leading to public development	*	*		*
Government support				
Assessment of support on basis of financial shortfall	*		*	

For example urban programmes cannot provide government assistance to the private sector, conversely the city grant is not a vehicle for general reclamation. Whilst derelict land grant can fulfill many objectives it is not, under current policy, able to tackle development packages nor provide government support for financial shortfalls. There are, of course, "hybrid" examples developing such as in the Birmingham Heartlands and Salford Quays where local authorities in conjunction with the private sector are blending derelict land grant with other forms of support to provide the complete package necessary for the comprehensive redevelopment of rundown areas.

Criticism of the urban development corporation approach from, for example, the House of Commons Employment Committee and the Centre for Local Economic Strategies has, whilst recognising the success of physical renewal, concentrated on the absence of policies and projects dealing with the deeper seated social and economic problems of inner cities. These are perceived to relate to local employment, housing and the regional economy. Specific examples are quoted in the CLES Report of potential conflicts of interest in respect of support for existing industry and lack of attention to social housing needs, the worst areas of which are often outside the tightly drawn UDC boundaries. Examples in Teesside demonstrate that this need not always be the case.

The proposition may be argued therefore that any failing in the "strategic approach" philosophy is to define the limit too narrowly. Application of the working methods over a wider area, with appropriate funding would give the flexibility to achieve both physical renewal and perhaps a more substantial impact on social deprivation. Practitioners in the field will be all too aware of the obvious edge of reclamation or redevelopment schemes where "marginal" industries or poor quality housing are left in stark contrast to the new or renovated land and property adjacent to it.

OPERATIONS

The actual operational methods within a strategically planned area are complex but come under four main headings:

1. Infrastructure, 2. Materials, 3. Relocations, 4. Impact and after use

Taking Teesside as a specific example, these plans are refined by consideration of eight sub areas within which there is an obvious ease of movement and integration of objectives, but without excluding wider opportunities for linking with other sub areas. The philosophy of the TDC, individually, is very much market led which allows reasonable flexibility in dealing with the private sector opportunities whilst working within a broad strategic framework defined by the eight sub areas.

Before proceeding to the main examples it is interesting to consider the position of Area 1, Hartlepool, where its isolated location from the rest of the TDC and its physical boundaries, constrained by the sea and the east coast railway have precluded solutions of the sort achieved elsewhere. Relocations have all been difficult to achieve with limited alternative land available and infrastructure improvements have been mainly limited to the immediate local needs of the derelict area.

Infrastructure

This is best exemplified by the road requirements for Area 4 which also impinge on Areas 5 and 6. The scale of development at the Teesdale and former Stockton Racecourse sites required a new diamond interchange on the A66. In addition, the link from here into Teesdale had to be supplemented by one improved and two additional accesses.

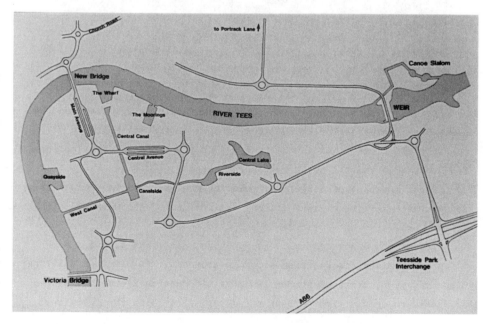

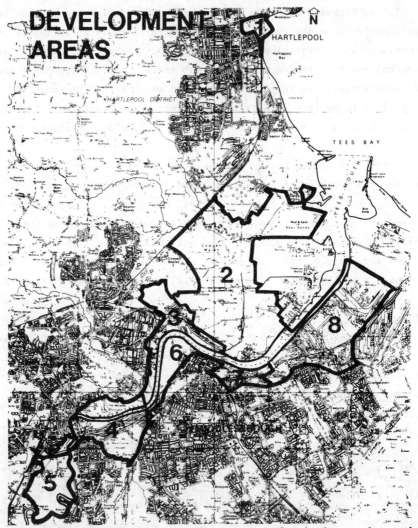

DEVELOPMENT AREAS

A further consequence is the improvement of existing roads on the north bank, which deal with the local growth in traffic but also complete the final links in the County Council planned highway network from the south and the north west/north east.

Other examples of beneficial road schemes are seen in Areas 5, 6 and 8 where existing highway schemes are integrated and enhanced with access roads serving developments and also acting as local distributors.

The second major component is sewerage. In Areas 3, 4 and 5 integrated extensions are planned or in progress. At Haverton Hill/the Clarences major areas of former industry and housing are under-sewered or unable to comply with current standards. Provision for individual sites would be expensive and inefficient and only substantial new mains and pumping can solve the problems. In Area 4 improvements to defective or substandard sewers are being carried out through a comprehensive masterplan which will, in the same process, create an acceptable water quality in the upstream river above the proposed Tees Barrage, suitable for the planned leisure and commercial use.

A third element concerns public utilities whose forward planning of major primary supplies can be more easily coordinated when the potential loads from major derelict area redevelopments can be assessed in one operation.

The last infrastructure example is the proposed LRT system for Cleveland, being promoted by the County Council after a feasibility study jointly funded with the Development Corporation. Routing to collect potential passenger growth areas is affected by redevelopment proposals and in specific locations corridors are being reserved to accommodate the track and stations.

Materials

Any urban area of dereliction is likely to contain contaminants and a major problem in reclamation is the disposal of these materials and the provision of a suitable replacement fill. Sites such as Teesdale in Area 4 which contains a mixture of organic pollutants, heavy metals and asbestos, from a range of previous industrial uses, present a design and development problem where containment on site will entail a sacrifice of development land but transport off site creates potential conflict between the developer and environmental health officers or the waste disposal licencing authority. This can be caused by the excavation, transport or disposal phases.

The traditional reclamation principle of balanced earthworks within a site is followed wherever possible, and has resulted at Teesdale in the use of landscape mounds to screen the new residential and commercial areas from the adjacent railway sidings which will remain in use. There remains a substantial surplus to be removed from site and, following rejection of a first site in Area 5, after public opposition, two alternative locations are under consideration both of which are Development Corporation sites requiring infill for landscape/screening. One is in Area 2 and the second in Area 4. In a similar vein, the old refuse tip in Area 4 adjacent to the retail development on the former

Stockton Racecourse has been removed to Area 3 as the core of a peripheral landscape treatment on a new industrial zone.

In Area 2 unsuitable expansive slags and chalk/gypsum wastes on an industrial site are designated to go into a fire bund at a nearby chemical plant.

When it comes to replacement fill, use is obviously made of in-situ material as much as possible. On Teesdale the best modern analysis of potentially difficult slag was carried out, allowing eventually some 400,000 cu. m. from old slag banks to be crushed and replaced in the voids created by the removal of contaminants. Old foundations were crushed and reused and supplemented by the same material from the former Whessoe site in Area 5. Surplus boulder clay from a greenfield site in Area 5, which was zoned for industry, was also imported to Teesdale. All this resulted in minimum use of crushed rock which had to be transported from quarries in Durham.

The final opportunity for materials interchange occurs with soils, where greenfield developments such as the proposed International Nature Reserve provide an opportunity to specify supply sources without recourse to expensive purchase.

Relocations

A centre of concern in the strategic approach to redevelopment of an area is the problem of often isolated small businesses whose continued presence would inhibit full comprehensive reclamation. Contrary to the belief sometimes expressed by opponents of the Development Corporation approach that these businesses are insensitively handled, it can in fact be the scope and flexibility on offer which can solve the problem. The concurrent or programmed reclamation of adjacent sites can offer plots for re-establishment as an alternative to compensation. There are many examples in Teesside where this has occurred. Business relocations have been arranged from Area 4 to Area 6, Area 5 to Area 3 and in Greater Teesdale, Area 4, but from the core of the Teesdale site to the north bank.

Similar exercises have been carried out for gypsy sites, allotments and a skinnery site on the river at Yarm where the Barrage requirements for water quality could require extensive treatment or amelioration works to deal with process effluents.

Impact

Perhaps the most striking aspect of strategic planning over a wide area is a change in circumstances which can be created, allowing the consideration of after uses which would otherwise be unthinkable. A combination of the Tees Barrage construction and the massive reclamation of Teesdale has opened up opportunities for housing, commerce and leisure on a area of previous derelict industry. The higher retained water level not only provides an enhanced development potential, as proved in similar schemes elsewhere, but also and perhaps more importantly in the long term allows leisure facilities based on water sports to be established. These include an artificial white water canoe slalom course, a more dependable rowing course and conditions for sailing, wind surfing and water skiing. Other areas adjacent to the upstream river which have traditionally turned their back on a despoiled stretch of water, will be open to a wide range of uses whereas at present they are limited to small and rundown industrial premises. Areas on the north bank, previously reclaimed for industry are now open to other options. In general there is a realisation that better zoning of uses can be achieved by strategic relocation and reclamation, leading to innovative development and a better environment.

One of the most despoiled parts of Teesside lay in Area 3, North Tees, where abandoned chemical sites, a derelict shipyard and a backcloth of heavy industry, including ICI and offshore construction yards, created an uninspiring atmosphere for the only block of existing housing in the UDC area. Given the planned reclamation of the derelict sites, the infrastructure improvements which have already been mentioned, environmental improvements to the remaining works and the proposed International Nature Reserve to the north the situation can be viewed very differently. The Clarences housing area has been transformed with the aid of city grant to affordable good quality housing and new build extensions are in progress.

Equally interesting is the south bank where again heavy industry and chemical plants were intermingled with derelict sites and with inadequate communications. The County Council's South Bank extension of Middlesbrough By-Pass has opened up access to the area and this is now being complemented by other infrastructure improvements and reclamation. Already the Tees Offshore Base for sub-sea technology has transformed the former Smith's Dock, and the Integral Corporation of America is expanding rapidly in the former Bay 16 building of the Teesside Engineering Works. Land values and demand have risen so much that the local Borough Council is now advocating more extensive reclamation to create further industrial land on difficult areas previously thought suitable only for landscaping. A final feather in the cap may be the creation of a multi-use motorsports park providing a new form of entertainment and excitement in a previously drab area.

CONCLUSIONS

Urban reclamation is difficult and expensive but has an enormous impact on the lives of many people. It is therefore a high priority and must be done in an efficient manner. Large scale strategic planning of an area and a rolling programme of integrated schemes can produce not only economies in joint infrastructure but release land use options of a greater variety and value than otherwise would be the case. It is a practice which should be confirmed and enhanced in all major areas of urban dereliction by appropriate use of the various agencies available.

Acknowledgements

The opinions expressed in the paper are the author's and do not necessarily reflect the view of Teesside Development Corporation. The author wishes to acknowledge the assistance given to him by the Department of the Environment and other members of staff at the Development Corporation.

REFERENCES

1. Department of the Environment., A Review of Derelict Land Policy. 1989. pp 42

2. Department of the Environment., A Review of Derelict Land Policy. 1989. pp 32

PLANNING FOR RECLAMATION

D M HOBSON B.Tech C.Eng MICE MIHT
W A Fairhurst & Partners, Environment Division
1 Arngrove Court, Barrack Road, Newcastle upon Tyne NE4 6DB

ABSTRACT

Land reclamation is increasingly becoming an integral part of major infrastructure projects which lead to the redevelopment of land.

The implementation process for such projects must be carefully planned to allow the various inter-related elements of the project to be addressed at the appropriate times whilst recognising the need to minimise early costs which may be at risk due to planning and other considerations.

The process is examined, step-by-step, with reference to real projects and the effects of conflicting interests highlighted.

INTRODUCTION

In the Government's White Paper on the Environment [1] the policy of re-using derelict land in preference to green field sites was clearly emphasised and most local planning policy decisions are also guided by this principle. Approval of new development on previously undisturbed land is therefore becoming less and less common.

The decline of older manufacturing industry and the changing balance of economic activity from heavy manufacturing to lighter "high-tech" uses, commerce, the service sector and housing has led to an increase in pressure for the redevelopment of disused industrial land.

Many new developments, therefore now involve the reclamation of land as part and parcel of site preparation. This is often treated as part of the infrastructure works which are undertaken as a precursor to built development. The research and design of the treatment of such land must therefore be carried out in the context of the whole development project.

FINANCIAL CONSIDERATIONS

Funding of development projects, including the treatment of land, may be from a variety of sources including Government Grant and capital investment, but almost without exception the value of the land created is of fundamental importance and often is the sole means of finance. Government Grant rules are formulated so that they can recover costs from the enhancement of land value [2] and therefore even where these are available land values will have a controlling effect.

The moves towards the principle of "the Polluter Pays" introduced by the EC Directive on Civil Liability for Damage Caused by Waste [3] and accepted as Government policy in their response to the Select Committee on the Environment [4] will increasingly create situations where funding must be made available irrespective of the value of land after treatment. Provisions are also contained within the Environmental Protection Act [5] for Local Authorities to take steps to enforce clean up of land. Nevertheless those responsible for implementing this work will need to recover whatever costs they can from the value added to the land by its treatment.

The consequence of the above is that, except in unusual circumstances, finance to pay for reclamation will not become available until the work has been carried out. There will therefore be a need for interim funding and the time between expenditure and realisation of assets will have a major impact on the viability of the project. The cost of such funding will be dependent upon its source, the guarantees which can be provided, the length of time for which the finance is required and the general level of interest rates. At the time of writing an indicative cost for one month's finance charges on £100,000 might be about £1,200.

The phasing of projects to minimise early costs will always be essential, however, this must be balanced with the need to avoid additional or abortive costs resulting from inadequate preparation, inefficient design or unrealistic contract periods. The balance of this compromise may be radically altered by after uses which generate high time based returns (eg retail uses).

THE FIRST STEPS

At the earliest stage the project will be identified. This may occur due to:

> An opportunity arises as a result of a specific requirement for land or its availability.

> A change in planning policy may create possibilities not previously existing.

> Land falls into disuse and there is a need to realise its value.

> An obligation to treat land (eg action under the Environmental Protection Act to prevent damage to the environment).

In most cases a specific use will be identified due to non-physical considerations such as planning policy, market forces, operational needs. Occasionally, however, opportunities may arise where no preconceived view has been formulated. In any event the first step should involve an initial feasibility study to identify the opportunities and constraints presented by the site.

An initial feasibility study should be a paper exercise based solely on information already available and involving relatively small costs. At this stage any expenditure will be at risk as there will be little certainty that any project is viable. these costs will also be outstanding for the longest period before they are recovered. Shrewd judgement is therefore required about the level of this work.

The study will need to consider:

>   Accessibility
>   Land use and market value
>   Land ownership and wayleaves
>   Legal issues
>   Site history
>   Topography
>   Geology and known ground conditions
>   Contamination Potential
>   Existing infrastructure including availability of services
>   Local environment
>   Costs

Ideally this should involve professionals from the appropriate disciplines such as planning, transportation, valuation, legal, engineering, landscape, environment, and quantity surveying. Local Authorities and other public institutions may well have all these available in-house. In the private sector it will often be necessary to assemble a small team of experts. A number of consultancies now offer a full range of professional services, but often a number of practices are brought together to work as a team.

The initial study should clarify the options which merit further consideration and will eliminate ideas which are not feasible. This important early step will save time and costs later on by avoiding more detailed work which is simply not feasible.

PROJECT FORMULATION

Armed with the results of an initial study it will then be possible to proceed with some confidence to evaluate a number of alternatives. This work will involve a number of sub-stages each of which will be pre-planned and costs estimated to establish the financial commitment necessary. At the start of this work the first cost plan should be prepared. This should identify **in general terms** overall costs, programme and phasing. It should consider the means of funding and set targets for achievement of returns.

Detailed evaluations should be avoided as they will result in unnecessary detailed preliminary design increasing costs and causing delay. They are also most unlikely to be any more realistic than a more general approach based on yardstick costs and experience.

The main objective of this stage is to provide sufficient information to confirm that a viable project probably exists and to progressively narrow down alternatives to produce a single set of outline proposals which can be worked up in more detail.

As in the initial stage a multi-disciplinary approach is essential.

The attitude of the various controlling authorities should be sounded out and wherever possible an open approach will be beneficial. Involvement at an early stage of authorities from whom approval will eventually be sought will enable their views to be fully understood and promote an atmosphere of trust. This will usually be rewarded by a genuine desire by all involved to see the project succeed and avoid the understandable tendency to look for problems and technical difficulties rather than their solutions. Of course, in the real world this will not always be possible, particularly where existing policies or established interests are challenged.

In one recent project an entire reclamation strategy together with decontamination standards were jointly prepared by a developer with local authority planning and environmental health officers, county waste disposal officers, the National Rivers Authority and other interested parties. Where disagreements did occur, these were openly discussed and solutions found to the satisfaction of all concerned.

During the initial stage of the project a picture of the site conditions will have been produced from data available on its history, topography, geology etc. It will almost always be necessary to confirm and refine this picture with additional investigative work. This will normally be of a preliminary nature and will be a compromise between the need for information and the need to minimise costs which will still be at risk at this stage and also will be incurred well in advance of returns. However, it will be essential that sufficient information is obtained to provide a sound basis for the decision making that will be necessary. For this reason site investigation and survey work is often undertaken in stages so that the relevant level of detail is available at each stage of the project without incurring excessive costs sooner than necessary.

During this stage a number of fundamental issues will need to be settled.

Land uses will be selected for compatibility with planning and physical requirements, the intended uses and also as a means of providing an acceptable return on investment.

The level of treatment necessary will be determined as a result of a complex relationship between planning, physical, environmental and financial issues. Where land is contaminated a decision will have to be made on the objectives of the treatment. Generally there are two main issues:

1. Suitability of the land for its intended use.

2. Treatment of the land to reduce its potential to harm the environment.

The first issue is normally a simple matter of separation of the user and hazard. Rational methods for the design of barriers are available [6] and a number of methods of treatment may be selected to achieve this end.

The second issue may be more difficult and is primarily concerned with mobile contamination. The Environmental Protection Act [5] contains provisions requiring treatment of land containing "noxious liquids or gases" which may cause harm to the environment. This requires an assessment of the integrity of containment or disposal systems and the implication of failure should it occur following development of the site.

An additional consideration will be the effect on the value of the land if contamination remains insitu. Local Authorities will be required to produce registers of contaminated land which will list sites even if they have been properly treated to separate the users from hazards and safely contain any contamination.

The density of development proposed will be affected by planning issues, accessibility and cost of infrastructure, but will also determine the end value of the project. For sites with especially difficult problems the cost of treatment may sometimes justify higher densities than would otherwise be acceptable. Architectural advice may well be needed at this stage if not already included in the team.

The extent of land required for a successful project must also be determined. This may well include areas under different ownerships and its availability and likely cost will have to be assessed. This will inevitably involve negotiations with third parties as will wayleaves for access, new services, drains etc. It will be essential to have a firm understanding of other project costs before undertaking such negotiations.

Finally a realistic and achievable programme must be selected against which an overall cost plan can be calculated. This must identify the timing of all costs that will be incurred including fees, surveys, site investigations and advance payments that might be needed for items such as service diversions etc. The financing of all costs must be identified in the cost plan so that they can be set against the anticipated returns. This problem is sometimes presented by discounting the value of income from the anticipated date of receipt to the base date for the estimate. This enables all estimated costs to be expressed as an amount discounted to the same base. This technique is used by the Department of Transport in its Cost Benefit Analysis of Trunk Road Schemes [7]

PLANNING APPLICATION

Once the fundamental decisions in formulation of the project have been taken and a sound programme/cost established the main business of developing the design and associated planning to produce a well supported planning application can be commenced. It should be remembered that until a planning consent is obtained together with other consents such as waste management licences, outfall consents and agreements on wayleaves, land deals etc there is still a risk that the project may not be capable of implementation. The main thrust of work should therefore be the preparation of details for submission of a planning application.

Very often a detailed approval is sought for the site preparation and infrastructure works whilst only an outline application to establish land uses for the proposed development is pursued. In order to minimise the design costs which are still at risk it is customary to leave much of the detail for resolution as reserved matters and planning authorities usually understand the necessity for this. However, sufficient detail must be produced to properly define the proposals and discussions with the planning and other authorities will give a good indication of the extent of their support or otherwise for the project.

Most major schemes will require an environmental assessment [8] in support of the planning application. The earlier environmental work will have ensured that environmental impact was taken into consideration in the formulation of the project and detailed work at this stage will be used to demonstrate that the impact of project on the environment are acceptable and that suitable mitigation measures can be taken.

As negotiations and discussions with interested parties progress during the preparation of the planning application and during the period when it is being considered, the degree of confidence that approval will be forthcoming will become clearer. If signs are positive then the opportunity should be taken to progress matters which might otherwise delay the project following its approval. This would include detailed investigations, surveys and studies necessary for completion of the detail design. Assessment of water quality and ecology work for example, frequently requires a full year to complete. As more information is acquired it will be possible to refine the cost plan in a rational manner consistent with the stage of the scheme and the detail available. This should allow the lead in time to commencement of the works following receipt of planning permission to be minimised and the financial viability of the project to be properly monitored and controlled.

CONCLUSIONS

The reclamation of derelict and contaminated land can be effectively achieved and financed as part of the site preparation and infrastructure provision for new development. Its implementation requires careful planning and programming and involves a compromise between the need to incur expenditure and acceptance of risk that the project cannot be sustained. The process requires a range of professional disciplines which must be properly co-ordinated and managed.

REFERENCES

1. This Common Inheritance, Britain's Environmental Strategy. HMSO, London, 1990.

2. D M Hobson and M T Foggon, Realising the Assets of Derelict Land - Self Financing Reclamation, Land Rec 88 Conference Papers, Durham County Council, 1988.

3. Proposal for a Council Directive on Civil Liability for Damage Caused by Waste and Certain Industrial Activities COM (89) 282 final - SYN 217, Commission of the European Communities, Brussels, 1989.

4. Department of the Environment. The Government's Response to the First Report for the House of Commons Select Committee on the Environment; Contaminated Land HMSO London, 1990.

5. The Environmental Protection Act 1990, HMSO London, 1990.

6. Dr I Anders, Clean Cover Reclamation - The Need for a Rational Design Methodology, Land Reclamation Conference, Cardiff, Elsevier Science Publishers Ltd, 1991.

7. COBA 9 Manual 1981 (as amended 1991), Department of Transport, London.

8. Town & Country Planning (Assessment of Environmental Effects) Regulations, 1988, HMSO London.

TREATING DERELICTION, THE LESSONS LEARNED

GERAINT DOUBLEDAY
RPS CLOUSTON
St Cuthberts House, Framwellgate Peth, Durham DH1 5SU

THE INITIAL STRATEGY

In the twenty five years since land reclamation began on a large scale in the UK, there has been ample time and accumulated wisdom to compile a number of guidelines. The first such rule is that successful reclamation demands flair, imagination, hardwork, utter professionalism and remains a minefield for the inexperienced. Each derelict land site is unique and usually needs treatment planned by a multidisciplinary team with skills in surveying, civil engineering, landscape design, environmental science and land management. If there is to be future hard development, involvement of the architect in even the early stages of planning reclamation has considerable merit.

Most sites contain problems, for example contamination or heavy foundations, but also at least some of the necessary resources for treatment. In larger schemes it may seem expedient to compartmentalise the area and treat each section separately. While implementation may have to be phased, planning must be integrated if all the resources are to be deployed effectively, and the best landscape solution is to be achieved.

No reclamation should be planned until there has been a full investigation to identify the location, nature and quantity of physical and chemical problems, services and resources on site. Surveys should also extend to the surrounding environment to assess the landscape context as well as social and ecological factors.

At this point, if the object is soft landscaping, a choice of strategy must be made. Where land values are low and there are no hazards on site, the landform may be left as it is. Revegetation may proceed by natural recolonisation or by the use of varieties of plants carefully selected for their tolerance to the specific conditions on site. Some Sites of Special Scientific Interest have been created this way, as well as much amenity woodland now of considerable landscape impact. For a variety of reasons most derelict sites need regrading and a range of landuses is demanded. Here, the type of vegetation to be established is dictated by the landuse proposed and the strategy must be to create a land surface capable of sustaining the chosen vegetation.

SCHEME DESIGN

The choice of afteruse is crucial both to the design of the reclamation scheme and to its future viability. Local demand must be ascertained and interests in peripheral land should be acquired to give any public and private access needed, and access to any resources necessary to carry out the scheme. Land uses should be restricted to what is feasible. For example there is no virtue in creating football pitches, however green the turf may look in summer, if in winter because the ground surface drains so poorly, the pitches become unplayable. Agricultural areas should not be created inless they have a cover of soil at least 200mm thick. Anything less and the fields will probably add to the growing UK surplus of unprofitable farmland.

Scheme design will be constrained by the capital budget available, but it should also have regard to the expertise and commitment to the aftermanagement which all reclaimed land needs. Intensive afteruses may have to be toned down if post reclamation site management is likely to be minimal. Alternatively, positive and constructive aftercare can make it unnecessary to achieve everything to perfection in the reclamation phase.

Land uses on reclaimed sites must change and adapt periodically. Many reclamation schemes should be designed with secondary as well as primary land use objectives, and thus provide flexibility.

IMPLEMENTATION ON SITE

Many good scheme designs have been spoiled by poor implementation on site. There is the same need for care and professionalism in supervision of works as in their design. Some contractors will take short cuts if not supervised, but equally most sites spring surprises during the contract works and it is in everyone's interest that amending decisions are made quickly, and given as instructions without delay. Whether for the purposes of contract management or for the benefit of the land owner, all works should be documented. The need to know of the existence of impenetrable foundations or toxic materials just below the formation surface is obvious. Making such information available, for what it is worth, must be a responsible policy and does not reduce a purchaser's responsibilities under caveat emptor. In any case such sites will probably have to be listed on the local contaminated land register required by the Environmental Protection Act (1990).

MANAGEMENT OF RECLAIMED SITES

All too often the high point for a site is the reclamation phase, from there on it deteriorates until it finds its equilibrium sometimes in a state not much better than the original dereliction. Reclamation should be only the starting point from which the site progresses to full development, ecological diversity and landscape maturity. Management of the site post restoration is the key and this requires imagination, professionalism and financial commitment, rather like the qualities that every stage needs!

THE RECLAMATION OF HARD ROCK QUARRIES: POLICY AND PLANNING IMPLICATIONS

BRIDGET SNAITH AND PETER GAGEN
Camlin Lonsdale, Landscape Architects
2 - 4 Oxford Road, Manchester M1 5QA

A quarry is a landscape archetype, a cultural mark that expresses the close relationship between man and the land he inhabits.

An increasing demand for stone and advances in extraction technology have led to an expansion in the scale of quarrying. There has been an adverse reaction, from both the public and mineral planners, to the impact of modern quarrying techniques on the existing landscape. Mutual benefits have become obscured to a point where conflict now exists between producers and consumers, and the political bodies who act as their intermediaries. The issue under contention is not whether quarrying should exist, but how quarries should interact with the landscapes of which they are part.

As land available for excavation becomes limited, due to refusal of planning permission on environmental grounds, so existing sites must be systematically and economically stripped of all their resources if prices are to remain competitive. This results in the permanent scarring of the landscape so abhorrent to public and planners alike. Simultaneously, operators are encouraged to view not only the planning system, but the landscape itself as a barrier to the development demanded by the marketplace.

The promotion of screening and greening as a planning strategy is ill conceived. As an attempt to maintain the myth of the rural idyll by concealing production processes, a green wall is built up between producer and consumer. The interface between production and consumption should be one of honest interaction if misunderstanding is to be avoided.

A clear strategy is needed which must combine an understanding of landscape processes and meanings, both scientific and cultural, with a thorough knowledge of the quarrying industry. Clear criteria and objectives must be determined and enacted, to counteract the present discrepancy between the formulation of national policy and the interpretation of local planning objectives. A prescriptive approach can be immediately discounted as operations differ in size, type and location. Future solutions must respond by mapping and modelling the extraction process within its parent landscape. Regionally based bodies such as R.A.W.P.s, with an extended remit to include not only mineral planners and operators, but landscape architects and landscape scientists, skilled in reading the cultural and physical make up of our environment, would be ideally suited to formulation of such strategies.

The creation of varied, stimulating and profitable quarried landscapes becomes possible through change in the unsympathetic methods of quarrying encouraged by present planning policy and by the realisation that the quarrying process is itself the best regeneration method.

The role of the operator as skilled craftsman harvesting the land, while acting as its custodian, within the framework of a clearly defined landscape strategy, known to the local population and responsive to its needs, will resolve the present conflict.

CONSIDERATIONS IN THE REDEVELOPMENT OF DERELICT LAND: EXPERIENCES OF AN URBAN DEVELOPMENT CORPORATION

J.S.Gahir [1], C.Dutton [2] & H.L.M.Jones [3]

(1) Gahir Consulting Services, 105 The Manor Drive, Worcester Park, Surrey KT4 7LN
(2) Black Country Development Corporation, Rounds Green Road, Oldbury, West Midlands B69 2DG
(3) Pinsent & Co., 26 Colmore Circus, Birmingham B4 6BH

The Black Country Development Corporation was established in May 1987 with a remit of regenerating its 10 square mile Urban Development Area (UDA) at the industrial core of the Black Country, of which derelict land comprises some 1500 acres. Exploitation of natural minerals over several centuries and more recently, chemical and heavy industries (the latter in the form of iron works and metal workings) and public utilities (in the form of sewage and gas works) occupied a large proportion of the area. The resultant decline of heavy industry led to widescale dereliction. Chemical industries which grew up in the area not only contaminated their operational sites but also led to the widespread disposal of waste products in adjacent areas. In addition the digging of large marl pits to supply the local brick industry left a legacy of deep holes which were used for dumping a wide range of materials.

In discharging its duties in bringing forward development proposals for the derelict sites for a residential, commercial or an industrial end user, the Corporation has had to carefully balance its regenerative role with its development control responsibilities; the Corporation being the local planning authority for its area. Because of its regenerative aims, the Corporation has been actively involved in finding appropriate engineering solutions to enable the safe and responsible grant of planning permission, rather than reacting to apparently adverse ground conditions by a simple refusal.

Regeneration of derelict sites within the UDA is generally development led. Developers expect a quick and trouble free planning permission from an urban development corporation. Instead, because the contamination problems are so extensive within the UDA they can face demands which are more onerous and on occasions time consuming, than they might have experienced elsewhere. In assessing planning applications for development on contaminated sites a detailed knowledge of ground conditions is necessary, so that the principles of a reclamation strategy can be addressed prior to grant of planning permission. However, the cost and time involved in undertaking a comprehensive site investigation before grant of planning permission is a risk which can deter developer interest. To improve the situation, the Corporation has had to move from a reactive stance of requiring ground conditions information along with the planning application to a pro-active stance of assisting developers through a pre-planning process. This has to date involved assisting the developer through joint discussions by alerting him to known data on ground conditions or if such information is lacking providing grant assistance for site investigation works in appropriate cases.

The need for a site investigation conducted in general compliance with ICRCL Guidance notes and British Standards: DD175, is impressed on the developer to define and quantify the contamination problems relating to the site. The different problems encountered are then overcome by devising appropriate engineering solutions on a site by site basis, with due respect to cost and emphasis on safety. Generally the lowest reclamation costs are for industrial or commercial end uses, where substantial areas will be hard standing. The reclamation in this case may include reinstatement of mine workings and mine shafts, the removal off site of the worst contaminants, the dilution and/or containment of the remainder and appropriate foundation protection and arrangements within the structure for the venting of any landfill gas where this is a problem. The redevelopment of derelict sites generating landfill gas is a special contamination problem and redevelopment of such sites, particularly for a sensitive end use, requires careful consideration and compliance with government guidance. Over costs necessary to incorporate remedial works in the development proposals are taken into consideration in the calculation of any City Grant for which the scheme might be eligible.

The objectives outlined above are being achieved through planning conditions with support through co-operation with the local Council's Building Regulations function. The highest reclamation specification is understandably for a housing end use where unresolved contamination issues can pose a potential hazard to the domestic gardener. The broad principles of reclamation are similar to those outlined above, however, in this case the Corporation has issued a clean cover specification for garden areas, which is utilised to ensure that the inert capping materials brought on site meet a stipulated guideline, both at source and in situ. A site monitoring mechanism has also been put in place on certain sites during reclamation and construction, to ensure and demonstrate that the design intent of decontamination of the site at the grant of the planning permission has been achieved in the field.

The nature and extent of the contamination on a few sites within the UDA is such that these sites are unlikely to be economically viable for any suitable form of built development. On these sites, the obvious solution of simply providing open space or general "greening" cannot be achieved with the degree of simplicity that might be supposed, not least because in some cases contamination is so severe as to prevent sustained plant growth. Such sites are being handled on case by case basis.

The Corporation is also a land owner within the UDA and since inception has been conscious of its duty of care to adjacent land owners. This duty of care has now been formalised with the advent of the Environmental Protection Act 1990, albeit Section 34 is not yet in force. In these instances the Corporation has commissioned the undertaking of a comprehensive site investigation, not only to quantify the contamination problems but also to identify means of stabilising and detoxifying the sites before enabling a built development. Rendering safe, contaminated sites close to residential areas is a complex and costly task and may need to be conducted in a manner to avoid exposing residents to substantive environmental impact in the course of the decontamination process itself. The non technical issues are being resolved through public information exercises where the community is kept abreast of emerging reclamation and development of sites close by.

Thus in its approach to the redevelopment of derelict and contaminated sites the UDC, whilst adhering closely to the Department of Environment and other guidelines, is seeking to be as proactive as possible and accordingly this stance has led to a significant number of sites being returned to a "wholesome" state to enable subsequent built development.

THE VIEWS EXPRESSED IN THIS NOTE ARE THOSE OF THE AUTHORS AND SHOULD NOT NECESSARILY BE CONSIDERED TO CONSTITUTE THE FORMAL VIEWS OF THE BLACK COUNTRY DEVELOPMENT CORPORATION.

ROUMANIAN CONCEPTS AND ACHIVEMENTS

Ramiro Sofronie
Professor of Civil Engineering

Lucia Otlăcan
Lecturer of Soil Erosion

Faculty of Land Reclamation and Environmental Engineering
Bd.Mărăşti No.59, 71331, Bucharest, Roumania

ABSTRACT

The paper deals first with the causes leading to soil degradation in Roumania as well as the experience gathered in all the fields of reclaiming the derelicted lands and their subsequent cultivation. One presents then the trends in degraded land treatment as the results obtained in soil erosion control, drainage and irrigation. The paper also discusses the works to be compleated in environomental engineering.

Roumania, with a total area of 23.75 million ha, has 14.93 million ha of agricultural land. 7.0 million ha of land with sufficient but unsuitably distributed annual rainfall, or lying on slopes without water-storage works, is subject to drought. The areas with temporary or permanent water excess extend over 5.5 million ha, particularly in low-lying flood plains and flat lands with heavy-texture soils. A major natural risk of degradation by water erosion of soil affects 6.37 million ha. During a year and particularly in the storm season, 106.6 million tons of solid material is washed off what means an average specific erosion of 16.28 t/ha/year. In most cases a given area may be affected simultaneously by the action of two or three of these degradation factors.

Although on certain areas the soil's properties were inadequate for farming, in Roumania there has been no trend or tactic of abandoning such areas. Even at the cost of great technical and economic efforts the aim has been to keep all farming land under culture considering the quota of only

0.43 ha of such land per inhabitant. In view of the many diverse causes and processes of degradation, the basic principle of reclamation and improvement is the complex management of the land. Land reclamation works have been correlated both internally and with activities in other fields such as water management, agriculture. Practically, parallel decisions are made in land-use planning to be subsequently brought together into a coordinated program, thereby achieving the compatibility, integration and correlation of solutions. Each decision is optimized in advance by an anticipative assessment of the effects of planning on the restoration of degraded ecosystems.

The area with anti-erosion works is 2.2 million ha. The results in this domain are as follows : the quantification of parameters concerning the spatial variability of the original natural conditions, in the search for solutions, is achived by pluridisciplinary research and by photogrammetry; mathematical models have been worked out for predicting alluvial effluence in small watersheds under natural conditions and alternatively under those modified by engineering works ; environmental protection by reduction the risk of alluviation in the river and/or accumulation basin network and by beautifying the landscape.

The area with drainage works is 3.1 million ha. Specialist efforts in this domain include : radical improvement of moisture-exces soils by staggering the execution of works, priority being given in the first phase to altering the original hydrogeologic conditions of the area, while the second phase aims at enhancing soil permeability and ensuring the drainage of excess water through the channel and drain network ; raising the productive potential of saline and alkaline soils and preventing the secondary salinization of irrigated soils by introducing drainage systems along with the irrigation systems.

The energetically economical irrigable potential is 5.0 million ha. At present the irrigation systems extend over 3.2 million ha. The following areas of interest may be evidenced : water saving by drip irrigation, by raising system efficiency through impermeabilization, by water recirculation and waste-water use ; introduction of advanced equipment such as automated devices and increasing system dependability ; monitoring of soil and ground-water quality.

We are interested in ecologically managing our soil and water resources, in extending of informatics and in producing equipment for the monitoring of degraded areas.

# SECTION 2 : ENGINEERING TECHNIQUES

KEYNOTE ADDRESS TO THE "ENGINEERING TECHNIQUES" SESSION OF THE THIRD INTERNATIONAL CONFERENCE ON LAND RECLAMATION

Land Reclamation as an applied science has evolved by the application and adoption of existing techniques and processes to new and challenging uses. It is a new science, and as such has had to evolve very rapidly. For it to succeed fully, it has to continue to evolve.

This evolutionary process is fuelled by three influences.

1. **New demands**

   Pressure on land for development coupled with a greater awareness of the potential offered by reclaimed land is bringing with it far more exacting demands from a wider range of developments. These demands include more exacting levels of decontamination and better bearing capacity.

   A growing acceptance of the need for better quality in design and finishes has resulted in laudably heightened aspirations regarding site layouts, site features such as water, and higher overall quality of hard and soft landscaping.

   While striving to meet these more refined demands, there is as ever a need to maximise the cost effectiveness of land reclamation monies, and technical innovations have a significant role to play.

2. **New problems**

   This description is shorthand for a variety of problems, some being newly identified but of ancient origin, others being genuinely newly created. In broad terms the majority of these new problems relate to site contamination, and to a lesser extent to elements of earlier reclamation activity which are proving to be less than adequate to withstand the ravages of time. Early identification of the latter category is best achieved by a programme of appraisal and reassessment of previously reclaimed sites, specifically aimed at the early identification of shortcomings. Early identification is important, not only to permit correction of the specific site inadequacy, but also to prevent a continued repetition of the same defect on other sites.

Contamination is an ever growing challenge for three main reasons. These are:-

- 2.1. Greater awareness of possible sources of contamination, coupled with more effective and thorough site investigation techniques.

- 2.2. Identification of previously unsuspected hazards. In the last twenty years there has been a steady procession of "new" threats, e.g. phenols, cadmium, dioxins, asbestos, P.C.B.s, landfill gas, and benzene.

- 2.3. Higher required standards of decontamination on previously known contaminants, and stringent limits on levels of new contaminant.

3. Technical Developments

Much of the progress in land reclamation techniques arises from developments in pure science or other allied fields. Research directly applicable to land reclamation has however also played an important role. Those advances of particular relevance are:-

- 3.1. Improved plant, machinery and equipment for site investigation, chemical analysis, earthmoving, drainage installation, soil preparation, tree planting, transplanting of trees and site maintenance.

- 3.2. New products such as geotextiles, geogrids, and tuley tubes.

- 3.3. Microbiological techniques for dealing with contamination, and for improving plant growth, e.g. rhizobium inoculation.

- 3.4. Chemical techniques for neutralising contaminants.

- 3.5. Methods of encapsulation of contamination.

- 3.6. Regimes for control and management of vegetation, including grazing regimes and biomass generation.

The field of land reclamation has attracted a wide range of technical talent. It is essential that this talent should be stimulated, and the fruits of its labours fully exploited. We must ensure that we monitor the effectiveness of past reclamation activity, and that we fully employ the lessons learned by that monitoring. We must also make every effort to anticipate the problems to be faced in future. Such anticipation will enable us to give adequate consideration to the techniques to be developed and refined.

It is of paramount importance that we share knowledge, so maximising the benefit to be gained from the fruits of everyones labours, so preventing us all making similar mistakes and each having to discover similar solutions to identical problems. This sharing of knowledge is essential if we are to achieve a uniform nationwide policy of good practice of acceptable standards.

In conclusion I must emphasise that as curers of a huge legacy, it is incumbent upon us to press for, and assist in more effective systems of prevention of contamination and dereliction.

D G GRIFFITHS
DIRECTOR, LAND RECLAMATION

LANDFORM REPLICATION AS AN APPROACH TO THE RECLAMATION OF LIMESTONE QUARRIES

DEBRA BAILEY AND JOHN GUNN
The Limestone Research Group,
Manchester Polytechnic,
Chester Street, Manchester, M1 5GD, UK

ABSTRACT

Increased environmental awareness has led mineral operators and mineral planners to seek practical, environmental and economic solutions to the reclamation of limestone quarries. One possible approach is *Landform Replication*, the construction by *Restoration Blasting* of varied slope sequences of rock screes, buttresses and headwalls which can be selectively vegetated to replicate natural dalesides. Trials to examine the potential of the technique have been undertaken in two Peak District quarries where three daleside landform sequences have been constructed. It is suggested that landform replication could be used to improve the visual appearance of those parts of large quarries no longer in production, or to increase the diversity of after-uses for quarries following final cessation of working. The implementation of landform replication as part of a progressive reclamation scheme could form part of a mineral operators planning application for a quarry extension or for a green field site.

LIMESTONE RECLAMATION IN THE UNITED KINGDOM

In recent years the environmental awareness of mineral operators, mineral planners and the general public has increased and with it has come the need for practical and economic solutions to limestone quarry reclamation which provide a more environmentally acceptable result. In contrast to extractive industries such as sand and gravel where reclamation techniques are well developed [1], the reclamation strategies available for hard rock quarries are surprisingly limited.

A study of reclamation conditions imposed on planning permissions for limestone quarries in the Derbyshire Peak District during 1951-1981 demonstrated a preoccupation with softening the margins of quarries in order to reduce the visual impact on the local amenities of the area [2]. Typical requirements were to reduce the visual impact of the pit by implementing a tree planting programme at quarry margins, the construction of mounds (often out of character with the surrounding landscape) to screen quarry plant and the seeding of waste materials produced during extraction. Quarry floors were to be left in a tidy condition.

There have been very few attempts to reclaim the void produced during extraction and even fewer attempts to treat the abandoned quarry faces. The commonest outcome following the cessation of the extraction operation has been for the company to leave the pit to evolve naturally. Quarrying operations in such sites may only be temporarily suspended and the quarry may form part of the company's long term stone reserves. If the quarry has been completely abandoned an attractive option for the mineral operator is to sell or rent the void. This may be for the tipping of waste materials or slurries, although this strategy is dependant upon the site being impermable, which in most limestone terrains can only be achieved by expensive engineering works.

Quarry voids may also be sold or rented for storage, industrial units or housing or the quarry floor may be reclaimed for agriculture or for amenity. These after-uses require some form of treatment to the quarry faces to reduce rock fall and to make them more visually attractive. Current techniques known to the authors involve wrecking quarry faces with a back-actor (a hydraulic rock breaker) followed by the tipping of waste either at the base of the face or over the top of the face. The resulting engineered landforms are then dressed (covered) with soils and revegetated. The major drawback of this approach is that tipping of material over the face or pushing material up to the foot of the face will not produce anything which resembles, or behaves like, a natural landform. Moreover, in order to ensure that the angle of the constructed rock pile does not exceed the safe working limits of quarry machinery only the lower part of a typical quarry face is treated using these methods. Hence, long sections of quarry face treated in this manner would have a very uniform and unnatural appearance and a great extent of the quarry face would still be able to liberate rock fall. In addition the use of back-actors to wreck quarry faces is likely to actively promote rock fall due to the disturbance of the upper face which is highly fractured during the quarries programme of production blasting. A recent variant of this approach leaves abuttments of quarry face around which materials are pushed, covered with soil and revegetated, the aim being to mimic the rock buttresses and vegetated screes of natural limestone landforms. However, the authors consider that even this leaves an essentially 'engineered' landform and that an alternative approach more fundamentally based on geological, geomorphological and ecological considerations is required.

LIMESTONE QUARRIES AS DYNAMIC LANDFORMS

Current reclamation techniques do not take account of the dynamic nature of the anthropogenic landforms created as a direct result of quarrying activities. Gunn and Gagen [3] identified three large scale anthropogenic landforms which result from quarrying activities. These comprise enclosed or semi-enclosed rock basins formed where quarries are worked downwards from the surface of a limestone plateau; valleys with one or both sides modified by quarrying and artificial 'dry valleys' produced by excavating into, as opposed to alongside a natural slope. These landforms are of sufficient scale so as not to change their form for a substantial time period. However, the quarry faces on the margins of the landform assemblages are out of equilibrium with their surrounding landscape and will evolve rapidly as a direct result of natural processes. Quarry face recession and the scale and extent of the suite of landforms which develop as the face recedes, is related to geological conditions, excavation methods and time since abandonment [3].

Gunn and Gagen [3] considered two major blasting techniques used in limestone quarries: black powder blasting and the use of Ammonium Nitrate and Fuel Oil (ANFO) and slurry explosives. Black powder blasting used slow action explosives which were comparatively inaccurate in small scale, single bench quarries up until 1949. The margins of these quarries were shown to evolve under the action of mechanical and solutional processes to produce irregular suites of landforms which are readily assimilated into the natural landscape. In contrast the precise and controlled drilling and blasting techniques of ANFO-slurry explosives produce only two principal landforms, blast fracture cones and rock buttresses. Blast fracture cones are produced by the concentration of explosive forces which occurs between adjacent shot holes. Rock buttresses which accord with the position of drilled shot-holes project out from the quarry face and increase in lateral extent towards the floor. Buttresses occur alternately between blast fracture cones which leads to the formation of a very regular suite of landforms. Anfo-explosive slurry blasted quarry faces are believed to degrade more slowly and are not as easily colonised by vegetation.

NATURAL COLONISATION OF LIMESTONE QUARRIES

An accidental by-product of mine and quarry operations over the past 2000 years has been the creation of habitats for natural colonisation by plants and animals [4]. The geomorphological form of the constructed landscape provides habitats with differing physical and chemical constraints. Colonisation will depend on land use in the area surrounding the quarry, the ability of a given species to disperse its seed and the length of time since abandonment [5,6]. The botanical importance of limestone quarries is attributed to the quarry floors whose harsh physical and chemical characteristics provide ideal refuges for species rich plant communities similar to those of existing semi-natural calcareous grasslands [4]. Quarry face communities comprise plants associated with high disturbance through rock fall, movement of bedrock clays and spalling of soils and materials from the plateau above the face.

The natural colonisation of limestone quarries by calcareous grassland species reduces as the scale of the quarrying operation and distance from limestone dalesides increases and as time since abandonment decreases. Studies of disused quarries in the Peak District by the authors and by Hodgson [6] and Wigglesworth [2] have shown that those in which semi-natural calcareous grasslands have established were generally worked by hand or by black powder and are in close proximity to semi-natural habitats. Modern quarries are unlikely to develop a species-rich flora due to their remoteness from a seed source and to current methods of reclamation which generally rely on the use of imported top soil because of the shallow nature of soils at quarry sites. Imported top soil is often more fertile than the well drained, infertile, lime-rich soils on steeper daleside slopes and weed species quickly out-compete limestone flora. In addition the hostile physical and chemical nature of quarry waste mediums reduces natural colonisation by individual plant species especially if they are of agricultural rather than semi-natural orgin and therefore not adapted to these conditions. Park [7] found that the seed rain of quarries was generally adequate and that plant loss was due to high seedling mortality through desiccation and failure of the radicle to penetrate the substrate. Hence, if a calcareous flora similar to that of a natural limestone daleside is required in a modern quarry as part of its overall reclamation strategy it will need to be actively promoted through a revegetation programme.

LANDFORM REPLICATION

The authors advocate an alternative approach to quarry reclamation which takes into account the geomorphological and ecological constraints discussed above and which should permit progressive reclamation of the dynamic limestone quarry margins. This is embedded in an on-going research programme entitled 'Landform replication as a technique to the reclamation of limestone quarries' commissioned by the Minerals and Land Reclamation Division of the Department of the Environment (DoE) in May 1988. The research is being undertaken by the Limestone Research Group at Manchester Polytechnic with the support of ICI General Chemicals and Blue Circle Industries Plc. (BCI). Landform replication involves the construction of a landform assemblage and associated habitats which mimics that of the nearby natural environment. The research project aimed to replicate natural limestone landforms and vegetation assemblages on quarried rock slopes in the English Peak District, a region dissected by valleys (dales) where the dominant natural limestone landform assemblage consists of soil covered rock slopes, bare and vegetated screes, buttresses and headwalls [3,8]. The limestones are of Dinantian age and generally are horizontally bedded, the bedding thickness varying from 30cm to 3m. The vegetation includes varying proportions of grassland/wildflower and woodland community types. Experiments were undertaken in two quarries, ICI Tunstead Quarry near Buxton and BCI Hope Quarry near Castleton, the objective being to construct three *daleside landform sequences* which function as self-sustaining ecosystems and replicate natural dales. The present paper briefly describes these experiments which are considered in greater detail elsewhere [9,10] and focuses on the general principles established.

Landform replication requires a knowledge of the geology, geomorphology (landforms and processes) and ecological communities present within the unexpected landscape in the vicinity of a quarry, together with an understanding of the ecological and geomorphological evolution of disused mineral workings. A geo-ecological restoration strategy is prepared which encompasses preparatory site investigation, geological, geomorphological and ecological survey and geo-ecological mapping. In the DoE project geo-ecological mapping was undertaken across the Peak District to identify the occurrence and characteristics of limestone landforms and the nature and extent of their associated vegetation. A more detailed survey was undertaken of a valley adjacent to Tunstead Quarry and a daleside model for replication was produced. Surveys were also undertaken in twelve abandoned quarries. The combined results of the surveys were used to produce a grassland seed mixture which reflects the typical limestone species of the northern Peak District.

Restoration blasting [11] was used to produce three skeletal daleside landform sequences whose scale, form and extent is visually attractive, inherently more stable than a production blasted quarry face and mimics their natural counterparts. The faces treated in the construction of the sequences were 10-30m high and 80-120m long. Drilling and blasting designs were determined in response to the geological and geomorphological condition of the quarry faces. If the technique becomes more widely used then it will also be necessary to consider the requirements of an overall quarry reclamation strategy in support of desired after-use(s). A restoration blasting design differs from a typical single row production blast by use of both single and multi-row, chevron pattern charged and vented shot-holes in strategically located charge zones. Rock fragmentation, distance of throw and degree of heave and swell generated by each charge zone can be altered to produce multi-faceted slope sequences which contrast with the uniform and rectilinear appearance of production blasts. The scale and extent of rock buttresses and headwalls and the volume of rock forming the screes is a function of the location, number and size of charge zones. The explosive loading and the timing and sequence of detonation of explosives in each charge zone determines the throw, heave and swell of the scree-blast pile, and, importantly for dressing operations, surface fragmentation and internal structure.

Due to the coarse nature and lack of soil materials at the surface of the rock screes generated by restoration blasting it is necessary to dress the piles with a seed bed prior to their revegetation. The materials used for the dressing operation were determined in a series of laboratory and field planting trials. The use of top soil was rejected due to its limited availabilty at quarry sites and as its chemical, physical and biological nature would allow the invasion of aggressive weed species which would out-compete the establishing limestone flora. Hence, its use would be an expensive waste of a valuable finite resource. As an alternative, quarry materials consisting of various grades of crushed limestone were used. Seven quarry materials were trialed to determine the level of physical and chemical amelioration necessary to support plant establishment, growth and survival.

Dressing of the quarry materials onto the skeletal landform was based on the position of grasslands on natural dalesides, as determined during geo-ecological mapping and upon the intrinsic conditions of the skeletal rock scree. In particular, the size and spacing of near surface voids was found to influence the receptivity of the blast piles to the dressing material. Dressing was undertaken by a dragline positioned at the base of the landform on or next to piles of the dressing materials. The material was positioned according to a pre-determined landscape design which retained the natural appearance of the pile (its heave and swell). A dragline was preferred to the other methods tried as it allows precise placing of material and avoids compaction. This contrasts with traditional approaches which may produce regular slope forms and can compact the material being used for reclamation. Following dressing, the daleside landform sequences were sown by hydraulic seeding. The subsequent germination and survival of the seed mixture's are being monitored.

The establishment of trees also forms part of the landform replication technique. Species selected reflect observations of local dalesides and abandoned quarries. Their location on the daleside landform sequence was determined partly by observations of where trees exist on natural dalesides and partly by the inherent conditions of each rock scree. It is considered that tree survival will be dependent upon the plants ability to root into finer material generated by the intermixing of rock scree from the restoration blasts. The tree transplants were pit planted either directly into the rock screes using an organic ameliorant or into the dressed material and provided with rabbit protection. An after-care management specification was prepared for the initial five year period of grassland and woodland establishment.

The objective of this geo-ecological restoration strategy is to enable a vegetation community which is similar to a semi-natural limestone flora to be established within 2-3 years. The grassland swards should remain open enough to allow the invasion of other beneficial limestone grasses and herbs.

PRACTICAL AND TECHNICAL IMPLICATIONS

It is considered that a main advantage of the Landform Replication technique is its adaptive use of existing equipment, materials and methods, most of which are employed by the extraction companies in their daily operation. The authors consider that rock skeletons may rapidly be constructed as part of an ongoing reclamation programme. The skeletons may subsequently be dressed, hydroseeded and planted with trees and shrubs to produce a series of daleside landform sequences. It is considered that dressing, hydroseeding and tree planting operation will not be necessary for every rock skeleton and that in areas of low visual impact it will be possible to produce skeletal daleside landform sequences which may be left to naturally revegetate. If a rock skeleton is produced which has a high percentage of soil/fine material at the surface it may be adequate to hydroseed directly onto this medium or to allow natural colonisation, particularly if the quarry is close to a semi-natural seed reservoir. The geo-ecological approach to landform replication also allows the evolution of the daleside landform sequences to be predicted, based on observations of the natural and excavated limestone landscape. Hence, a management specification for vegetation after-care may be implemented and built into a planned after-use for the site.

In contrast to existing methods, the rock screes constructed by restoration blasting are considered to have a more natural placement in relation to their rock headwalls and buttresses. In addition, coarse screes wrapped around rock buttresses and screes underneath headwalls which have been constructed so that they slope backwards into the quarry face can be left undressed to provide a natural capture zone for any subsequent rock fall. It may also be possible to predict the evolution of the rock landforms in order to isolate other areas which may liberate rock fall.

Quarry faces to be reclaimed by the landform replication technique normally require a bench width of face height and a half, although rock skeletons can be constructed on bench widths of face height. Concern has been expressed by some mineral operators that the bench width required for this technique represents a large loss in their stone reserves. The technique will sterilise the stone out of which the rock skeleton is constructed and all the stone beneath the pile whereas existing methods of quarry reclamation aim to sterilise the minimum amount of stone. However, compared to the total tonnage available within a major quarry, particularly on new sites, this will be relatively small. Nevertheless this represents an additional cost to mineral operators which must be set against the perceived environmental benefits. The amount of stone sterilised may be reduced by selective use of landform replication in areas of high visual interest.

CONCLUSIONS

Few effective techniques for the reclamation of large-scale limestone quarries are currently available and there have been no previous attempts to treat the pit as a dynamic anthropogenic landform. The new landform replication technique may enable the construction of a predictable suite of landforms (bare and vegetated screes, buttresses and headwalls) which should improve the visual appearance of those parts of large quarries no longer in production and increase the diversity of after-uses for quarries following final cessation of working. The combination of landform replication with existing methods of reclamation as part of a progressive scheme could form part of a mineral operators planning application for a quarry extension or for a green field site.

ACKNOWLEDGEMENTS

The research on which this paper is based was undertaken as part of a Department of the Environment (Minerals & Land Reclamation Division) Research Contract. The views are those of the authors and are not necessarily those of the Department. The authors would like to thank ICI General Chemicals and Blue Circle Industries Plc. for providing sites and support for the trials. Dr P J Gagen (Camlin Lonsdale Landscape Architects) was responsible for the Restoration Blasting Trials and Technical advice was provided by The Groundwork Trust (St. Helens), ICI Nobel's Explosives, Explosives and Chemicals Products Ltd., Rock Environmental Ltd. and Dr. S. Shaw (Dept. of Biological Sciences, Manchester Polytechnic.

REFERENCES

1. Hawksworth, P.M., Restoration of Sand and Gravel Workings, 1982, MAFF (Publications), Booklet 2377, 18pp.

2. Wigglesworth, P., <u>Limestone Quarrying and Nature Conservation</u>,MSc. thesis, CNAA (Manchester Polytechnic), 1990.

3. Gunn, J. and Gagen, P.J., Limestone Quarrying as an agency of landform change. In <u>Resource Management in Limestone Landscapes: International Perspectives</u>. eds. Gillieson, D.S. & Smith D.I., Special Publ. No.2, Dept. of Geography & Oceanography, University College, Australian Defence Force Academy, Canberra, 1989 pp. 173-181.

4. Davis, B.N.K., Wildlife, Urbanisation and Industry. <u>Biological Conservation</u>, 1976, **10**, 249-291.

5. Usher, M.B., Natural communities of plants and animals in disused quarries. <u>Journal of Environmental Management</u>, 1979, **8**, 223-236.

6. Hodgson, J.G., The botanical interest and value of quarries. In <u>Ecology of Quarries</u> ed. Davis, B.N.K., Institute of Terrestrial Ecology, Cambridge, 1982, 3-11.

7. Park, D.G., Seedling demography in quarry habitats. In <u>Ecology of Quarries</u>, ed. Davis, B.N.K., Institute of Terrestrial Ecology, Cambridge, 1982, 32-40.

8. Gunn, J., Pennine karst areas and their Quaternary history. <u>In The Geomorphology of Northwest England</u>. ed. Johnson, R.H., Manchester University Press, 1985, pp. 263-281.

9. Gagen, P., Gunn, J. and Bailey, D., Landform replication experiments on quarried limestone rock slopes in the English Peak District. <u>Zeitschrift fur Geomorphologie</u>, in press.

10. Bailey, D.E., Gunn, J., Handley, J.F. and Shaw, S., Construction of limestone ecosystems on quarried rock slopes (In prep). Australian Mining Industry Council, The Second International and Sixteenth Annual Environmental Workshop, 7-11 October, 1991, Perth, Australia.

11. Gagen, P and Gunn, J. Restoration blasting in limestone quarries. <u>Explosives Engineering</u>., 1987, **1(1)** 14-15.

GARDEN FESTIVAL WALES - EBBW VALE 1992

C. D. GRAY
Keltecs (Consulting Architects and Engineers) Limited
Keltecs House, 5 Pascal Close, St Mellons, Cardiff CF3 0LW

ABSTRACT

During the period between 1978 and 1982 British Steel progressively reduced operations at their works in Ebbw Vale, Gwent, eventually leaving an 80 hectare site derelict. In 1986, Blaenau Gwent Borough Council were awarded the 1992 National Garden Festival to be located on this site. Keltecs were given the task of planning and supervising the reclamation of the site for use by the Festival prior to its use for industrial development. Within the next 3 years the problems of the site had to be identified and overcome and the site made ready for Festival development. This paper describes the process of reclaiming the site and some of the difficulties that had to be resolved including the relocation of 12 km of rail sidings; the strengthening of 950 metres of live river culvert; the treatment of toxic materials; the capping of eleven mineshafts and the movement of some 2 million cubic metres of slag and shale materials.

SITE DESCRIPTION

The Festival site is situated in the Borough of Blaenau Gwent immediately south of Ebbw Vale town centre, some 25 miles north of Newport in Gwent. It is near the head of the Ebbw Fawr river valley two to three miles south of the main A465 "Heads of the Valleys" road.

The site is divided into two by the route of the A4046 Newport to Ebbw Vale road. The road runs along the eastern side of the site at its southern end and then crosses the river to the western side of the valley almost at the midpoint of the site. It then runs northwards into Ebbw Vale itself.

This northern part of the 2.5 kilometre long site was overshadowed, in its northwestern corner, by the former Prince of Wales Colliery. All the colliery headworks and buildings had been dismantled but eight old mineshafts remained uncapped. The remainder of the northern site had been left derelict as the Steelworks operations had retrenched further to the North. Most of the furnaces and buildings had been demolished to ground level but several huge structures remained intact on the periphery of the Festival area.

South of the A4046 road crossing, the western part of the site was dominated by the large slag waste tip associated with the Steelworks and the eastern part by the spoil tip generated by the old Waunlwyd and Victoria collieries. These steep sided waste tips were up to 40 metres high and covered most of the southern half of the site. Figure 1 shows the extent to which these tips dominated the site.

Figure 1. Aerial view from the southern end of the Festival site showing the slag and shale tips in the foreground and the Steelworks in the background.

The Ebbw Fawr river used to flow through the middle of the valley but had been culverted at the northern end of the site to enable development of the Steelworks. To provide space for the discard of the slag and colliery wastes the river was also culverted over the southern section of the site. The culvert was extended many times and in many different forms as the demands for more tipping space grew. Control over the tipping appears to have been fairly limited as large mountains of waste were stockpiled above the culvert. These tips led to considerable overburden pressures on the culvert, far in excess of the design loads. As a result there was serious cracking of the culvert. There was also considerable erosion of the culvert invert.

The tops of the shale tips had been levelled to accommodate the extensive network of railway sidings that served the Steelworks. These sidings occupied much of the southern part of the site with the rail lines into the Steelworks snaking across the site in several directions.

To the East of the southern part of the proposed festival site is the main British Rail line supplying the Steelworks and further East the site of the old Waunlwyd Colliery. As at the Prince of Wales Colliery further north, all the headworks had been demolished to ground level, but three untreated mine shafts still remained.

In 1984, Blaenau Gwent Borough Council commissioned Keltecs (Consulting Architects and Engineers) Limited to undertake a feasibility study for the reclamation of the site. However, this study was overtaken in November 1986 when the site was chosen to host the 1992 National Garden Festival. This is to be the last of the series of five festivals in Britain and the first to be held in Wales.

SITE INVESTIGATIONS

Following an initial feasibility study of the site, two detailed site investigation contracts were undertaken between January and August, 1987.

General Investigation

The first was a general investigation of the old Steelworks site and had various aims. These included :

i   Determination of the general geology of both the superficial layers and underlying strata.
ii Assessment of the extent and nature of the slag and shale tips.
iii Testing of the potential fill materials to ascertain their physical and chemical properties.
iv Obtaining further information on the extent of landslips on the western side of the valley overlooking the festival site.
v   Determining levels of contamination and toxic residues. In particular, the flue dusts generated by steelmaking processes were a cause for concern. These had been deposited at the southern end of the site.
vi Locating the extent of underground structures and obstructions. Figure 2 shows a typical structure encountered.

The investigation was one of the largest ever funded by the Welsh Development Agency. The contract required a comprehensive series of trial pits, boreholes, trenches and laboratory and on-site testing. Over 100 boreholes were drilled but, for a site area of 200 acres, this represented an average of only one borehole in every two acres. In view of the extensive former use and heavy previous development of the site, this was, in hindsight, barely adequate.

Mine Shaft Investigations

Two main areas of mining activity were known to have existed within the site and its environs, but nothing was left on site to indicate the positions of the eleven shafts recorded as having existed at the two collieries.

Using information gathered from British Coal's archives, trial trenches and probe holes eventually located all the shafts.

Figure 2. One of the many underground structures. This was part of one of the old furnaces.

Further Investigations

An interesting feature of the site was the existence of two huge crags of fused slag material standing proud of the main slag tips at the southern part of the site and in the middle of the proposed Festival area. The Festival Company believed that these two imposing "slag bluffs", standing over 30 metres high, would make an impressive feature; and a waterfall was planned to cascade down between them.

Standing nearly vertical in places, but with material spalling away from the faces, the bluffs warranted a full scale investigation. This was completed in March 1989 following a full assessment of their history, composition and stability. Whilst adequate drainage measures could ensure their overall stability, localised remedial measures were recommended including dowelling, bolting, dentition, geogrids or shotcrete together with minor scaling works.

Made Ground

The general investigation showed up to 45 metres of made ground overlying the natural superficial deposits throughout the site. This made ground was largely composed of industrial waste materials produced by the iron and steel making processes and coal mining activities. The three main categories of waste identified were :

i   Blast furnace and steel slags, appearing typically as grey, fine to coarse gravel, with cobble and boulder sized fragments of blast furnace and metal slag.

ii Colliery shale, appearing typically as colliery waste material comprising black, medium to very dense, fine to medium gravel-sized fragments of coal, ash, shale and mudstone.

iii Flue and open hearth dusts from the blast furnaces.

In addition to these, demolition rubble and refractory materials existed in abundance.

The greatest thickness of made ground occurred in the area known as the Victoria Tip, where up to 45 m of blast furnace and steel slags were encountered. Blast furnace slag was also abundant throughout the northern part of the site. The red flue and open hearth dusts which had been located were essentially confined to the southern end of the Victoria Tip. A view of the southern end of the tips is shown in Figure 3.

Figure 3. The two tips dominating the southern end of the site. The slag tip is on the left and the shale tip to the right.

Glacial Deposits
Between 1.4 m and 27 m thickness of sandy, silty clay with gravels and cobbles underlies the made ground over much of the site. The deposit is generally orangey/brown and can be described essentially as a glacial till. Angular gravels comprising fine to coarse fragments of sandstone, with associated horizons of sand and silt were evident at the extreme southern end of the site. These deposits were also assumed to be of glacial origin.

Extensive thicknesses of variably coloured silty clays and sands, enclosing angular fragments of shale, sandstone and occasionally coal, were encountered in boreholes to the North of Victoria Pond on the western perimeter of the site. These deposits were believed to represent material which had slipped from a higher level on the steep valley side, where a series of landslips were known to exist.

Bedrock
The solid strata comprised mudrocks, including shales, mudstones and siltstones, with sandstone and occasional thin seams of coal.

In the southern part of the site sandstone was encountered immediately underlying the superficial rocks. This forms part of the upper sequence in the Upper Coal Measure rocks.

In the northern part of the site the solid strata essentially consisted of siltstones with interbedded mudstones.

CONTAMINATION

Samples were taken from boreholes and trial pits undertaken throughout the site. The area of principal concern was to the southern end of the site where concentrations of flue dusts were known to exist. Samples from throughout the site were tested for sulphides, phenols, cyanides, tars and total and available metal contents.

Sulphides
A number of locations examined contained quite high levels of sulphides (in excess of 25,000 ppm in one case). The problems associated with sulphides are twofold. Firstly, the presence of significant amounts of sulphides may give rise to hydrogen sulphide gas being evolved in acid conditions. This is a flammable gas and high concentrations can give rise to toxic effects. Secondly, the sulphides present may be oxidised by chemical or bacterial processes to produce sulphates which will increase the aggressive nature of the soil towards concrete.

Fortunately, the site is alkaline in nature throughout and, as long as it remains so, no problem should arise. The formation of sulphates is a possibility, and concrete mix designs were amended to provide adequate resistance to attack.

Phenols and Total Cyanides
The concentrations present were low and insignificant.

Tars (Toluene extractables)
A maximum of 53,000 ppm (5.3%) was found in the soils and at concentrations greater than 20,000 ppm (2%) staining of, for example, concrete could occur. Coal tar is a very complex mixture of organic compounds, some of which can be aggressive.

However, concentrations throughout the site were generally well within acceptable levels, and the high concentrations were in localised areas which could be dealt with individually.

Metallic constituents

All samples taken throughout the site were low in terms of levels of Arsenic, Cadmium, Chromium and Nickel.

Most of the soils were relatively low in total Copper but up to 1,240 ppm were present in one of the red dust samples. The available Copper content of this particular soil was found to be 167 ppm. The usual total Copper content of soil varies from 2 to 100 ppm with available Copper levels at about one tenth of these concentrations.

Normal soils have Lead levels of between 2 and 200 ppm with no more than one tenth normally being available for plant uptake. Most soils were relatively low in Lead but the six red dusts tested contained up to 23,600 ppm.

The majority of the soils examined were also relatively low in Zinc content, normal soils might contain from 10 to 300 ppm. However, the red dusts produced high available Zinc levels, and two in particular exhibited significant Zinc concentrations. Surprisingly, the water leachable Zinc concentrations were not unduly high, and suggested that the Zinc present in the soil was in the insoluble form.

The possible uptake of metals from the red dusts was, therefore, a matter of concern and a principal design consideration.

AIMS OF THE RECLAMATION SCHEME

The general philosophy of the land reclamation was to convert the existing derelict site with its tips, shafts, culverts and railway sidings into an interesting landform suitable both for a Garden Festival and for later development of the site. This reclamation included :-

i Reprofiling of the slag and shale tips as part of a major earthworks operation. This involved the movement of approximately 2 million cu.m. of material.
ii. Treatment or removal of toxic materials.
iii Strengthening of the southern section of the overstressed Ebbw Fawr river culvert over a one kilometre length. Replacement of 170 metres of the northern section beneath a new railway embankment constructed to serve a new rail viaduct.
iv Releasing large areas of the site by relocating the B.S.C. railway sidings from the South to the Northeast of the A4046. Twelve kilometres of active rail track were to be relocated and a new 2 span 50m. long railway viaduct constructed, all without disrupting the normal operation of the Steelworks.
v Preparation of the site for the realignment of 1.8 kilometres of main highway together with 1 kilometre of new road infrastructure.
vi Demolishing old buildings and removal of foundations or filling over them to a depth at which future development could proceed unhindered. Approximately 80,000 cu.m. of reinforced concrete were broken out and removed.
vii Capping of eleven old mine shafts.
viii Installing over 14 kilometres of new and replacement foul and storm water drainage systems.

ix   Providing new buildings for British Rail, British Steel and the Festival Company itself.
x   Advance structure planting and surface treatment of other prepared areas.

TIMETABLE OF CONTRACTS

A series of thirteen contracts designed by Keltecs and relating to the reclamation of the site were awarded between February, 1987 and October, 1989.   Their combined value was in excess of £15 million and included four different clients :-

Blaenau Gwent Borough Council   Garden Festival Wales
Welsh Development Agency   British Steel Corporation

The principal elements of the reclamation included:-

Planting Contracts
The Ebbw Vale Festival had a potential period for planning of just over 5 years, considerably longer than earlier Festivals. On a site with no substantial vegetation, hostile ground conditions and an elevated, exposed aspect, it was essential that full use was made of this lead-in period if the planting was to appear established. The reclamation, therefore, included a substantial amount of on-site and off-site structure planting. The first of six separate planting contracts was awarded in February, 1987 to plant over 40,000 whips and transplants and 1500 larger trees on areas of peripheral land. Alder, Birch and Poplar proved particularly successful in providing rapid growth for screening and shelter.   Over 100,000 trees were planted as part of the planting contracts.

As described earlier, the surface consisted principally of slag and colliery waste.   To import sufficient topsoil to cover the site to a reasonable depth was impractical, and the regraded slag was therefore covered with colliery shale as part of the earthworks contract. The shale, after improvement with peat, fertilisers and water-retaining polymer, produced a suitable planting medium. Some peat was brought from a development site in the area, whence it had been removed as waste.   Mulching, using wood chip or mushroom compost, has been concentrated on the most visible areas to repress weed growth and retain moisture.

Culvert Strengthening
As part of the feasibility study, carried out in 1986, the Ebbw Fawr river culvert was inspected in detail. Whilst the culvert under the northern part of the site was in relatively good condition, the southern culvert, which passes directly below the main Festival area, was in far worse condition. Considerable erosion had taken place to the culvert invert and extensive cracking was evident in the crown. Reinforcement was exposed in parts of the culvert and the concrete was spalling.

As a result of the distress evident over most of the length of the 950m culvert, a contract was awarded in March 1987 for remedial and strengthening works to be undertaken.

This comprised two basic parts :-

i   Strengthening of the culvert arch by means of an internal bolted segmental steel lining.

ii   Reconstruction of the culvert invert.

Figure 4 shows these works shortly after completion in October, 1988.

Figure 4:   The Southern culvert on completion of strengthening work

Mine Shaft Treatment Works
Two separate contracts were awarded for the treatment and subsequent capping of the disused mine shafts. The first in June, 1988 for the 8 shafts of the former Prince of Wales Colliery and the second in April, 1989 for the 3 shafts of the old Waunlwyd Colliery.

Of particular interest was Prince of Wales Shaft 7. This had been positively located during the site investigation but could not be located for the actual capping operation. It eventually transpired that Shaft 7 was in fact a staple shaft that had been incorrectly transferred from the original mine records onto the more formal records normally available. Unfortunately, at exactly the same location on site, a large bricklined sump had been constructed under British Steel's railway system and subsequently backfilled and it was this that had been located during the site investigation.

Site Preparation Contract
This was the main reclamation contract awarded in November 1987 to Davies, Middleton and Davies Limited. The scheme involved the majority of the items listed previously and was completed in October 1989. The conclusion of the Contract marked the handover of the reclaimed site to the Festival Company for its own development. Figures 5 and 6 show the

site at commencement of work on the reclamation scheme and two years later at handover to the Festival Company.

Figure 5: View of the southern part of the site prior to reclamation

Figure 6: View of the southern part of the site after reclamation

To a large extent, the red flue dusts discussed previously were removed from site. However, some dusts remained at depth together with contaminated clays. These materials, with their elevated metal concentrations, were capped with a layer of slag 1000 mm thick and then covered with a 750 mm thick layer of shale for tree planting or a 300 mm thick layer in areas to be grassed.

The slag forms an effective capillary break layer to prevent the upward movement of heavy metals, whilst rainfall leaches downwards any soluble material. The inhospitable nature of the slag also acts as a barrier to plant root penetration.

Relocation of Rail Sidings

A contract for the relocation of British Steel's railway sidings was awarded in September, 1988 and completed in February, 1989. Twelve kilometres of active rail track were removed from their original location in the middle of the proposed Festival area and relocated further north along the eastern boundary of the site. The sidings were in use for the whole period and this presented major planning problems.

The only non-operational period available was a four day break over Christmas 1988 when switchover from the old to the new system was achieved. This was the most critical period of the whole of the reclamation works as a failure to meet this closedown period would have delayed the whole reclamation works by six months until the next available shutdown in the following July.

AFTERUSE

The site has been reclaimed with the design considerations of the afteruse a foremost priority. An infrastructure was therefore constructed which could initially serve the Festival but would eventually form the basis of any future industrial development. This work included roads, foul and storm drainage, electrical, gas, water and telecom services.

The first "end-user" has in fact been on site since October 1989 when Bass Wales and West took possession of a two hectare parcel of land for their new regional administration offices and bottling depot.

CONCLUSIONS

A scheme that commenced as a normal reclamation exercise with interesting but not overwhelming technical problems, became a major planning and logistical exercise with many contracts running in parallel and numerous contractors occupying the site. Deadlines and key dates were interdependent. However, in just three years the 80 hectare site has been transformed by the largest current derelict land reclamation project in Wales. A sombre example of industrial dereliction has been converted into probably the most dramatic and exciting landscape of all the Garden Festival sites.

ACKNOWLEDGEMENTS

The author wishes to acknowledge the assistance and valuable role played by the Welsh Development Agency, Blaenau Gwent Borough Council and his colleagues at Keltecs.

ALBION LOWER TIP LANDSLIDE - CONTROL OF GROUNDWATER LEVELS USING WELL POINTS AND BORED DRAINS

J D MADDISON
Sir William Halcrow & Partners Ltd
25, Windsor Place, Cardiff, CF1 3BZ

ABSTRACT

Albion Lower Tip is situated on the east side of the Taff valley at Cilfynydd, Mid Glamorgan. There is a history of ground movements at the site predominantly associated with an ancient landslide upon which the tip was constructed. The paper describes the installation of well points and a trial of five 150m long bored drains to regulate groundwater levels in the landslide deposits below the tip and thereby control ground movements.

The findings of a desk study and morphological mapping of the tip and adjacent hillslope are given and the results of an investigation of the ground and groundwater conditions at the site are presented. The construction of well points installed in the lower part of the site to control artesian water pressures is described and the options which were considered for deep drainage measures to regulate groundwater levels in the upper parts of the site, are discussed. A trial of five bored drains was undertaken to ascertain their viability as a means of deep drainage. The drains are described with respect to their installation and construction. Substantial reductions in piezometric levels within the landslide deposits were observed following the installation of the well points and bored drains. Subsequent monitoring has revealed piezometric levels to have been maintained at these reduced levels showing only minor fluctuations to significant rainfall events.

INTRODUCTION

Albion Lower Tip lies on the west facing hillside of the Taff Valley at Cilfynydd, Mid Glamorgan approximately 20km north west of Cardiff, ref Figure 1. The tip comprises colliery spoil and was constructed between circa 1886 and circa 1920. It covers an area of about 8 hectares lying between 95 and 170m OD. In 1975 Albion Lower Tip was reprofiled as part of a reclamation scheme carried out under the direction of Mid Glamorgan County Council (MGCC). The reprofiled tip face slopes at about 1 in 3 and is drained by two 5m wide berms which extend the full width of the site at elevations of 128 and 148m OD. A general plan of the site is shown on Figure 1.

Ground movements had occurred at the site prior to the 1975 reclamation scheme. In about 1980 some 5 years after completion of the reclamation scheme MGCC recorded movement in the reprofiled tip. Subsequent monitoring by MGCC showed continued intermittent ground movements. Halcrow

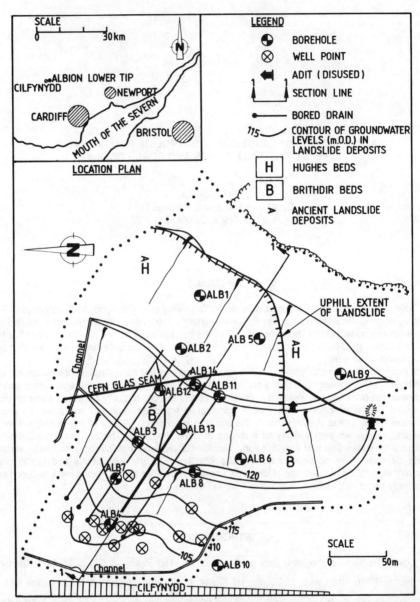

Figure 1. Site location and general plan

were engaged by MGCC in August 1989 to investigate those movements and to report on the stability of Albion Lower Tip. Between September 1989 and January 1990 survey markers installed across the site showed extremely slow to very slow movements, defined in accordance with Ref [1]. In early February 1990 rates of movement significantly increased following prolonged winter rainfall which coincided with high groundwater levels. The well points and bored drains described below were subsequently installed to regulate groundwater levels and thereby control the ground movements.

GROUND CONDITIONS

A desk study and morphological mapping of the tip and adjacent hillslope were performed to establish the general geological conditions at the site. These studies indicated Hughes and Brithdir Beds of the Upper Coal Measures to crop below the tip with sandstone of the Hughes Beds forming the steep hillside rising to the east above the site. The junction between the Hughes and Brithdir Beds is marked by the Cefn Glas coal seam. The Cefn Glas seam was worked from two levels at the site, ref Figure 1, between circa 1890 and 1933. Coal was also extracted below the site from a further five seams at between 470 and 613m depth prior to 1930. Subsidence associated with mining will almost certainly have influenced the hydrogeological conditions of the Hughes and Brithdir Beds at the site and surrounding areas.

Between September and December 1989 boreholes were sunk at six locations (ALB1 to 6 inclusive) on the tip to investigate the ground and groundwater conditions. Following increased ground movements at the site in February 1990 boreholes were sunk at a further eight locations (ALB7 to 14 inclusive) predominantly for monitoring groundwater levels. The investigations proved Hughes Beds comprising predominantly sandstones to crop below the upper part of the tip with the Cefn Glas seam cropping across the central part of the site, ref Figure 1. The underlying Brithdir Beds were found to comprise interbedded argillaceous and sandstone strata. Correlation of the borehole data and plans of workings in the Cefn Glas seam showed the strata dip to be about 7 degrees north north east.

Bedrock below the tip was found to be overlain by ancient landslide deposits of between about 9 and 15m depth (ref. Figure 2). These deposits comprised some 5m depth of ablation till as defined in Ref [2] overlying deposits grading with depth from sand and gravel into gravel, cobbles and boulders in some areas overlying blocks of displaced bedrock. The ablation till comprised silty becoming clayey sand and gravel intermixed with subangular to subrounded cobbles and boulders overlain below the tip

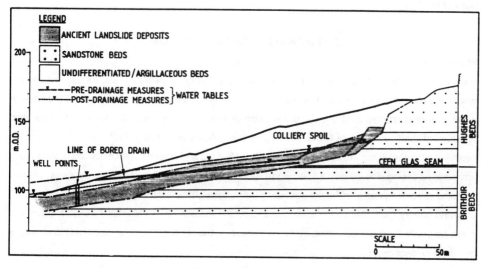

Figure 2. Section 1-1

lower slopes by a sandy clay containing many highly weathered fragments of siltstone and sandstone possibly representing weathered till or colluvium. The observed ground movements at the site were predominantly associated with this ancient landslide.

As part of the 1975 reclamation scheme some 10m depth of spoil was removed from the tip. The reprofiled tip varies in thickness from about 5m depth at the toe to some 20m depth below the upper slope. The spoil is a well graded, slightly clayey to silty, sandy gravel containing occasional cobbles and boulders.

GROUNDWATER CONDITIONS

The predominantly sandstone strata of the Hughes Beds possesses an extensive groundwater catchment on the high ground east and south east of the site. Much of the catchment lies up dip of the crops beneath the tip. The presence of south east - north west faulting accompanied by a parallel joint trend (strike 121 to 136 degrees north) and open fissures resulting from mining subsidence would be expected to give enhanced north westerly flow of groundwater towards the site. Piezometric data confirmed an unconfined groundwater table in the Hughes Beds cropping at the site with a series of differential water tables existing under confined conditions in the underlying Brithdir Beds strata with discharge from bedrock into the landslide deposits. Drainage from old mineworkings beneath the south of the site is also likely to augment recharge of the landslide deposits from bedrock.

Groundwater in the landslide deposits was found to move from unconfined conditions below the tip upper slope to confined conditions downslope, the confining layer being the zone of weathered glacial till. A contour plan of groundwater levels, ref Figure 1, reveals a distinct lobate form with flow in a north west direction and relatively higher piezometric pressures in the north west of the site. Artesian water pressures of up to 4m above ground surface were observed in the north west of the site in February 1990. Groundwater levels within the colliery spoil were generally less than 1m above the base of tip.

GROUNDWATER CONTROL MEASURES

Well Points

Following the significant increase in ground movements at the site in February 1990 fourteen well points were installed on the tip lower slope, ref Figure 1, during February and March 1990 to relieve the artesian water pressures referred to above and thereby reduce ground movements. No movements have been recorded at the site since late March 1990. The well points were installed in 200mm diameter cable percussion boreholes sunk to bedrock. Demco Terrafilter 130 well screen (100mm diameter) comprising 3mm slotted casing sleeved with a 250 micron filter was installed within the landslide deposits. Demco Terraline 130 casing was used for the well riser passing through the confining weathered till deposits and overlying colliery spoil. The borehole annulus around the well screen was backfilled with pea gravel and a 3m bentonite seal placed above. The remainder of the borehole annulus was backfilled with a bentonite cement grout.

Following installation of the well points a series of dewatering tests were performed to determine the hydraulic properties of the landslide deposits. The average transmissivity and storativity were $10m^2$/day and 10^4 respectively, typical of confined groundwater conditions. Observations during the tests and subsequent groundwater modelling indicated that substantial lowering of piezometric pressures in the landslide deposits below the tip lower slope might be achieved by allowing the well points to drain at some 5m depth. These measures however would have little effect on groundwater levels upslope of the tip lower berm. In July 1990 works were undertaken to drain the well points at some 5m depth to discharge into the open channel at the toe of the tip. Flow from individual well points of up to $58m^3$/day has been recorded and a maximum total flow from the well points of $92.5m^3$/day was observed in early January 1991 coincident with a period of intense rainfall. The performance of the well points in regulating groundwater levels within the landslide deposits is described below.

Bored Drains

The groundwater table in the landslide deposits upslope of the tip lower berm lies at between 8 and 20m depth. Excavation through disturbed landslide deposits to such depths was considered impractical and thus precluded deep conventional drains and well points for regulating groundwater levels in this part of the site. An adit drainage gallery was also unsuitable to achieve widespread regulation of groundwater levels as there was no evidence of a discrete source of strata water recharging the landslide deposits. It was considered that regulation of groundwater levels in the upper part of the site might be achieved by bored drains. However such drains have not been used extensively in circumstances similar to those prevailing at Albion Lower Tip and the shallow tip slopes and the depth of the water table necessitated drains some 150m long, significantly longer than generally constructed in the UK. A trial of five bored drains was therefore undertaken to establish their viability as a means of deep drainage.

A contract for installation of the trial bored drains was let to Wimpey Geotech Limited and the works were undertaken between October and mid December 1990. The contract final cost was £280,000. The design alignment of the drains located them within the zone of relatively higher piezometric pressures at an inclination of 3 degrees above horizontal, ref Figures 1 & 2. This inclination allowed for some possible upward deviation of the bore and hence reduced the potential risk of the bore emerging above the groundwater table.

Deviation of the bore from the design alignment was considered to be potentially a major problem given the predominantly coarse granular and blocky nature of the deposits which had to be penetrated. To ensure that drilling commenced on correct line a 15m long, 200mm diameter steel guide tube was installed in a trench at each drain location as part of preliminary works at the site. The guide tube was aligned using ground survey techniques and it was bedded in concrete to prevent movement prior to backfilling the trench. Immediately downslope of each guide tube a concrete slab and plinth to carry the horizontal drill mast and drill head assembly were constructed. At the uphill end of each guide tube a 1.5m diameter manhole was built for the collection and removal of the borehole cuttings.

The bores were constructed using a G. Klemm K806 power pack and HDK800 internal rod double head drill feed unit, operating a combination of drag and rock roller bits depending on the ground conditions. The two independent rotation devices on the equipment facilitated the casing being drilled in immediately behind the drill bit to minimise deviation from the design alignment. A series of stabilisers to centralise the drill string within the casing were also used to reduce deflection of the bore. A foam flush was used to remove cuttings and to ensure maximum stability of the bore. Heavy casings of 178, 133 and 101mm outside diameter were used in constructing the bore. The two largest casing sizes were advanced some 60 and 100m respectively. The advance of the smallest casing was terminated at 150m.

On completion of each bore a drain comprising 3m approximately screw fit lengths of 75mm outside diameter Demco polypropylene pipe was installed. The 90m length of drain within the landslide deposits had 3mm side slots cutting approximately 20 per cent of the pipe circumference at 75mm spacing staggered to maintain the structural integrity of the pipe. The slotted pipe was sleeved by a 250 micron filter mesh bonded to the pipe to maintain it in position during installation in the bore. Blank casing was used for the 60m length of drain immediately above the bore outfall to minimise loss of groundwater into the weathered till and colliery spoil deposits.

It was originally intended to seal the initial 60m of each drain within its bore using a polymer grout. Grouting of the first bore, however, was severely disrupted by groundwater flow from the annulus of disturbed ground around the drain upslope. The initial 60m of subsequent drains were sleeved with blank Demco casing of 127mm diameter in order to collect most of the flow from the disturbed annulus and to prevent its loss into the colliery spoil.

Upon completion, each drain was surveyed for inclination and azimuth at 30m intervals using a camera single shot system supplied by Sperry-Sun UK Ltd. The surveys revealed predominantly small variations in inclination and azimuth of individual bores of between 1.5 and 2 and 1.5 and 4 degrees respectively. The average inclination of the bores was between 1.5 and 2.5 degrees above horizontal and four of the five bores were within 1 degree of the design alignment of 3 degrees above horizontal. The average azimuth of the bores was generally within 1 degree of the initial alignment.

Flow of between 2 and 109m^3/day from individual bored drains has been recorded with a maximum total flow of 170m^3/day observed in early January 1991 coincident with a period of intense rainfall.

GROUNDWATER LEVEL OBSERVATIONS

Typical groundwater levels in the landslide deposits at the site recorded between October 1989 and February 1991 are presented on Figure 3. The records show a sustained rise in piezometric levels over the winter period 1989-90 reaching peak levels in February 1990 following intense rainfall in the preceding two months. A typical water table in the landslide deposits in February 1990 is shown on Figure 2.

Following the installation of the well points observations of piezometers 3 and 4 (borehole locations ALB 3 and 4 respectively) installed in the landslide deposits below the tip lower slope showed abrupt reductions in water levels of some 3 and 5m respectively from February 1990 peak levels. The reduced piezometric levels were approximately 0.5 and 3m respectively below pre winter minimum levels.

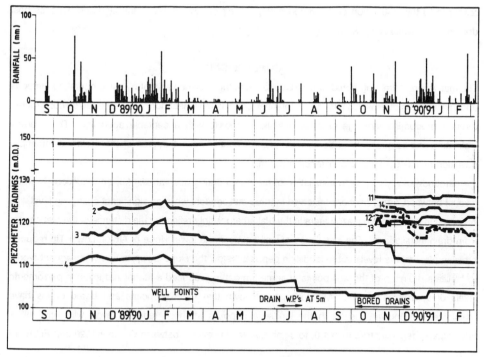

Figure 3. Piezometer readings in landslide deposits

Observed reductions in water levels in the landslide deposits below the upper part of the site (piezometers 1 and 2) were small and were considered to merely reflect reduced rainfall.

In July 1990 the works to facilitate drainage of the well points at 5m depth produced a further sharp reduction of approximately 3m in groundwater level at piezometer 4. No significant changes in groundwater levels within the landslide deposits elsewhere at the site were observed. Piezometric levels below the lower slope have remained at these reduced levels showing only minor fluctuations (less than 1.0m) to significant rainfall events between October 1990 and February 1991.

During the installation of the five trial bored drains between October and mid December 1990 observations of piezometers 3, 12 and 14 located in the area of the trial showed abrupt reductions in piezometric levels of between 3 and 7m coincident with the construction of adjacent drains. Piezometers 11 and 13 also located within landslide deposits in the trial area however showed no significant change in water levels. Subsequent monitoring to February 1991, including a period of intense rainfall in late December 1990 and early January 1991 has revealed only small fluctuations, (less than 1.0m), in water levels in piezometers 3, 11, 12 and 13. These fluctuations are much smaller than those observed during the preceeding (1989 - 1990) winter period prior to drains installation, ref piezometer 3 on Figure 3. A rise of 2.5m was observed in piezometer 14. This water level however is still some 4.5m below the maximum level recorded prior to installation of the bored drains. Water levels in piezometers 1 and 2 installed in the upper part of the site and piezometers outside the trial area have shown no significant response to the installation of the bored drains. A typical watertable in the landslide deposits post drainage measures is shown on Figure 2.

CONCLUSIONS

The trial of five 150m long bored drains undertaken at the site showed that by using operating methods and drilling techniques which minimised potential for deviation a near horizontal bore can be constructed through predominantly large granular landslide deposits with acceptable accuracy. The average inclination and azimuth of the bores were predominantly within 1 degree of the design alignment, a deviation of less than 3m at a distance of 150m.

Groundwater response to installation of the bored drains was varied. Abrupt reductions in water level of between 3 and 7m were observed in three piezometers while no significant change in level was observed in a further two piezometers in the trial area possibly reflecting the distance of the piezometer from drain installations. The installation of well points in the lower slope of Albion Lower Tip reduced artesian piezometric levels in the landslide deposits below the slope by up to 5m. Subsequent works to facilitate drainage of the well points at some 5m depth produced a further reduction in piezometric levels of approximately 3m. Monitoring has shown that the well points and bored drains have maintained groundwater levels in the landslide deposits at these reduced levels with only minor fluctuations, generally less than 1.0, to significant rainfall events between October 1990 and February 1991.

ACKNOWLEDGEMENTS

The Client for this project was Mid Glamorgan County Council, the Engineer was Sir William Halcrow & Partners Ltd and the Contractor for the installation of the bored drains was Wimpey Geotech Limited. The Author's thanks are given to colleagues in all three organisations who have assisted on the project and in particular to Mr P E Wright, County Land Reclamation Officer, Mid Glamorgan County Council. The bored drains trial was funded by the Welsh Development Agency.

REFERENCES

1. Varnes, D J. Slope movement types and processes. In Landslides : Analysis and Control Transportation Research Board, National Academy of Sciences, Washington D.C. Special Report 176, 1978, Chapter 2.
2. Fookes, P.G., Gordon, D.L. and Higginbottom, I.E., Glacial landforms, their deposits and engineering characteristics. Symposium on glacial materials. Midland Soil Mechanics and Foundation Society, Birmingham 1975.

GAS CONTROL MEASURES FOR DEVELOPMENT OF LAND AFFECTED BY LANDFILL GAS

DAVID P ROCHE
MSc BSc CEng MIMM MIGeol FGS
Associate Director
Frank Graham Geotechnical Limited, Consulting Engineers
22 Waterbeer Street, Guildhall Centre, Exeter, Devon, EX4 3EH

and

GEOFFREY B CARD
PhD BSc CEng MICE EurIng FGS
Associate Director
Frank Graham Geotechnical Limited, Consulting Engineers
Shinfield House, School Green, Shinfield, Reading, Berkshire, RG2 9EW

ABSTRACT

Based on case study examples this paper illustrates how effective gas control measures have been designed and implemented to enable building development on sites affected by landfill gas. A comprehensive assessment and design process is required to identify the hazards posed by landfill gas and to fully understand and evaluate the risks to the location and type of building development. An integrated system of gas control measures is preferred. The principal types of control measures are illustrated, including encapsulation, gas venting trenches and wells, gas resistant membranes and barriers, natural ventilation to structures, and gas detection and alarm systems.

INTRODUCTION

Current demand for building land in the United Kingdom has increased the pressures for urban regeneration and the redevelopment of former industrial and derelict land. Land within and adjacent to landfill waste sites is increasingly being used for industrial, commercial and residential purposes. Rigorous interpretation of recent guidelines to preclude development within 250m of a gassing landfill could sterilise vast areas of land, unless the risks are countered by effective engineering solutions.

The risks of development on or adjacent to landfill waste sites are well documented. This paper seeks to illustrate how effective gas control measures have been designed and implemented with reference to case study examples.

LANDFILL GAS HAZARDS

The nature, properties and hazards of landfill gas are well documented (1, 2 and 3). The gas can vary considerably in composition. Its most common constituents are methane and carbon dioxide, with other trace gases including hydrogen sulphide and complex organic compounds. The gas is produced by the decomposition of putrescible material within the waste.

Understanding the hazards and evaluating the risks posed to building development is fundamental to arrive at a successful and safe design strategy. This involves interpretation of investigation and monitoring data, prediction of future trends, selection of appropriate design parameters, evaluation of gas control measures, safety during construction, and long term servicability and monitoring.

The main hazards from landfill gas with respect to building development are:

- Flammability and Explosivity
- Asphyxiation and Toxicity
- Odour

Migration of landfill gas from its source of generation and through the ground to nearby building structures poses the principal risk to development on or adjacent to gassing sites. It is necessary, therefore, to undertake detailed investigation and monitoring to enable a full assessment of gas migration and the consequent risks to development. Typically the investigation would comprise conventional trial pits or boreholes with standpipes inserted for a subsequent programme of gas monitoring. The frequency, duration and scope of monitoring at each site will depend upon the concentrations of gas actually measured and the scale of risk anticipated. The monitoring programme must be adequate to establish the gassing nature of the site.

The degree of risk to the development depends upon:

- the gassing nature of the landfill, especially rate of generation and emission
- the nature of the surrounding ground conditions and potential routes for migration
- the composition and concentration of the gas
- the sensitivity of existing and proposed developments.

The implication of current guidelines published by Her Majesty's Inspectorate of Pollution (1) is that any development within 250m of gassing landfill should be fully assessed to identify the potential risk prior to development. The recommended 250m limit is however considered to be somewhat arbitrary and takes no account of specific ground conditions, for example if the local geology is intact clay of low permeability or fissured limestone of high permeability.

INTEGRATED GAS CONTROL SYSTEMS

Having assessed the degree of hazard and risk at a site, it is possible to consider the nature and extent of gas control measures which will be necessary.

Measures to control landfill gas migration fall into two broad categories:

- controls for preventing or regulating gas emissions and migration from the landfill source, both from surface and subsurface boundaries

- controls for preventing migration of gas into confined spaces within building structures.

For building development, an integrated control system combining both categories and more than one individual method is generally preferable. Thus a staged system of preventative and precautionary controls is designed and specifically tailored to the nature and scale of the problem. In the event of a malfunction or in-service failure of one part of the system, there should be at least one other measure to provide control and therefore minimise risk with a large margin of safety.

Engineering design control measures should be based on worst credible values of gas concentration and emission rate so that the system has adequate capacity and margin of safety at all times and under variable climatic conditions. These values are derived from careful evaluation and weighting of the monitoring data with particular reference to the maximum recorded levels.

Based on current guidelines with respect to safety limits and safety margins, the key design concentrations are considered to be:

- Methane - less than 0.25% by volume in air (<5% LEL)
- Carbon dioxide - less than 0.25% by volume in air.

The design of gas control measures has been discussed previously by Card and Roche 1991 (4). The principal methods to control gas emission and migration from source include encapsulation, interception and cut-off, ventilation and abstraction. The principal methods to control gas migration into buildings include gas resistant membranes and barriers, underfloor ventilation, and gas detection and alarm systems. Key design features are illustrated in Figures 1 and 2.

Examples of integrated gas control systems which have been designed and implemented to provide effective control are illustrated by the following case studies.

CASE STUDIES

Site A - New Residential Development

Landfill gas was detected in the course of a ground investigation at a housing development site in Torbay. Gas levels ranged between negligible and 5% methane, with a sporadic distribution of high concentration 'hot spots'. The gas was measured in a variety of monitoring points including boreholes, trial pits and specially constructed monitoring wells. The site itself included an area of made ground comprising mainly inert materials in which gas generation was extremely unlikely. It was suspected that the gas was migrating into the site from adjacent filled ground where former domestic refuse waste disposal was recorded.

The implications of landfill gas to the development proposals were highly significant. Unless effective remedial measures could be implemented, two extreme possibilities could have jeopardised the development: either abandonment of the site or wholesale removal of the made ground. Furthermore there was a serious risk of blight to adjoining areas of existing housing.

Following detailed investigation and monitoring, and evaluation of the implications for the development proposals, an integrated system of gas control, preventative and precautionary measures was designed to facilitate safe development.

The measures included:

- keeping made ground areas as open space with no house construction; this was achieved by modifying the building plot layout and allocation of open space.

- removing from site some of the fill in slope regrading works.

- installing a series of passive ventilation cut-off trenches within and around the periphery of the made ground as a primary control measure to isolate the source of gas migration into the site and prevent gas migration towards the development areas.

- providing suspended ground floors to dwellings and ventilated underfloor void spaces as a second stage precaution to preclude any build up of gas should it migrate beneath the dwellings.

- including gas resistant membranes and seals to floors and service entries as a third stage precaution to form a positive barrier to gas entry into buildings.

- providing special precautions to domestic garden areas, involving placement of a granular underblanket connected to the peripheral ventilation trench, and a low permeability clay subsoil beneath the garden topsoil. Garden sheds, greenhouses etc were prohibited.

- monitoring of gas levels on a longterm basis.

The problems at this site have been effectively solved by the implementation of the gas control measures. Installation of the initial trenches and regrading works substantially reduced the gas measured on site to insignificant levels (less than 0.1% methane), and appear to have been extremely effective in this instance.

Site B - Existing Residential and Commercial Development

In Exmouth a landfill gas migration assessment was carried out in the area around a major closed waste disposal site in a former claypit. The investigation involved installing a series of monitoring wells in trial pits and boreholes to various depths and distances from the landfill boundary, and measurement of gas levels using portable gas detection instruments.

Significant levels of landfill gas migration off-site was detected. Methane measurements in the natural ground beyond the landfill boundary were over 1% in many monitoring points, and ranged locally to over 50% in places. In particular, significant gas migration was detected in close proximity to houses and commercial premises. Although no gas has been detected during internal checks to these buildings, the problems of danger and blight to the existing developments necessitated remedial action. In consequence, a series of gas control measures have been implemented including:

- passive ventilation trenches

- gas pumping trial to assess potential for active ventilation by pumped extraction from wells to a flarestack

- passive gas ventilation wells

- continuation of monitoring of gas levels on a regular longterm basis

The problems at this site have proved very difficult to remedy. Despite its age, most of the landfill waste has been in place for more than twenty years, high concentrations of gas are still recorded, although the rate of generation is not sufficient to sustain active abstraction to a flarestack.

Site C - Existing Industrial and Leisure Development

In Barnstaple a landfill gas investigation was carried out in a developed area of landfill reclamation including some zones of domestic refuse waste. The development included existing industrial buildings and a leisure centre. The investigation involved establishing a series of monitoring points in exploratory holes and within buildings.

Significant levels of landfill gas were detected in parts of the estate. Whereas very high and potentially explosive levels of gas were measured in monitoring wells in the ground, in general the buildings were naturally well ventilated with no measurable gas. However slight traces of gas were detected inside some buildings, and particularly high levels were emitted via a service duct in one building. A series of gas control precautionary and detection measures were implemented at this site including:

- passive ventilation/cut-off trenches to minimise the potential for gas migration and accumulation beneath buildings, and to isolate the more highly gassing areas

- permanent gas detection and automatic alarm systems to buildings to afford an effective continuous safeguard to warn of any future build up of gas

- continuation of monitoring of gas levels

Gas detection and alarm systems were adjudged particularly appropriate at this site, where the existing buildings were at risk from gas migration and remedial methods of ventilation and cut-off could not be engineered effectively. The systems were specifically designed for each individual building taking account of its use, its form of construction and possible entry points for gas migration. Sensors were located in identified key locations and poorly ventilated spaces within and beneath buildings, and linked to a central display and alarm panel. A two stage alarm activation was incorporated: a low level warning at 0.5% methane, and a high level alarm at 1% methane for immediate evacuation of personnel.

The use of permanent gas detection and alarm systems in the existing buildings at this site has proved to be extremely effective. It has enabled continued use of the site with enhanced security and confidence and with reduced reliance on periodic monitoring.

Site D - New Retail Development

At Thurrock, Essex an infilled old chalk quarry was included within a site for coach and car parking facilities, with associated management buildings, for an 'out of town' retail development.

Landfill gas was detected in monitoring wells on the site and a distinct pattern emerged. High levels of methane up to some 13% by volume in air and carbon dioxide up to some 6.6% volume in air were recorded in the backfilled chalk quarry. Beyond the zone of the quarry much lower levels of methane were detected although levels of carbon dioxide indicated migration over greater distances through natural ground.

In view of development programme constraints there was limited scope to undertake a detailed monitoring exercise to establish gas generation and behavioural patterns with time. For this reason 'worst credible' parameters were adopted to take account of time and climatic variations on possible fluctuations in the landfill gas regime.

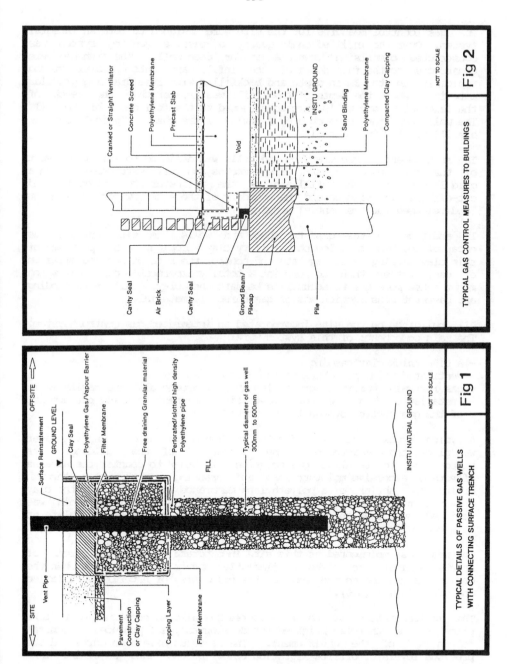

Figures 1 and 2. Typical details of gas control measures.

The gas control measures for the site were designed to form an integral system. Over the infilled chalk quarry a passive venting system was constructed comprising trenches and/or deep wells. The trenches were constructed to the full depth of the infill and into natural ground typically up to 3m in depth and backfilled with free draining granular material. Where the depth of the infill was greater than the reach of the excavator gas wells were constructed to the full depth of the fill. The wells were interconnected by a series of trenches. Typical details are illustrated in Figure 1.

The arrangement of gas trenches and wells was designed to suit the layout of the car and coach park. The trenches and wells were generally on a grid layout across the infill and on the perimeter to intercept and cut-off gas migration beyond the boundary of the site towards development land released for residential housing.

In addition to the gas trenches and wells the site surface was encapsulated beneath a low permeability clay capping. The purpose of this clay capping is to regulate and control the rate of surface emission of gas, rather than to provide a totally impermeable clay seal which might cause more gas to accumulate beneath the clay. All hardstanding and pavement construction was of gas permeable material.

All buildings on the site incorporated an integrated gas control system. The main components of this system were:

- a 0.5m thick clay capping
- ventilated void space beneath the ground slab
- gas restraint membrane across plan area of ground slab and cavity walls
- installation of gas monitoring/detection equipment to allow performance of the gas control system to be checked.

A typical cross section is indicated in Figure 2. The clay capping layer was placed and compacted over the 'footprint' of the building. Its purpose is to regulate gas emission. In order to counter the adverse effects of excessive moisture loss and cracking of the clay through drying and shrinkage a polyethylene membrane was placed over the clay and in turn protected by a sand blinding. The ventilated void was designed such that the gas design concentrations could not accumulate or be exceeded in any 72 hour period.

The measures implemented at this site incorporate an adequate margin of safety against the risks of flammable vapours and explosions when the design was checked on the basis of the principles of natural ventilation and gas dilution (5,6).

The implementation of these measures has allowed construction to take place on land otherwise blighted for development, and produced a general 'clean up' of the environment. The measures have also curtailed the previous levels of offsite migration below ground into adjoining land where planned residential development could also have been blighted.

Site E - New Office Development

The uncontrolled tipping of domestic waste over a number of periods between 1930 and 1950 provided engineering problems for the development of a site at Shinfield in Berkshire for prestigious offices.

Low but variable levels of landfill gas were monitored during investigations, and gas control measures were therefore recommended as a precautionary measure. A passive venting system was adopted to ensure no build up of gas beneath the development. The ground floor was designed as a suspended slab with an underfloor void. Air vents through the external walls were carefully designed to disperse any gas, with ample safety margin to allow for any subsequent increase in methane generation or partial blockage of the venting system.

Monitoring of gas concentrations continued during and subsequent to the development and showed persistent but low levels of gas. No gas however has been detected in the ventilated underfloor void space. The building was occupied at the beginning of 1989, by Frank Graham.

CONCLUDING REMARKS

The purpose of this paper is to illustrate how effective engineering solutions have been applied to overcome problems on sites affected by landfill gas, and thereby safeguard existing developments and permit redevelopment through new construction and an end to dereliction.

The key requirements are:

- detailed investigation and monitoring of the gassing regime to allow the full hazards and potential risks to development, either existing or proposed, to be evaluated

- selection of appropriate gas concentrations and emission rates for use in design

- the design of integrated gas control systems to meet the specific problems of each particular site with adequate margins of safety

REFERENCES

1. Her Majesty's Inspectorate of Pollution, Waste Management Paper No 27, The Control of Landfill Gas, Her Majesty's Stationery Office, London, 1989.

2. Health and Safety Executive, Guidance Note EH40, Occupational Exposure Limits, 1989.

3. Edwards, J.S., Gases - Their Basic Properties. Methane Facing the Problems Symposium, Nottingham, 1989.

4. Card, G.B. and Roche, D.P., The Design of Gas Control Measures for Development Affected by Landfill Gas. Methane Facing the Problems Symposium, Nottingham, 1991.

5. British Standards Institution, BS 5925 : 1980, Code of Practice for Design of Buildings : Ventilation Principles and Designing for Natural Ventilation.

6. Building Research Establishment, Ventilation in Relation to Toxic and Flammable Gases in Buildings. Current Paper CP 36/74, 1974.

REDEVELOPMENT OF CYMMER COLLIERY PORTH

N J TAYLOR
L G Mouchel & Partners Ltd
Harcourt House, 15 St Andrews Crescent,
Cardiff, South Glamorgan CF1 3DB, UK

ABSTRACT

In the reclamation of any derelict site the problems need to be identified at an earlier stage, the future use of the site must be determined, and a study undertaken to compare the cost of reclamation with the value of the site and the cost of development knowing the proposed end use.
 The site under consideration was previously a colliery which had been left to fall into a state of dereliction. The reclamation works were considerable but despite this, and with the aid of grant money, the project proceeded to the satisfaction of all concerned including the property developer.

HISTORY OF THE COLLIERY

The area known as Cymmer Yard must have been known for the coal which outcropped just on the western edge of the site. There are no records but we can only assume that from earliest times our ancestors dug into the shallow seams that outcrop in Cymmer Yard.

 On 25 September 1844 George Insole, a coal dealer, and his son took a lease on 375 acres of land at Cymmer for a period of 70 years. In 1847 George Insole sank the Number One shaft in Cymmer Yard to work the number three Rhondda seam. The coal from this seam proved to be fine coking coal and by 1836, thirty six coke ovens had been erected on Cymmer Yard.

 In December 1987 a pair of shafts 70 yards apart were sunk to exploit lower coal seams. The downcast shaft was 16 feet in

feasibility study to identify the specific abnormal aspects, and make an estimate of the cost of treating these abnormal aspects, to achieve a site which would equate to a green field location.

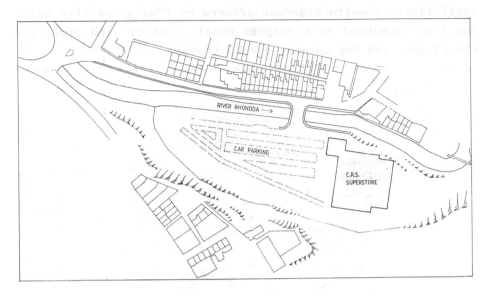

Figure 3. Plan for Food Superstore.

ASPECTS OF THE RECLAMATION

The study identified a number of aspects which required engineered solutions and in turn these solutions attracted abnormal costs which had to be estimated.

Demolition

An advance contract was let for the demolition of the colliery buildings - the headframes and steam winding engines had already been removed. This included, of course, the demolition of the winding engine house chimney.

The specification included for the demolition of all basement walls to a metre below ground and the filling of the basements with rubble. The rubble was tipped into the basements to be compacted at a later date by vibro-compaction.

A number of tunnels and culverts were discovered during the demolition work and later during construction. Where possible these were broken out and filled. Small culverts were pumped full of pulverised fuel ash and cement grout.

The Demolition Contract was completed in four weeks at a cost of £28,500. This closely matched the estimate for the work set out in the initial study.

Diversion of Surface Water Culvert and Gas Main

A stone built arch culvert crossed the site taking surface water run off from the mountains on the west of the valley into the River Rhondda. The absence of records made this difficult to locate, but it became apparent that it crossed the proposed superstore site. Immediately following the demolition contract a new concrete culvert 1.2m x 2.15m in size was constructed around the new building position. This was a structure with manholes at regular intervals with vented covers to aid methane escape. At the same time the trunk gas main which served this valley was diverted to the edge of the site.

Shallow Workings

As part of our initial study we consulted historical records, geological memoirs and maps. These indicated that the upper coal seams, in particular the Rhondda Number 2 was within 12m of the surface and outcropped at the boundary of the site. It was therefore most likely that shallow mining had been carried out to remove some of this coal. A contract was let immediately following demolition, to investigate ground conditions by means of shell and auger drilling and rotary coring to establish the extent of these workings. Twenty-nine holes were sunk both by shell and auger and rotary cored methods. The investigation confirmed the existence of voids which corresponded in depth to the Rhondda Number 2 seam. An estimate of the extent of the removal of this seam by mining was made from the boreholes taken.

The Specification for grouting shallow workings required a primary grid of holes at 12m centres to a depth of 13.6m. This grid was then infilled at 6m centres and then at 3m centres. In the event the grout take was substantially completed during the drilling and grouting of the second grid. This work progressed well, although it soon became apparent that yet more underground waterways and culverts not discovered during the demolition contract existed. A sharp look out was kept for grout leakage into the River Rhondda and any wide scale contamination was avoided.

This contract was completed within budget, despite these difficulties, at a total cost of £360,350 inclusive of mine shaft capping.

Mine Shaft Capping

It is known from the history of Cymmer Colliery that three shafts existed on the site. Insoles first shaft (Cymmer Old Pit) was actually outside the development site. The two further shafts which were sunk in 1875 were just within the site and had been sunk to a depth of 397m. There had been no effective shaft capping - only timber at ground level. It was necessary to effectively cap the shafts at rock head with a 1 metre thick reinforced concrete cap. The rock was some 10m down. Consideration was given to the best way of constructing the shaft caps and it was decided that by doing this work prior to the necessary ground improvement work the rock head could be exposed in a battered excavation. This required an excavation 10m deep and almost 30m in diameter. By suitable selection of plant this exercise progressed very speedily and was completed ahead of programme.

Ground Conditions

The desk study and subsequent site investigation showed that the site was overlain with up to 7.5m of fill. This fill was predominantly colliery spoil of varying density. The surface had been contaminated by both colliery debris and spoil from the non-ferrous metal recovery operation.

Below this loose fill a 5m thick band of dense sand and gravel overlaid the coal bearing sandstones. In view of the variable density of the fill we opted for vibro-consolidation of the fill to a depth of 5m using pressure points at 2.2 m centres beneath the building. This considerably improved the density of the granular fill. In the knowledge that we had grouted up the old shallow workings and that there would be no further deep mining beneath the site we were able to design conventional pad foundations at a permissable ground bearing pressure of $200KN/m^2$.

High methane concentrations were measured in a number of locations. Whilst initially it had been assumed that the most significant methane emissions were from the shafts, subsequent work revealed that the fill was the major contributor to the methane. Whilst the reinforced concrete shaft cappings had provision for a vent which would have discharged above ground the levels measured at rock head were found to be very low and therefore high level venting was not deemed to be necessary. Nevertheless to counteract the general methane level, the under slab membrane was specified as two layers of 1200 gauge polythene taped and sealed at all joints. Drainage was designed with vents to discharge any methane to atmosphere. There have been no complaints in recent years of the Porth Pong and we wonder whether the source of this was the organic putrescible material which had been disposed of down the redundant shafts, some remains of which were excavated during our shaft capping.

Outside the building area most of the site was to be paved for car parking and in view of the inhospitable ground conditions, it was agreed that there would be no soft landscaping. This would have required substantial importation of top soil. Instead locally quarried stone was used to form scree slopes.

The ground improvement was let as a separate contract to Bauer Foundations at a cost of £42,857. The additional polythene membrane, the special detailing of the drainage, and the stone scree as opposed to the planting, was not considered to have any significant increase on the total development cost.

Access

The access into the original colliery site was from the north at a junction immediately adjacent to the A4058 river crossing (See Figure 1.). This arrangement was awkward and inadequate for modern vehicles and was neither acceptable to the highway authority or from the retailers viewpoint. The adjacent public road the A4058 (Pontypridd Road) which passed through Porth and continued up the Rhondda Valley is in fact sited on the opposite side of the river Rhondda to the site. Furthermore this road is narrow. It is bounded on one side by terrace houses and the other by the river. Pontypridd Road is one way and a mere 6.0 metres wide. The earliest intention was to provide a skew bridge crossing the river at the point where Pontypridd Road approaches the River Rhondda. It is apparent from early maps that there had once been a bridge in this position (See Figure 1). However, mainly because an entrance into the site at this point would make the internal site planning awkward, and because of the need to re-locate the Public Conveniences, this bridging site was abandoned in favour of a bridge further north on the straight section of Pontypridd Road.

This site gave access to the development site at its widest point which gave greatest flexibility of planning. It must be remembered that major food retailers have specific requirements on the standardisation of both size and shape of new stores. This precludes long narrow plan forms notwithstanding the site configuration.

The chosen bridge site soon brought further difficulties in terms of level. The requirements of Welsh Water was for one metre clearance above highest known flood levels to the underside of the new bridge structure. The top of the bridge was dictated by the existing level of Pontypridd Road. Furthermore the existence of the terrace houses on the opposite side of the road did not permit a substantial revision to vertical road alignment.

The occupants of these houses had been able to park on Pontypridd Road outside their houses. The chosen bridge site required that parking would be prohibited. Whilst the residents of the houses were delighted to think that the decrepit and decaying colliery buildings were soon to be replaced by a clean new development which would also provide convenient shopping, they were very concerned about losing car parking outside their homes. It fell to the architect and myself to visit all the affected residents to re-assure them that our client would do his best to replace their loss of amenity. In the event every house that required the facility was provided with a hardstanding in their small front gardens for one car with an access on to the highway.

We finally adopted a bridge deck of only 1030mm structural depth formed of precast concrete bridge beams transversely stressed with Macalloy bars. To avoid disruption to the traffic in Pontypridd Road we elected to use the existing river wall, adjacent to the road, as an abutment. This wall was constructed of random rubble and in poor condition. Careful probing revealed that if this could be adequately stabilised the thickness of stone was sufficient. A scheme incorporating tension piles down through the wall, ground anchors back under the wall and grouting to preserve the homogeneity of the wall was derived. In practice the work was made more difficult by the presence of a 12 inch diameter cast iron sewer laid on the river bed and a large number of services beneath the existing road. The work despite being difficult to execute was in fact carried out to budget and to programme. A total of 13 vertical tension piles were installed and a considerable number of ground anchors. Our worst fears concerning the collapse of the river wall were not realised although difficulties were encountered in the grouting.

The western bridge abutment was designed as a reinforced concrete retaining wall founded on rock in the river bed. This was constructed by forming a rock fill bund down the centre of the river and pumping out the site for the bridge abutment.

The bridge access which was the result of much discussion and agreement with the highway authority, the water authority and the residents did in fact cause further aggravation when the transverse stressing bars failed during stressing as a result of hydrogen embrittlement. This was overcome by re-stressing above and below the deck. This, of course, required a revised vertical alignment and our Client is grateful to the realistic approach taken by both the highway and water authorities in permitting some relaxation of standards.

River Walls

The river at this point was not only polluted with debris but the river walls were in a state of collapse. Those supporting the Pontypridd Road were in the best condition. Notwithstanding this, in addition to the work required to strengthen the wall to make it adequate as a bridge abutment, considerable work was required in replacing and repointing stonework. The development site river bank was partly supported by existing walls containing outfalls from culverts which discharged water pumped from the old colliery. Furthermore the river was restricted in width by a redundant concrete flume. The walls required making good and the concrete structure had to be demolished.

In areas where the river walls were irreparable or non-existent, Macafferi Gabions were specified. The specification called for the stone in the front face to be hand placed to resemble dry stone walling. The batter above the new wall was a stone embankment "scree". This was considered to be in keeping with the surrounding landscape and would be relatively maintenance free.

The work to the river was considerable and cost £131,500.

THE COST OF RECLAMATION

The total abnormal costs for the development were made up as follows:-

Site Investigation	£ 15,280
Demolition	£ 28,500
River Works and New Surface Water content	£131,500
Ground Improvement	£ 42,857
Shaft Capping and Grouting Shallow Workings	£360,330
New Road Bridge	£311,313

The above work was partly paid for by the Urban Development Grant made by the Welsh Development Agency of three quarters of a million pounds.

This compared with the overall cost of the development of £4.3m and the site value, which although not known in this case, would typically be in the order of £3m.

CONCLUSION

It can be concluded that the most unlikely sites can be reclaimed for relatively small cost compared with the total cost of development. Furthermore a study undertaken early enough can give prospective clients the assurances they need to proceed with such a venture and enable grant money to be sought which will make the prospect attractive enough to proceed with.

REFERENCES

1. Rhondda Collieries Volume 1 by John Cornwell published 1987 D Brown & Sons Ltd, Cowbridge. ISBN 0905926822.

2. Institute of Geological Sciences map - Sheet 248
    - Glamorgan sheet XXVII SE

THE APPLICATION OF STRATA 3 TO THE ASSESSMENT OF CONTAMINATED LAND -
A CASE STUDY

B.R. THOMAS and L.M. GREENSHAW
Wallace Evans Ltd.,
Penarth, South Glamorgan. CF6 2YF, U.K.

S.P. BENTLEY and A.S. STENNING
University of Wales College of Cardiff,
Cardiff, South Glamorgan, CF2 1XH, U.K.

ABSTRACT

Strata 3 is a PC-based computer program for producing and interrogating 3-D models of the subsurface environment. The application of Strata 3 to the assessment of contaminated land is discussed with reference to investigations carried out at the Grangetown Gasworks site in Cardiff. The proposed impoundment by Cardiff Barrage will cause a groundwater rise of varying amounts within South Cardiff, including the the Grangetown Gasworks site. This paper demonstrates the use of Strata 3 as a tool for the assessment of existing pollution and contamination hazards at the site and evaluates the subsequent implied risk resulting from modified groundwater levels.

INTRODUCTION

The proposed construction of and impoundment by Cardiff Bay Barrage will cause a groundwater rise of varying amounts within South Cardiff including the Grangetown Gasworks site. From previous studies (1,2) the predicted groundwater rise at the site is approximately 1.5 to 2.5m to a maximum level of +5.0m OD; this compares to existing ground levels of between +7.2 to +10.1m OD.

A survey of pollution and contaminated land in south Cardiff (3) identified deep fill and potential contaminants at the Grangetown Gasworks site. Although no detailed information existed, expected contaminants included tar emulsions and spent oxides. Concern as to the nature, extent and level of contamination of the ground and groundwater and the potential for the mobilisation of contaminants resulting from a change in groundwater levels at the site indicated that further site investigation was necessary to provide more precise information.

The paper outlines the site investigation procedures adopted, and in particular focuses on the use of the data management system, Strata 3, as a tool for the collation, presentation and assessment of the geological and chemical data. Considerations of possible risk to the site operational works are outside the scope of this paper.

Contaminants at Gasworks

Gasworks sites may be expected to be seriously contaminated from the raw materials used and the products and wastes of town gas manufacture, including ammoniacal liquors, coal tars and other hydrocarbons, cyanides, heavy metals, sulphur and sulphates.

The potential contaminants and the principal hazards presented by them are summarised by the Interdepartmental Committee on the Redevelopment of Contaminated Land (ICRCL) (4). Given the range of possible contaminants, which may be present as solids, leachates and gases, the chemical assessment and determination of risk from contaminants is problematic. The ICRCL notes on the redevelopment of gasworks sites (5) and ICRCL (4) provide guidance for site assessment using trigger concentrations for specific chemicals. The range of contaminants covered is incomplete and is due to be extended together with further guidance on interpretation. These guidelines must be interpreted with care and informed professional judgement is a prerequisite of their use in site assessment.

The EEC Groundwater Directive 80/68/EC (6) together with Welsh Office Circular 34/90 (7) provide further guidelines on the protection of groundwater against pollution.

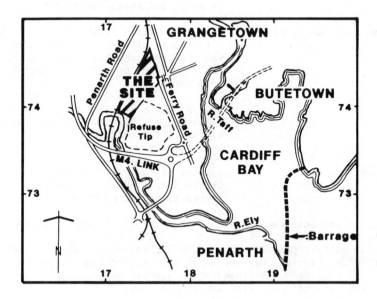

Figure 1. Location of Grangetown Gasworks Site

Site Location

Grangetown Gasworks comprises a 15Ha site in South Grangetown, to the west of Cardiff. The site is bounded to the south by the tidal River Ely, to the north and west by the Penarth-Barry railway lines and to the east by Ferry Road landfill tip, Figure 1.

The existing site complex consists of numerous above and below ground structures, with shallow and deep foundations, and large areas of concrete/tarmacadum hardstanding overlying filled-in ground and old foundations.

SITE INVESTIGATION

The general lack of chemical and pollution data indicated the need for a comprehensive site investigation. This was generally undertaken in accordance with the British Standard Draft for Development DD175 (8) and BS5930:Code of Practise for Site Investigations (9), as follows:

Desk Study — involved collection and assessment of all relevant published data, previous site investigation reports, aerial photographs and discussions with on-site and retired Gasworks staff.

Identification of priority — an understanding of the history of the site areas for investigation development (especially infilled areas) and past manufacturing processes enabled site zoning into areas of unlikely, possible, probable and known contamination.

Design of Site Investigation — based upon all available data and the site zoning plan, a site investigation was designed comprising six boreholes to 20m depth and forty-three trial pits to 3-5m depth.

Sampling — approximately 200 soil and groundwater samples were collected for testing and examination during the site investigation.

Laboratory Testing — a total of 127 soil and water samples obtained from the site investigation were tested as indicated below:-

Ninety one soil samples were tested for the following parameters:- pH, total cyanide, acid soluble sulphate, sulphide, toluene extractable matter, phenols, lead, arsenic, selenium, copper, nickel, zinc, chromium, cadmium, barium, beryllium, molybdenum, strontium, vanadium and manganese. Representative samples were also examined for water soluble sulphate and free and simple cyanide.

Thirty six water samples, twenty four from trial pits and twelve from boreholes, were tested for the following parameters:-
pH, electrical conductivity, cyanide, sulphate, sulphide, total organic carbon and polycyclic aromatic hydrocarbons.

More details of the site investigation and laboratory test results are presented elsewhere (10).

GEOLOGY AND GROUNDWATER

The geological succession is shown in Table 1. Information from the site investigation was input into Strata 3 and a 3-D geological model of the site was produced. Typical outputs from the Strata 3 system in the form of a geological cross-section, contour map and perspective view are given in Figures 2,3 and 4.

TABLE 1 Generalised geological succession at the Grangetown Gasworks site

Strata	Description	Thickness (m)	O.D. Levels at Base of Strata
Made Ground	Clay, silt, sand, gravel, ash, bricks, building rubble	0.8 to 4.2	+3.9 to +6.85 (generally > 5.5+)
Estuarine Alluvium	Blue/grey/brown silt-clay with peat and organic matter	2.9 to 9.1	-3.1 to +2.5
Fluvioglacial Gravels	Grey brown very sandy fine to coarse gravel	4.5 to 13.7	-7.6 to -15.0
Mercia Mudstone	Red-brown silt-clay with mudstone lithorelicts	2.6m (proved)	-

Long term groundwater observations from existing piezometers indicated an ambient piezometric surface at approximately +2.2m OD i.e. within the Estuarine Alluvium. The readings showed that the groundwater is not significantly tidal. Groundwater levels recorded during the site investigation indicated the presence of two groundwater regimes which are not in direct hydraulic continuity:-

a) a perched water table in the Made Ground observed as strikes and seepages especially in trial pits. The perched water table phreatic surface is shown in Figure 2 representing the upper groundwater surface contour at the site, generally at levels of between +6.0 to +7.0m OD.

b) a sub-artesian groundwater table within the gravels, confined by the relatively impermeable Estuarine Alluvium, Figure 2. It was observed as strikes near the alluvium/gravel interface which rise slowly to levels of +1.0 to +2.2m OD in agreement with previous observations.

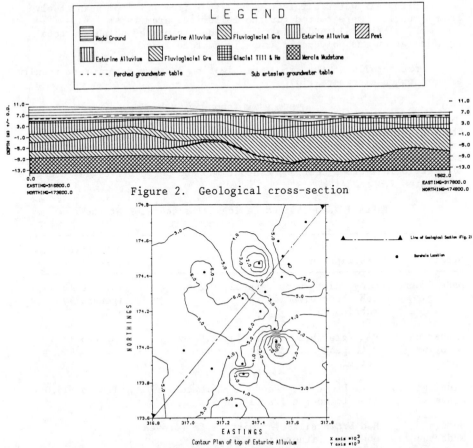

Figure 2. Geological cross-section

Figure 3. Contour map of top of Estuarine Alluvium (m. O.D.)

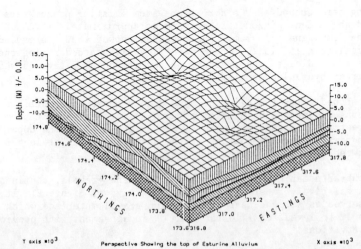

Figure 4. Perspective view showing top of the Estuarine Alluvium

DATA MANAGEMENT USING STRATA 3

Strata 3 is a PC-based computer program for producing and interrogating 3-dimensional models of the subsurface geological environment. The use of geological information, including borehole records, to contour and hence define geological surfaces is well-documented (11,12). However, resulting graphics packages tend to be of a general nature or have been tailored for certain applications, notably in petroleum and mining geology. Wager (13) gives an overview of some of the more recent packages. Strata 3 is one of the few computer programs designed by geotechnical engineers for use by geotechnical engineers.

Strata 3 system development has been undertaken in the School of Engineering at the University of Wales College of Cardiff, and has been sponsored by Wallace Evans Limited, Consulting Engineers, Penarth.

In its present form Strata 3 provides a variety of tools for interrogating geological models, these include functions to give numerical information, cross-sections, perspective views, isopachyte maps and geological maps. The 3-D subsurface model is produced by interpolation between data points. A number of different interpolation methods are available within Strata 3, including simple triangulation, kriging and weighted average. The aim of the interpolation is to fit the data points precisely and this is achieved using shaped surfaces which are geologically acceptable.

In this study Strata 3 was used in conjunction with an ORACLE database which was written to provide a general-purpose data store for site investigation information. Relevant data can be downloaded from the database directly to Strata 3.

Inputs to Strata 3 are geological information and geological knowledge/experience. Any item of geological information that can be defined spatially may be input. In this application boreholes and trial pits are the data source but information from geological maps, cross-sections and interpreted geophysical and remote-sensing surveys can also be incorporated. The second major input is geological knowledge or experience. Modifications to, say, geological sections can be made quickly and easily by transferring the section to a drafting package; one of Strata 3's functions allows automatic downloading to Autocad. An alternative approach is for geological knowledge to be introduced using artificial or knowledge-based boreholes or data points. This approach has the benefit of creating an "approved" model which can then be used by junior staff or non-specialists.

There are a number of ways in which the geological model can be interrogated. What they have in common is that the on-screen operations are very rapid and they are very simple to request, for example, numerous cross-sections in different directions can be inspected in a few moments.

Qualitative methods of interrogation using the graphics include

- Cross-sections showing strata layers and groundwater.

- Geological maps of the ground surface or maps with the upper layers stripped away to reveal either the surfaces of lower layers or the strata present at a particular elevation, for example, in an excavation.

- Perspective views of 3-D geology

Graphics can be plotted out at sizes from A0 to A4: cross-sections can be produced at working scales.

Numerical information can be obtained from:

- Contour maps, for example, base of Made Ground or top of rock.

- Isopachyte maps.

- Precis of any borehole log or other record, can be called up on-screen.

- x, y and z co-ordinates for points on any surface selected by the cursor.

Graphics depicting the geology of the Gasworks site have had to be simplified to black and white line figures: Figure 2 is a geological cross-section showing the positions of the two water tables underlying the site, Figure 3 is a contour map (in m.O.D.) for the top of the Estuarine Alluvium and Figure 4 is a perspective view of the site chosen to shown the deeper pockets of Made Ground above the Estuarine Alluvium.

Geological information from the site investigation was entered into the geotechnical ORACLE database, in the absence of a tailor-made chemical database the chemical data was entered directly into Strata 3 as ASCII files. The construction of a linked database would provide obvious advantages and tailor made systems in dBase, Lotus, Foxpro can easily be linked to Strata 3.

A number of simple enhancements to the functionality of Strata 3 were necessary to enable the processing and presentation of chemical data. Firstly, a function was needed to provide a continuous definition of concentration values down the sampling profile. A simple function connecting known values with straight lines within the limits of determination was used for this purpose. This function permitted interpolated chemical concentration data to be obtained for any depth for all data locations. The data obtained from this function was then used by existing Strata 3 routines to provide contour maps of horizontal slices through the site. Contour map showing concentrations of sulphates and polycyclic aromatic hydrocarbons (PAH) at an elevation of 5m O.D. within the ground are shown is Figure 5. The contour intervals used in each chemical map can be easily modified within the system to coincide with the ICRCL "threshold and action trigger" levels corresponding to each contaminant. Contour maps zoned with respect to the trigger levels

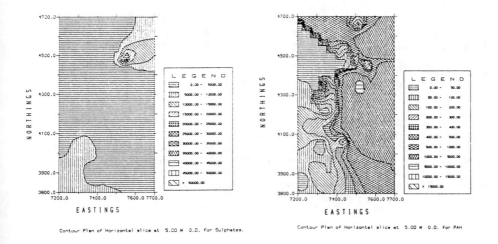

Figure 5. Contour maps of sulphate and PAH concentrations (mg/kg)

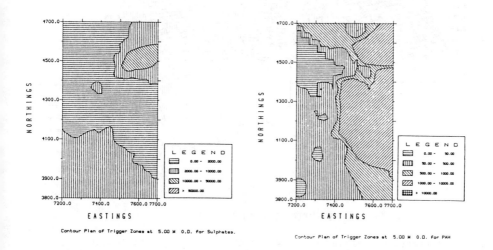

Figure 6. Contour maps showing trigger levels for sulphate and PAH (mg/kg)

for sulphates and PAH are illustrated in Figure 6. This form of
graphical output was achieved by coding the sampling profiles at each
location to coincide with appropriate ICRRL trigger values e.g.

a - < Threshold Trigger
b - > Threshold Trigger but < Action Trigger
c - > Action Trigger

These coded profiles were used to establish a 'layer sequence' for
the whole site which accorded with that for each individual sampling
location. The coded profiles and agreed 'layered sequence' where then
used by Strata 3 to create representations of the required information in
3-D. A surface perspective view of trigger zones for cadmium in soil is
shown in Figure 7 as an example of this form of output.

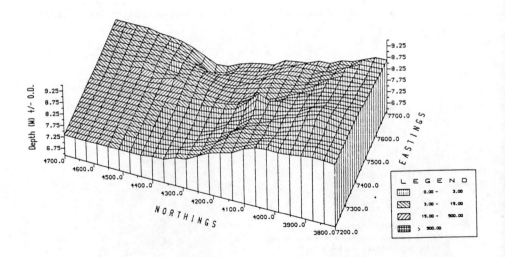

Figure 7 Perspective view showing trigger levels for cadmium

IDENTIFIED CONTAMINATION

The investigations carried out have identified and quantified the
pollutant loadings at the site indicating contamination typical of its
former usage as a town gas manufacturing site (10). Principal
contaminants were identified on the basis of past experience in the
assessment of contamination at gasworks sites, and have been mapped at
the site using Strata 3. Principal contaminants exist both in the Made
Ground and to a lesser extent in the underlying Estuarine Alluvium,
together with contaminants in both the perched and confined aquifers -
They include significant concentrations of the following chemicals:-

Chemical	Soils	Waters
Sulphate	X	X
Chloride		X
Phenols	X	X
Cyanide	X	
Polycyclic aromatic hydrocarbons		X
Ammoniacal Nitrogen		X
Toluene extractable matter	X	
Cadmium	X	
Copper	X	
Lead	X	

The use of Strata 3 allowed a rapid visual and quantitative assessment of the spatial distribution of the principal identified water and soil contaminants at the site. Example contour plan outputs from the contaminant mapping exercise are shown in Figures 5 and 6 for sulphates and PAH.

The contour plan for sulphates indicates highly variable concentrations in the Made Ground, which ranged from less that 50 ppm to 116620 ppm, which is above the ICRCL 'action concentration of 50000 ppm for buildings.

The levels of PAH ranged from 1-9700 ug/ℓ indicating very poor water quality with significant localised contamination of both the confined and perched aquifer.

The information obtained from the contaminant mapping exercise was used to assess the ground and groundwater within the zone of potential inundation associated with the barrage impoundment. This, together with consideration of possible pollutant transfer mechanisms at the site including displacement due to groundwater flow, inundation and capillary rise, allowed identification of the implied risk resulting from the predicted increase in groundwater levels.

CONCLUSIONS

The large quantities of chemical data generated in contaminated site studies and environmental audits are increasingly being collated and interpreted using computed-based information systems. The Strata 3 system, which was designed to build geolgical models, was adapted to handle chemical data and build 3-D models of the subsurface chemical environment. The resulting graphics provided easy-to-interpret visual images in the form of maps, cross-sections and perspective views, and these were complimented by the numerical functions of the Strata 3 system.

The principal findings at the Grangetown Gasworks site were:-

a) The perched water table within the made ground is contaminated with substances which under EEC Directive 80/68/EEC(6) should not be allowed to be discharged into natural groundwaters.

b) the potential exists for pollutant dispersion into the Made Ground through the perched water table and through very localised connections with, the underlying confined aquifer at 'low-spots' in the base of the Made Ground (see Figure 4). This potential is a function of the hydraulic gradient of the confined aquifer which will be substantially reduced by the impoundment. Pollutant mobilisation by this mechanism is therefore unlikely to constitute a significant or measurable hazard.

ACKNOWLEDGEMENTS

The authors wish to thank James Rendle, and Marcus Cato, UWCC, for their help in data processing and Cardiff Bay Development Corporation for permitting information from the site investigation to be used in this paper. The opinions expressed in the paper are those of the authors.

REFERENCES

1. Wallace Evans and Partners, Cardiff Bay Barrage Feasibility Study, Groundwater Investigations Stage III, April, 1989.

2. Wallace Evans and Partners, Cardiff Bay Barrage Feasibility Study, Report on Computer Modelling in North Cardiff and review of Groundwater data, December 1989.

3. Wallace Evans and Partners, Cardiff Bay Barrage Feasibility Study, Groundwater Investigations, Report 10. Survey of Pollution and Contaminated Land, November, 1988.

4. Interdepartmental Committee on the Redevelopment of Contaminated Land, Guidance on the assessment and redevelopment of contaminated land. ICRCL 59/83, Second Edition, 1987.

5. Interdepartmental Committee on the Redevelopment of Contaminated Land, Notes on the redevelopment of gasworks sites. ICRCL 18/79, Fifth Edition, 1986.

6. EEC-Directive 80/68, The protection of groundwater against pollution caused by certain dangerous substances, EEC, Brussels, 1979.

7. Welsh Office, EC Directive on protection of groundwater against pollution caused by certain dangerous substances, (80/68 EEC): Classification of listed substances, Circular 34/90, HMSO, London, 1990.

8. British Standard Draft for Development 175:1988, Code of Practice for the identification of potentially contaminated land and its investigation, BSI, London, 1988.

9. British Standard 5930:1981, Code of Practice for Site Investigations, BSI, London, 1981.

10. Wallace Evans and Partners, Cardiff Bay Barrage Feasibility Study, Groundwater Investigations, Contaminated Land Study at Grangetown Gasworks, January, 1990.

11. Davies, J.C., Statistics and Data Analysis in Geology. John Wiley, 1986.

12. Jones, T.A., Hamilton, D.E. and Johnson, C.R., Contouring geologic surfaces with the computer. Van Nostrand Rheinhold, 1986.

13. Wager, F.J., Geobyte Index and Reference - 1985 Through 1990. Geobyte, 1991, 5 (6), 20-27.

WINSOR PARK HOUSING SCHEME
THE RECLAMATION OF AN EX-GASWORKS SITE IN BECKTON

PETER TREADGOLD
Partner, Flynn & Rothwell
Charrington House, The Causeway, Bishop's Stortford, Herts, CM23 2ER
formerly: Area Engineer, London Docklands Development Corporation

SIMON TILLOTSON
Senior Consultant, Environmental Resources Ltd
1-2 The Enterprise Park, Boughton Green Road, Northampton, NN2 7AH

SOCIAL & POLITICAL ISSUES

The Demise and Regeneration of London's Docklands:

The Decline of the London Docks: During the 1970s, the slow decline of London's Docklands continued. In the relative affluence of the post-war years, dockers' increased militancy for higher rates of pay led to several major strikes. The increase in labour charges and disruption caused the employers to seek cheaper and faster methods of handling cargo. Containers fulfilled this need and the Port of London Authority (PLA) invested heavily in the latest technology, both in cargo handling and ships, and built the new docks at Tilbury.

The Formation of the London Docklands Development Corporation: In the late 1970s, the local London Boroughs worked together through the Docklands Joint Committee (DJC) to develop a strategy for the regeneration of London's Docklands. This was summarised in the London Docklands Strategic Plan. Inevitably, the aspirations of the member Authorities occasionally conflicted, and their philosophy tended to be 'needs-led', rather than responding to the market. More than anything, the performance of the Committee was restricted by the shortage of funds. However, and despite these problems, a significant amount was achieved, particularly by the London Borough of Newham (LBN) where extensive infrastructure provision was made in preparation for the new Beckton community, via Urban Aid funding packages.

The limited achievements of the DJC (for whatever reason) and the need for the Government to be seen to be responding to the problems of the 'Inner City', prompted action by Central Government. They stepped in to break the impasse and through the Planning & Land Act established Urban Development Corporations, in particular London

Docklands Development Corporation (LDDC), to promote the regeneration of derelict inner city areas.

Redevelopment of the Royal Docks: The Royal Docks Area Team of LDDC was established in 1984, though it was not until May 1986 that it had an on-site presence north of Royal Victoria Dock. The Corporation made an early decision to promote the Royals' development through a 'Consortia Approach', to attract a diversified range of co-ordinated proposals. In particular, this route encouraged offers to be made incorporating significant community and social facilities by off-setting these costly items against the more lucrative elements of a particular development scheme.

In the eastern regions of the Royal Docks, the Corporation chose to enter into negotiations with Rosehaugh Stanhope to redevelop the Albert Basin, the Royal Albert Dock, the Harland & Wolfe Site and an area of land to the south of Winsor Terrace and west of the Beckton Gas Works site, that became known as Winsor Park. See Figure 1.

Social Housing in Newham

The Housing Shortage in East London: The London Borough of Newham has a long standing housing shortage. Estimates suggest that in excess of 3,000 dwellings would be required to satisfy this need. In addition, the Borough is the second most disadvantaged, when measured against a range of social deprivation factors. There are currently over 9,000 households on the Council's housing waiting list. The problem has been exacerbated in recent years by the 'right to buy' legislation and national trends towards smaller households.

Public Sector Housing in Beckton: The Beckton area of Newham's Docklands has represented one of the Borough's limited number of remaining opportunities for major housing development. Although much of the land in public ownership was vested in the Corporation, the subsequent land assembly and re-distribution has resulted in new-build Council housing in North Beckton, where approximately 2.5 ha of public rented accommodation has been integrated with development that includes much private and shared-equity houses. Other public sites in North Beckton and West Beckton have been proposed but further new-build has been made difficult by recent government legislation requiring the majority of income from capital receipts to be used for debt redemption.

Social Housing as an Alternative to Council Housing: Central Government, whilst wishing to discourage the provision of rented council accommodation, has actively promoted methods of funding rented and co-ownership housing by alternative mechanisms, sponsored primarily by the Housing Corporation. However, Government grants to the Housing Corporation are limited, and it is generally necessary for 'Social Housing' alternatives to be linked to capital projects such as large-scale redevelopments, to achieve 'subsidy' through infrastructure or some other provision.

Beckton District Plans and the Royals Development Strategy

The Beckton District Plans: In the early days of the Royal Docks regeneration, the London Borough of Newham, which was party to the London Docklands Strategic

Plan, produced its own proposals for the Newham Docklands. The South Docks and

Beckton Local Plans were promoted at a Public Enquiry in 1986 (1). Objections raised to these plans by the Corporation were perhaps indicative of the lack of discussion between the Council and the Corporation and also that the Council was out of step with the feelings of the local community who generally supported the Corporation's proposals for regeneration.

Even though Newham Council has statutory responsibility to produce local plans, the Inspector supported the broad strategy of the Corporation and the principle that the Docks themselves remain sensibly as existing, for use by the Community. The Corporation has never produced firm plans to parallel the local plans of the Council and has chosen to rely on development frameworks to encourage market-led developer responses.

The Royals Development Strategy: The absence of a structure plan and its substitution with a framework should not be seen as the Corporation throwing open the Royals to ill-defined regeneration. Coupled with the disposal strategy defined in the 1985 framework (2), there has been intensive work to provide strategic services, transport and engineering work to ensure that sites can be properly marketed at the time of their disposal.

This work includes:

1. Complementing the existing North London Link (British Rail) with an eastern extension of the Docklands Light Railway to link Beckton through the Royals to the Isle of Dogs and on to the City of London. See Figure 1.

2. Provide a strategic highways network from the southern end of the Mll Link Road (A406) through the north side of the Albert Dock, and on to the Lower Lea Crossing via the Connaught Crossing and North Woolwich. See Figure 1.

3. Construction of a strategic drainage network costing some £50 million including new surface water and foul water pumping stations and the installation of strategic services.

4. Reclamation of polluted sites and the provision of on-site infrastructure to advance individual development packages.

5. Negotiate for the early building of a STOL (Short Take-Off and Landing) Airport between the Royal Albert and King George V Docks. See Figure 1.

This rolling programme of works has allowed the Corporation to offer serviced sites, in a prime position, linked to the national and international transport infrastructure.

Strategic Planning and the Winsor Park Development: Early proposals published by Newham Council had included infilling much of the Royal Victoria Dock to permit the development of public sector housing. This was opposed implicitly in the Royals Development Strategy (2), and in the negotiations that followed the publishing of the Draft Revised Beckton District Plan (1), it was proposed by the Corporation that the Council be approached on the basis that the land to the south of Winsor Terrace (Winsor

Park Site) could be used for public rented housing, that would otherwise have been created by the filling of part of the Royal Victoria Dock.

This idea was discussed during 1986/87 as Rosehaugh Stanhope were formulating proposals for the regional shopping centre site and business park development in the east of the Royals. Newham Council were concerned about the mainly commercial nature of the development, and in response to these concerns the Developers proposed to integrate a diversified social housing package on Winsor Park, into the consortium proposals. A Memorandum of Agreement between the Corporation and LBN in September 1987 identified the Winsor Park site and recognised that social housing needs in the Royal Docks could only be met by the reclamation of substantial areas of derelict land.

Marketing

Public Perception: The Winsor Park site had been previously abused, being subject to occasional indiscriminate dumping of polluted material from the operational gasworks to the east. Whilst the level of toxicity was known to be not extreme, local people recognised the site to be polluted generally with random distribution of wastes. The closely knit local community had lived in the shadow of the "largest gasworks in the world". It is fair to say that as scientific awareness of the dangers associated with ingestion of these pollutants has increased, public concern has been heightened, sometimes out of proportion to the real risks.

Political Perception: The politics of the redevelopment of East London are complex and many years of neglect and decay have left residents cautious of the new initiatives proposed by the Corporation. Whilst the Council was keen to obtain housing for people in need, the philosophies of the Corporation and the Council had to be reconciled. Objections to the use of this site for housing were raised early in the design process and pre-dated engineering proposals. Dialogue continued throughout the design and construction processes, although the real level of adverse feeling was difficult to gauge, and appeared to be founded on political rather than technical imperatives. Probably, much of the opposition resulted from a belief that too little land had been provided too late for social housing, and that whilst private housing had been built on clean marsh land, public housing was proposed on the remaining polluted site, apparently almost as an afterthought. This view was expressed at several public and council meetings.

Purchase and Disposal of the Site: Prior to redevelopment, the Winsor Park Site was owned by the Gas Board and in the early 1980s lengthy negotiations secured its purchase by the LDDC under the land assembly programme. The purchase price reflected the market value of the land and the need for reclamation works.

The disposal of the site became an intrinsic part of the Rosehaugh Consortium proposals and it was originally proposed that the Developer, as part of the Royal Albert Dock Development would fund the reclamation and infrastructure on the site. However, when these proposals were withdrawn in 1990, it became necessary to consider alternatives.

After negotiations, the Housing Corporation was able to offer a grant to fund the purchase of the land and LDDC offered an additional £20m to the Housing Associations towards the costs of house construction. The latter subsidy was determined at a level to secure affordable rents. Additionally, some of the revenue from the land sale was used to fund a management company dealing with the long-term monitoring and maintenance of the 'structure' of the reclamation scheme.

PHYSICAL CONDITION OF THE SITE PRIOR TO RECLAMATION

History

The site in question originally formed part of the East Ham Marshes, a low lying area drained by natural and man-made channels, for agricultural use. It was purchased by the Gas, Coke and Light company in 1868 and gas production started in 1870. The Beckton Gas Works eventually became the largest in the world.

The Winsor Park site covers an area of approximately 22ha and was not part of the main gas producing area, nor was it used for any of the later chemical processes, such as production of ammonia, inks, dyes and a range of chemicals. These processes were concentrated to the north of Winsor Terrace. Its major use was for stockpiling of coal, coke and breeze, either for, or from, the gas works. See Figure 2. However the site was subject to widespread dumping of waste materials.

There was much visual evidence of contamination on the site including large quantities of coal, coke and ashy material; staining of large areas by blue-billy (associated with complex chemicals in spent oxide); areas of brown 'bog ore' (used to clean coal gas); areas where tars had been disposed of on the surface and detectable tarry and phenolic odour from some of the material. Vegetation was poor or absent over much of the site.

Assessment of Conditions

Surface Topography: Running east-west across the site was an elevated area with isolated mounds. This area lay largely between 4 and 6m AOD, falling away at the western end to around 2 and 3m AOD. At its eastern end this elevated area extended to the southern and almost to the northern boundary of the site.

The southern edge of this area was characterised by a relatively severe bank down to a level of approximately 2m (west) to 3m (east) AOD. On its northern boundary, another bank sloped steeply to a level of 1 to 1.5m AOD (Area B). This area was covered seasonally by standing water.

To the north of this area, ground levels rose again to around 2m (immediately behind Winsor Terrace properties). In the north-west corner of the site, ground levels were variable but typically around 1 to 1.2m AOD.

Underlying Geology: Almost the whole of Winsor Park site was covered by made ground with thicknesses varying between 0.5 and 6m. Alluvial clay and peat deposits underlie the site. These deposits show significant variations of thickness, ranging between 1 and 5.5m. The peat is rarely greater than 1 metre thick and is often absent.

Floodplain gravels associated with the River Thames underlie the clay and peat deposits. These dense, grey brown, very sandy gravels range in thickness from 3 to 10 metres. The gravels overlie the London Clay.

Groundwater: There were perched water tables above the alluvial clay, the clay acting as an aquiclude. Water levels were generally encountered between 1.4m and 1.8m (AOD). A further aquifer within the floodplain gravels is encountered immediately below the base of the alluvial clay. The groundwater is partially confined by the overlying clays. A hydrogeological study was carried out (3) and discussions with Thames Water confirmed the long-term shallow water table to be at a level of 0.4m AOD.

Site Investigations: A number of chemical and geotechnical site investigations were carried out between 1982 and 1987 (4 to 8).

Supplementary investigations were carried out by the Corporation prior to and during reclamation. These investigations related primarily to gas and lagoon water analysis.

Summary Contamination

In many cases the contaminants present exceeded the appropriate trigger and action levels defined in the ICRCL guidance notes. For review purposes the site area was divided into 5 sub-areas. See Figure 3.

Area A:
Slightly elevated toluene extractable materials, sulphate, arsenic and boron levels, not normally requiring major action as they lie within the ranges experienced in many areas of London. Area A was considered to be relatively clean.

Area B:
This area was seasonally occupied by standing water. No analysis of soils was carried out here. Analysis of the lagoon waters indicated a pH of 2-3, concentrations of sulphates up to 6000 ppm and metals, in particular boron, copper, zinc and nickel. It was probable that contamination resulted from the leaching of contaminants from adjacent areas.

Area C:
The material was largely ashy with coal, coke, clinker, dust, rubble and bricks. There was mention of tar, bog ore and spent oxide, together with unidentified deposits of coloured materials. This was the most contaminated area of the site, with coal tars consistently exceeding 5000mg/Kg and concentrations of toluene extractable materials reaching values of 1 - 4%. Cyanide concentrations consistently exceeded 1000 mg/Kg, sulphate concentrations were regularly above 1% and there were high concentrations of elemental sulphur. There was patchy, but in some cases severe contamination by a range of heavy materials including cadmium (up to 80 mg/Kg), mercury (up to 26 mg/Kg) and arsenic (up to 194 mg/Kg).

Area D:
Fill was a mix of soil, rubble, ash, clinker and coke etc. Reference was made to the presence of tar in a number of the trial pits. Contamination was considered

patchy in this area, however to the northern edge of the area, contamination was severe with cyanide, coal tars, sulphates and arsenic significantly exceeding trigger levels.

Area E:
Beneath the existing floor slab, there was 0.4 - 1m of ashy fill passing into natural peats and clays below. Contamination was patchy and generally less severe, but there were elevated concentrations of toluene extractable materials, mineral oil, cyanide, phenols. Sulphate concentrations were very high in places.

RECLAMATION OPTIONS AND SCHEME SELECTION

General Considerations
The nature, concentration and distribution of contaminants present at the site, was not considered severe enough to render the site non-developable. However, given the preferred end-use of housing with gardens, remedial action was necessary for the following a number of reasons,reasons in particular:

- The variability of contamination was likely to lead to 'hot spots' not being detected by spot sampling. Remedial action had to take this potential into consideration

- Contamination in the surrounding areas could lead to contaminant transfer to the site in the future

- Contaminated shallow groundwater, seasonal fluctuations in the groundwater levels and the potential for capillary rise of contaminants during drought conditions were a concern and had to be taken into account

Options for Remedial Action
There were four broad options available for remediation.

- Removal of contaminated soil to a landfill site. This had been used extensively in the past since it leaves a low residual risk to the site in question

- Excavation and burial on-site. Such a scheme relies on an area of the site being relatively uncontaminated, allowing clean material from the excavation to replace the buried contaminated material. Dilution of remaining material could also be achieved by the mixing of clean materials with relatively uncontaminated material

- On-site soil treatment, by physical, chemical or biological means either in-situ or following primary excavation, including: addition of micro-organisms to materials containing high levels of coal tars and phenols, and chemical treatment of oils with lime

- Covering and isolating the contaminated materials from the end-users by a

> properly designed, multi-component layer reducing residual risk to buildings and site users to a low and acceptable level. Care is needed with such a scheme if the water table is at a shallow level

The preferred scheme involved the installation of a covering system designed to achieve the long-term safety of site occupants and minimise environmental impact both on and off-site.

The other options were dismissed for the reasons listed below:

> Removal of contaminated soil:
> The site was contaminated to considerable depths. The cost of excavation and earth moving such volumes was considered prohibitive. In addition the soil would have required disposal at specific licensed sites at a cost exceeding £30-35 per m^3 for 'special' waste. Area C alone represented 150,000 m^3 of material. The option would have required excavation below groundwater level. In addition, a soil cover would still be required.
>
> The transport operation could cause considerable environmental impacts, including odours, contaminated dusts, pollution of waters, noise etc, and extra risk to workers and the public. Removal to another site was not considered the best environmental option by the DoE and other regulatory authorities.
>
> Excavation and burial on-site:
> Whilst some improvement could be made by the transfer of the worst areas to a greater depth, importing clean cover material from off-site would have been necessary. The impacts of excavation and disposal below groundwater was unpredictable.
>
> In-Situ Treatment:
> Such techniques are not proven over the range of site contaminants. The performance of particular organisms could have been impaired by hostile environments. It was considered that a significant depth of cover material would still be required to provide a 'garden soil' of suitable consistency over the existing made ground. The metal concentrations on the site would not be reduced by treatment techniques, and further remedial work would have been required.

The intended site development also called for the significant raising of the site to form a dome, and soil import for this would be necessary.

Design Philosophy

The primary purpose of the cover scheme was to isolate the end-users of the site from the contaminants. Routes by which contamination could come into contact with future end-users were considered in the design. These are discussed below:

- Deep excavation into contaminated material by gardeners or during service installation or repair. LBN preferred a 900mm barrier to guard against over-zealous gardeners.

- Take-up by vegetation roots. The majority of garden plants can successfully grow on less than 500mm of soil. The LDDC commissioned a study to identify suitable species and those restricted by deeper rooting (9).

- Passage of contaminants vertically with groundwater fluctuations. Regrading and compacting of the original ground level achieved a 'starting profile' above the maximum recorded level of the groundwater in all areas except along the southern boundary. Here the thickness of break layer was increased to place the soil cover above the water table.

- Upward movement of contaminants by capillary action during dry conditions. This may cause significant transfer of contaminants. Fine grained materials are much more susceptible. The risk can be eliminated by a very coarse break layer. See Figure 4.

Cover Scheme Specifications

The outline design was refined after discussion with the relevant statutory bodies including: The London Borough of Newham, London Waste Regulation Authority and Thames Water Authority. Figure 4 shows the selected scheme. The sub-soil was imported from approved sites throughout London and was composed predominantly of London Clay, originating from deep pile arisings, basement and tunnel excavations.

To achieve doming of the site, additional sub-soil material was placed during the project. This resulted in cover depths in the order of 5-6m in the central area reducing to approximately 1.6m around the periphery of the site. The material of the break layer was greater than 25mm dia. with a minimum layer thickness of 400mm. A geotextile membrane was also placed above the break layer to prevent silting of the voids by fines from above. The break layer was crushed concrete imported from approved sites. Services placed beneath the break layer were laid in clean trenches with butyl HDPE protection for the pipe joints.

The break layer was also designed to intercept infiltrating surface waters and act as a precaution against a rising water table. The domed profile of the site encourages drainage to a perimeter drain and hence to a series of capped manholes. Direct discharge from these manholes to foul or surface sewer was considered inappropriate. It was agreed with TWA that the monitoring of the manholes be carried out periodically. Contaminant levels would determine whether the waters could be pumped to sewer or removed to a landfill or waste treatment facility. See Figure 5 for a typical cross-section through the site.

Gas Mitigation Measures

Spot surveys were carried out prior to the commencement of reclamation. Only negligible amounts of methane were shown as being generated within the site. Measurable levels of flammable gas were identified in two (out of 40) positions on-site. These were associated with volatiles from contamination 2m beneath original surface.

Generation rates were estimated to be low with a rapid dissipation of gas. Due to the sensitive end-use, overriding safety considerations and the timescale of reclamation, it was decided to include gas venting precautions and monitoring programmes within the remediation.

A series of measures were accepted by LDDC and LBN in late 1987. The break layer formed the basis of a passive venting system, venting to surface and was specified coarser than originally proposed to increase its effectiveness as a gas vent. The revised specification allowed for 10% fines with a minimum of 25mm diameter for the remainder. At the perimeter of the site, the break layer was keyed into a continuous venting trench 750mm wide. Gas vents, provisionally one per hectare, were provided and it was recommended at this stage that future building design should consider the potential for gas generation, and be subject to gas monitoring.

Implementing the Reclamation Scheme

Removal of Contaminants from Site: Reclamation required the minimum disturbance to the existing ground. However, local 'hot spots' were removed from site. Approximately 5,000 m^3 of a tar/gravel mix were removed to a licensed hazardous waste landfill during the reclamation, under procedures in-line with LWRA regulations.
The lagoon contained approximately 6,000 m^3 of contaminated water at commencement of reclamation. Subject to the addition of neutralising agents and a limit on the discharge rate to reduce suspended solids, permission was granted by TWA for the discharge of the water to foul sewer during February/March 1988.

Quality Control of Imported Fill: The cover material was imported from over 50 sources throughout London. The following control procedures were adopted:

- Each prospective source was subject to an inspection visit, and rejected if history or inspection suggested the material may be contaminated. If there was any doubt, analysis of representative samples was carried out. Approximately half the sites visited were rejected

- For approved sources, consignment tickets were issued by the Corporation for that site

- On the arrival of each load, the consignment ticket was collected. The load was then examined by the contractor and site inspectors to ensure compliance with the material expected from that source, and the acceptability of that particular load for use on-site

- At the point of deposition, the load was again inspected whilst being tipped. Material rejected was loaded directly back into the incoming transport for removal from site

- LBN Building Control had a Clerk of Works on-site throughout the contract

SHORT-TERM AND LONG-TERM SITE MANAGEMENT

Introduction
A site management document was prepared in late 1988 for the information of statutory

bodies, developers and housing associations. It describes the monitoring strategies to enable the long-term effectiveness of the remedial measures to be assessed, and to provide an early warning of any potential problems. Advice was also included on some practical precautions to prevent future damage to the reclamation scheme either during or after the development.

Monitoring Strategies

Gas Monitoring Strategy: Following discussion and approval by LBN. Environmental Health, it was thought appropriate to consider the monitoring in three phases:

- 'short-term', commencement to completion of the reclamation (January to September 1988)

- 'medium-term', end of reclamation to end of development

- 'long-term', post development

Short-term monitoring featured instantaneous gas readings taken using flammable gas meters at a set depth of 250mm, taken at 10m centres on grid layouts at between six and eight locations across the site. Random sampling was also carried out. Post-reclamation permanent monitoring stations were installed in key areas of the site. Their location was determined by preceding work and the findings of the short-term programme. At each station probes were installed within:

- the capping layer

- the break layer, and

- to a depth of 5m within the underlying contaminated material

Medium-term monthly visits were recommended to record gas concentrations at each probe, monitoring being carried out for carbon dioxide, oxygen and flammable gas. Trigger levels were set for specific gas concentrations and contingency measures defined for scenarios when the trigger concentrations were exceeded.

Long-term, post development, the frequency of monitoring will depend on the results of the preceding monitoring programmes. As an initial recommendation it is proposed that a minimum of four visits per year be undertaken. After ten years, a review of the monitoring results will determine subsequent monitoring frequencies.

Perimeter Drainage: It is possible that the water entering the break layer (and hence perimeter drain) from the polluted ground may be contaminated. Consequently the quality of any collected waters would determine its disposal route.

The following procedure has been agreed with TWA, now Thames Water Utilities Limited (TWUL):

- Monitor the quantity of water collected in the discharge manholes, initially monthly. Subsequently monitoring would be subject to a review of the results. Results to be submitted to TWUL

- When sufficient quantity for sampling has been collected, TWUL are to be informed

- TWUL to collect and test. If the sample is unacceptable for discharge to public sewers, necessary arrangements must be made for disposal at a licensed site

<u>Groundwater Levels Within Original Made Ground:</u> Water levels within the contaminated ground are of importance with regard to the potential for contamination of the site cover scheme. Monitoring of the probes installed within the contaminated material was recommended in line with the gas monitoring programme. Monthly monitoring in line with the envisaged gas programme was proposed. Such monitoring will enable an early warning of potentially damaging rises in water level to be recorded and remedial actions taken at an advanced stage.

Annual sampling of waters within the probes was recommended in accordance with guidelines specified by TWUL's Trade Effluent Division.

<u>Settlement:</u> Due to the nature of the underlying strata and the imported fill, LBN required the settlement of the site to be monitored during 1989.

PREVENTION OF FUTURE DAMAGE TO THE RECLAMATION SCHEME

A Summary of the Management Document

<u>Development Control:</u> The scheme design ensures that the majority of future construction (with the exception of deep sewers and piling) will be above the break layer. Development will be controlled by the LBN Building Control Department or other designated Building Control Authorities. Due to the presence of the capping scheme and the nature of made ground, full liaison with LBN is considered essential to ascertain the special construction details required. Above the break layer, standard construction techniques for a made ground site apply.

<u>Foundations:</u> Differential settlement renders shallow building foundations inappropriate. It is anticipated that all structures will be on piles founded in the floodplain gravels. Ground displacement piling techniques are preferred thereby avoiding the transport of potentially contaminated material to the surface. Any construction below the break layer must be resistant to aggressive ground conditions and allow for the possible settlement of the made ground.

<u>Services and Drainage:</u> Services positioned in the cover layer will not require protection but should be designed for probable settlement on made ground sites. Services in contaminated material require special protection. Gas monitoring should also be carried out in accordance with HSE guidance during any such trenching work.

Horticulture: Normal gardening activity by owner/occupiers can be undertaken. The work by Carol Coulson identified trees that will thrive in the 1.2m of fill above the break layer (9).

Penetration of the Break Layer and Filter Fabric (other than for piling operations): The party involved must immediately inform the LBN and the LWRA. The break layer and filter fabric must be reinstated such that no contamination material exists in or above the break layer.

Maintenance of Soil Depth: A minimum thickness of 1.2m of soil cover above the break layer is to be maintained under all future road or building construction.

REFERENCES

1. London Borough of Newham, The Revised Beckton District Plan. 1986.

2. London Docklands Development Corporation, Royal Docks a draft development framework for the Royal Docks Area. 1985.

3. Howard Humphries & Partners, East Beckton Housing Site hydrological appraisal. 1988.

4. Bostock, Hill & Rigby Ltd., Chemical & Environmental Investigation of Gaslands north of the Royal Docks. 1987.

5. Caleb Brett Laboratories Ltd., Site Investigation Area South of Winsor Terrace. 29 March 1985.

6. Johnson, Poole & Bloomer, Site No. 6 South Winsor Terrace. 14 January 1983.

7. Peter Reason Associates, Site Investigation at south of Winsor Terrace. December 1982.

8. Soils Engineering, Site Investigation of Gaslands North of Royal Docks, Beckton. April 1987.

9. Carol Coulson, An Assessment of the reclamation proposals for the East Beckton Housing Site in respect of species selection, establishment and effect of tree and shrub planting, landscape architecture, management and ecology. 1987.

Note: **Opinions expressed or inferred from this paper are those of the Authors and may not be those of LBN, LDDC or Site Developers.**

FIGURE 1 CONTEXT OF WINSOR PARK IN NEWHAM DOCKLANDS

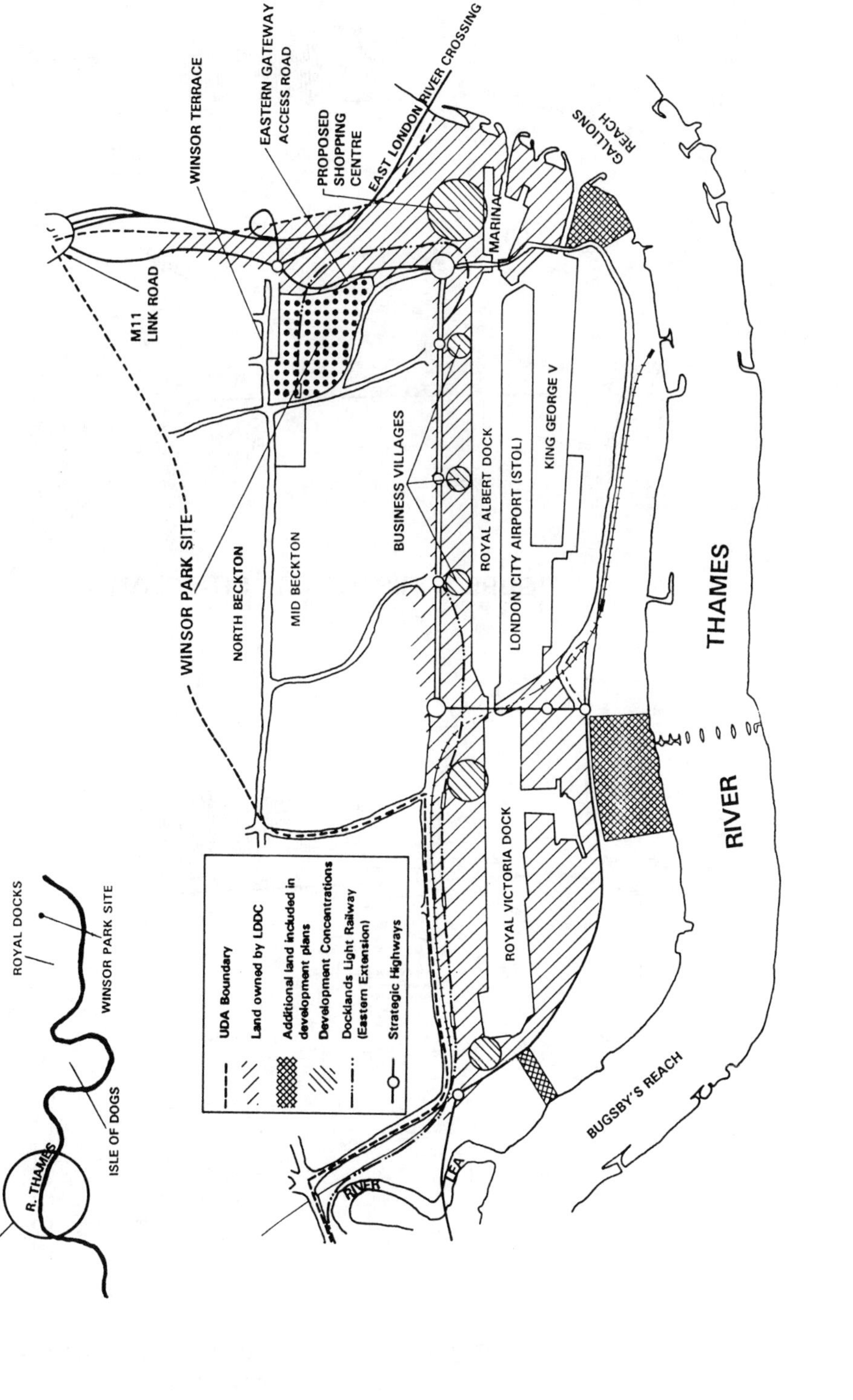

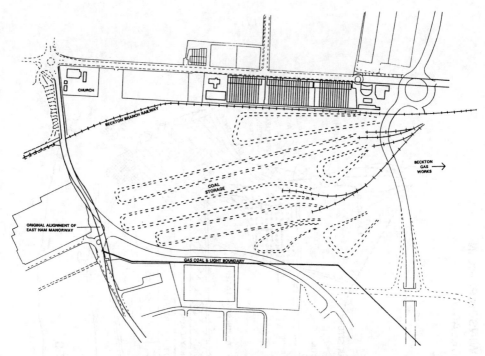

FIGURE 2 HISTORICAL SITE PLAN

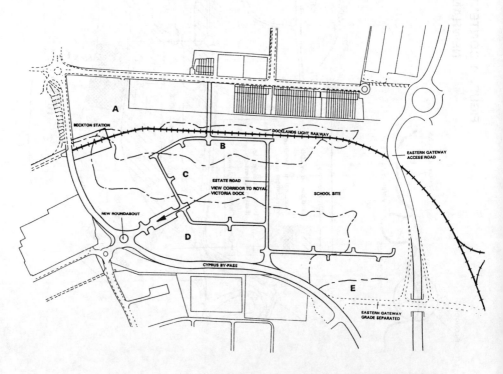

FIGURE 3 PLAN OF RECLAIMED SITE SHOWING LOCAL INFRASTRUCTURE

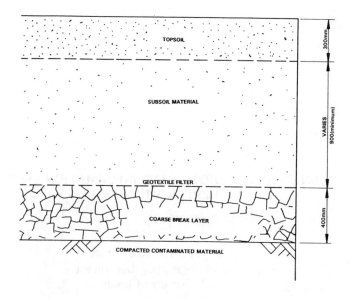

FIGURE 4 DESIGN OF THE CAPPING LAYER

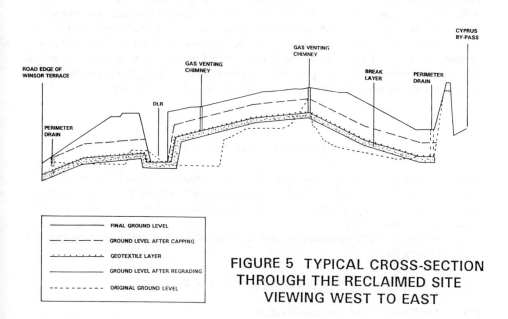

FIGURE 5 TYPICAL CROSS-SECTION THROUGH THE RECLAIMED SITE VIEWING WEST TO EAST

ADLIS - AN AIR PHOTO LAND INFORMATION SYSTEM FOR DERELICT LAND STUDIES

C.T.YONG and J.UREN
Civil Engineering Department
University of Leeds

ABSTRACT

This paper describes an Airphoto Derelict Land Information System (*ADLIS*) currently being developed in the Department of Civil Engineering at the University of Leeds. Aerial photographs have been used very successfully in the past for derelict land identification. *ADLIS* couples these with a powerful desk-top microcomputer to produce a complete and economical package which can be applied to any derelict land evaluation scheme. The article begins with a discussion of the various factors involved in derelict land studies and reviews the role of Land Information Systems in this type of work. It concludes with a detailed description of the emerging *ADLIS* system which is being developed and tested in a pilot study of derelict land evaluation in Keighley, West Yorkshire.

DERELICT LAND STUDIES

Derelict sites are generally caused by neglect and by improper or unorganised land use. Their study invariably involves the following sequence of events: planning, surveying, decision making and implementation, the latter usually taking the form of either reclamation or monitoring.

To ensure that this sequence can proceed smoothly, two essential items are required at the planning stage:

(1) A clear definition of the needs and objectives of the study.

(2) A reliable derelict land classification.

Only when these are available can suitable sources of land use data be identified and the collection of relevant information begin. This in turn enables potential problems to be foreseen and reliable predictions to be made about the extent of reclamation that is possible and the time required to complete the project.

Needs and Objectives

These may not be easy to define. Increasingly, derelict land is being considered on a regional or national basis rather than simply as a local issue. Other than for the most straightforward of schemes, the study of derelict land cannot be considered in isolation. The long term effects of any reclamation must be carefully considered and included in the overall plan.

In addition, the use of previous studies as precedents for future work is not always possible and, due to the demands made on land nowadays, derelict land information obtained from earlier studies may not be relevant to the current situation. Land use patterns change over time in response to economical, social and environmental forces [1]. Consequently, careful research is necessary before the needs and objectives can be formulated.

Classification

Attempts to classify derelict land have always been faced with many difficulties and uncertainties. The current Department of the Environment (DoE) definition of derelict land is :

Land so damaged by industrial and other development that it is incapable of beneficial use without treatment.

However, there is still much dispute concerning the extent and degree of dereliction, for example, areas defined as 'partial derelict', 'active derelict', 'potential derelict', 'waste land'. and 'marginal land' share many similar classification features [2], but differ considerably in priority when reclamation grants are allocated.

Inevitably, the availability of money tends to influence the objectives of the study with the classification being tailored to those categories for which the possibility of financial backing exists. This can lead to a variation in the interpretation of classifications for what are in effect similar studies and can cause difficulties when bringing together data from differently funded projects.

This is unfortunate because the crucial factor in determining the success of any type of land use mapping lies in the choice of an appropriate classification scheme designed specifically for an intended purpose [3]. In other words, the classification should be fitted to the purpose once the purpose has been defined and the interpretation of the classification must be consistent. This ensures that similar studies have similar classifications giving comparable results and was the approach adopted by the Department of the Environment for their "1988 Derelict Land Survey". The classification scheme they devised is shown in Figure 1.

- Colliery Spoil heaps
- Metalliferous spoil heaps
- Other spoil heaps
- Excavations and pits
- Derelict railway land
- Mining subsidence
- General industrial dereliction
- Other forms of dereliction

Figure 1 . DoE derelict land classification

Although the classification shown in Figure 1 is suitable for many derelict areas, some sites may require a combination of these categories, for example, a disused railway line passing through mining subsidence. In such a case, a more intricate classification system is necessary to fully describe the type and extent of the dereliction. This can only be provided if the classification allows itself to be extended into several levels in which each level contains a more detailed description of the category specified in the previous level. Such a multi-level classification system has been adopted by the United States Geological Survey (USGS) in its Land Use Classification [4].

LAND INFORMATION SYSTEMS

For very comprehensive derelict land studies, additional data are required. In particular, geographical information such as Ordnance Survey grid references, topographical features, vegetation, land areas, relationships with the adjacent environment and so on may need to be incorporated in the classification.

Such a requirement leads to the involvement of many diverse sources of data in the surveying stage of the study, ranging from textual information such as site investigation reports and feasibility studies to graphical data such as that provided by the interpretation of aerial photographs. Information from each source must be fed into the appropriate section of the classification within the system. Taken as whole, the various levels of information within the database form a complete package known as a 'Land Information System (LIS)' which itself comes under the general title of a 'Geographical Information System (GIS)'. Several projects have shown the capability of such systems and a recent report dealing with rural land planning [5] strongly recommended the advantages of a GIS approach.

It can be concluded, therefore, that a well-defined LIS is essential for a comprehensive derelict land study. However, it must be designed in such a way that it can not only accept data from a wide variety of sources but also output information in a form suitable to a wide variety of users.

Once a decision has been made to adopt an LIS approach, further commitment is required by the potential user in the form of financial outlay and staffing. It may be necessary to follow the example of several local authorities who have undertaken feasibility studies in the search for a suitable GIS/LIS system. GIS/LIS software packages will need to be developed/obtained and adapted as necessary to the requirements of the user. Appropriate hardware on which to run the system will be necessary and it is essential to allocate suitably trained personnel to look after the updating and day-to-day running of the system.

Most users, not having the time to develop their own systems, will opt to buy one of the commercially available LIS software and, if necessary, arrange to have it adapted for their own requirements. Although excellent, the cost of such packages can run into several thousands of pounds. This may deter potential users and, if expense is a limiting factor, they may be willing to accept a less sophisticated but more economical package that can still meet their requirements.

Just such a system is currently being developed in the Civil Engineering Department at the University of Leeds [6]. This is targeted specifically at derelict land studies and has been given the acronym of '*ADLIS*' - Airphoto Derelict Land Information System.

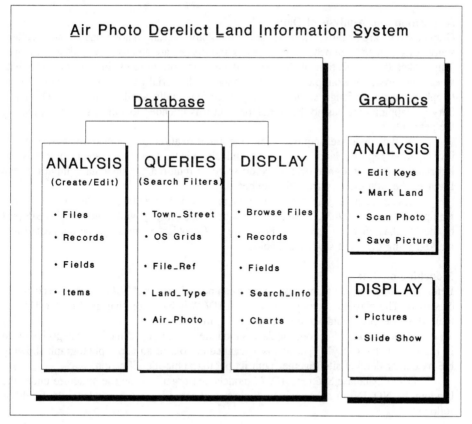

Figure 2. Layout of *ADLIS*

ADLIS-AIRPHOTO DERELICT LAND INFORMATION SYSTEM

ADLIS is a PC based system designed to run on any IBM AT compatible machine. It is currently being developed on an IBM PS/2 Model 70 microcomputer in conjunction with a commercially available database (dBase IV) and a hand-held scanner (Logitech).

The purpose of *ADLIS* is to create a multi-dimensional but readily accessible computerised derelict land data base which can store data from a range of sources and supply information to a variety of users in both graphical and textual form. The layout of the *ADLIS* is shown in Figure 2. It consists of two elements: a Database Section and a Graphics Section. Within each of these, steps are provided to enable the following to be performed:

- Acquisition and Analysis of Data
- Modelling of Data
- Storage of Data
- Retrieval of Data

Each of these components is briefly described in the following sections.

Acquisition and Analysis of Data

Graphical data are acquired from aerial photography. Aerial photographs are an excellent source of graphical data because their scales and images are comparable to those of maps. This makes the acquisition of data from them much more accessible to non-specialists when compared, for example, to satellite images. The aerial photographs are input to the Graphics Section through scanning. As yet, commercially available digitised Ordnance Survey mapping data has not been included but it is intended to add this to the system in the near future.

Textual data are acquired from the interpretation of scanned images, the site investigation reports, land use records and site visits. Generally, such sources yield information without a great deal of expertise being required and the data is input into the Database Section directly from the keyboard.

Having acquired the data, a facility is provided to enable the data to be further manipulated within *ADLIS*. Textual data can be edited directly within the ANALYSIS part of the Database Section as shown in Figure 2. Graphical images can be delineated and marked using different shades and colours.

Modelling of Data

Before any data derived from these sources can be included in *ADLIS*, it must be correctly classified. The classification system used in *ADLIS* is based on that used by the DoE in their 1988 Derelict Land Survey shown in Figure 1.

In the Graphics Section, the data is modelled by giving each of the categories listed in Figure 1 a 'fixed key' [7] which is a scanned cut-out of an aerial photograph showing the particular derelict site together with its photographic identification.

In the Database Section, *ADLIS* models and organises the file structure under the heading LAND_INFO as shown in Figure 3. Each file is sub-divided into fields as follows:

The fields in the MASTER file contain general information about each site including its site reference, O.S. reference, photograph reference, adjacent sites, ownership and so on.

The fields in the LAND USE file contain detailed information about the past and present use of each site together with its site reference.

The fields in the RECLAMATION file contain details of the site investigation reports, estimates of the costs involved and the nature of the reclamation that is required for each site together with its site reference.

The OTHERS file is provided to allow users to store any fields that are unique to the study in question. Each of the files within LAND_INFO is linked by a common site reference code which enables easy access to all the information relevant to a particular site.

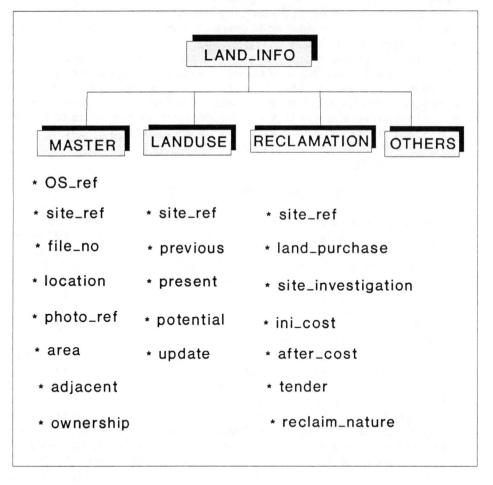

Figure 3. Database file structure of *ADLIS*

Storage of Data
Scanned photographic images are stored in the Graphics Section in a raster format called TIFF. Derelict land data are stored in the Database Section in Dbase IV which is running under DOS in the IBM PS/2 Model 70 computer.

These hardware and software components of *ADLIS* represent some of the most commonly available and widely used systems. They have been deliberately chosen in order to make *ADLIS* highly compatible and readily accessible/ adaptable to most of the commercially available GIS software packages.

Retrieval of Data
This is the final component of the system and is essentially an interface between the user and the Database and Graphics Sections within *ADLIS*. Fundamentally, it consists of a user-friendly 'dialogue box' which incorporates a facility for the user to select particular information before any retrieval is instigated.

Subsequently, it retrieves the necessary information from the storage component and presents this to the user either on the computer screen or as hard copy in the form of scanned images, statistical graphs or written statements.

To retrieve textual information, with reference to Figure 2, the user activates the QUERIES file within the Database Section and a list of query fields appears on the screen. On choosing one or more of these, all the relevant data is shown on the screen and can be output as hard copy if required by accessing the DISPLAY file. In cases where the required information is already known and no queries are necessary, it can be accessed directly through the DISPLAY file which can be browsed through as required to find the relevant information to be output. Graphical information can be retrieved and displayed via the Graphics Section.

PRACTICAL APPLICATION OF ADLIS

The concept and subsequent development of *ADLIS* arose as a result of approaches made to the Civil Engineering Department at the University of Leeds in 1988 by the Planning Department of Bradford City Council. They, in common with all the other metropolitan district councils were requested by the DoE to assess the amount of derelict land within their boundaries. Although the classification system required was specified by the DoE, the method of collection of the necessary data was left to the discretion of the Planning Department concerned.

Bradford, after researching suitable methods, decided that aerial photographs would be an ideal source of derelict land data [8] and opted for this method. Subsequent approaches by them to the Civil Engineering Department at the University of Leeds for assistance with the preparation of a flight specification led to the whole of the centre of Bradford and the surrounding areas being photographed using large format near-vertical colour aerial photography at two scales: 1:3000 and 1:6000.

Unfortunately, poor weather conditions delayed the necessary flights and the photographs arrived too late to be used in the derelict land study. However, the clarity of the photographic images and the amount of detail they contained generated considerable interest in their use in this type of work and the *ADLIS* project began. To date, the technique has been concentrated on a small area covering Keighley in West Yorkshire.

The Keighley Pilot Study

Keighley is confined within the O.S. National Grid kilometre square from 403 000 E to 410 000 E and from 443 000 N to 439 000 N. It takes about 40 1:6000 scale aerial photographs to cover the entire area. Each of the colour photographs is assigned a reference code (photo_ref in Figure 3) and is subsequently examined by eye for derelict and waste sites. Information from the 1988 Vacant Land Survey and Land Use records provides clues to the interpretation of the photographs.

Initially, any 'suspected' derelict site is given a site reference code (site_ref in Figure 3). Then, all known relevant information about the site is input to the MASTER and LANDUSE files for later editing as required. This results in each site record in the MASTER file containing some or all of the following information: OS_ref (numerical data), site_ref (numerical), file_no (numerical), location (characters), area (numerical), adjacent (another OS_ref) and ownership (characters). In addition, the 'PRESENT' field in the LANDUSE file is given one of the classification categories shown in Figure 1.

The database is progressively built up as more information becomes available through site visits and interpretation of aerial photographs. The photographic 'fixed-keys' provide a useful visual aid in this process.

The scanned images which are stored in the Graphics Section are viewed, delineated and marked with the appropriate site reference codes.

At the time of writing this article the Keighley Study is still at an early stage. However, four derelict sites have been identified for analysis by *ADLIS* which is coping well with the storage and retrieval of data from a number of different sources. Further details will be given at a later date.

SUMMARY

This article has discussed the on-going development of a multi-dimensional Airphoto Derelict Land Information System (*ADLIS*). The use of this multi-dimensional approach enables data about even the most complex of sites to be stored in a flexible and readily accessible form for subsequent analysis, editing or output.

Preliminary studies involving the use of *ADLIS* on a pilot study area in Keighley, West Yorkshire are very encouraging and illustrate the capability of the system to handle data from a variety of different sources.

The initial identification of derelict sites is done using aerial photographs with the relevant images being quickly scanned into the system for subsequent storage and retrieval. It must be stressed that this reliance on photographic images does not necessarily imply a considerable outlay on specially commissioned photographic sorties. There is a wide availability of recently taken aerial photographic coverage of the UK at various scales which may be obtained from a number of different photographic libraries and commercial organisations at a reasonable cost. Providing they are still relevant to present site conditions, such slightly out-of-date photographs can represent a very cheap source of invaluable information.

One of the main objectives of the *ADLIS* project was to develop a system that could be used on readily available IBM AT compatibles without the need for considerable additional expense. Given that in addition to the computer and the *ADLIS* software, the only other requirement is a small hand-held scanner costing a few hundred pounds then it is fair to say that this objective has been achieved.

ADLIS does, of course, have its limitations, the main one being that, at present, the Database Section and the Graphics Section of the system are two separate entities. To change from one to the other it is necessary to download one before loading the other. However, work is continuing on the development of a suitable interface between the two.

In addition, *ADLIS* is not as sophisticated as some of the commercially available GIS packages such as 'GS-M.A.P.' and 'MapInfo'. However, once it has been fully developed it is likely to be considerably cheaper than these and, as such, will represent a viable and economical alternative for any individual or organisation concerned with the study of derelict land sites.

REFERENCES

1. Mather, A.S., Land Use, Longman Inc., New York, 1986, pp.26-64.

2. Bridges, E.M., Surveying Derelict Land, Oxford Science Publications, Oxford, 1987, pp.7-36.

3. Lo, C.P., Applied Remote sensing, Longman Inc., New York, 1986, pp. 227-281.

4. Anderson, J.R., Hardy, E.E., Roach, J.T. and Witmer, R.E., A Land Use and Land Cover Classification system for Use with Remote Sensor Data. U.S. Geological Survey Professional Paper 964, 1976.

5. Bracken, I. and Higgs, G., The role of GIS in the data integration for rural environments. Mapping Awareness, Vol.4, No.8, October 1990, pp. 51-56.

6. Uren, J. and Yong, C.T., The development of an air photo based land information system. Mapping Awareness, Vol.4, No.3, April 1990, pp. 22-26.

7. Curran, P.J., Principles of Remote Sensing, Longman Scientific & Technical, England, 1985, pp. 56-99.

8. Bush, P.W., Derelict land in the West Riding of Yorkshire - an air photo study. Ph.D. Thesis, Department of Civil Engineering, University of Leeds, 1970.

RECLAMATION OF A BACKFILLED SAND PIT

P.J.WITHERINGTON, BSc, CEng, MICE
Wimpey Environmental Limited
Beaconsfield Road, Hayes, Middlesex, UB7 7ST, UK

ABSTRACT

This paper describes the redevelopment of a 25 m deep backfilled sand pit, the investigation of which has been previously reported. Stability of the site to permit construction of houses and associated infrastructure was achieved by staged surcharging. The procedures adopted, the monitoring systems installed and the results obtained are detailed.

INTRODUCTION

A site in south-east England was proposed for redevelopment with low rise housing. Investigations by Wimpey Laboratories Limited, predecessors to Wimpey Environmental Limited, indicated that the site comprised a backfilled sand pit. Extraction of the sand had commenced in 1934 and continued until 1970 when a maximum depth of about 25 m was attained. The site was subsequently progressively filled until 1985 when the development was proposed. Investigations were then instigated to assess the viability or redevelopment and the results of this work have previously been published[1].

In summary, the investigations indicated that the site had been backfilled from a variety of sources and contained principally cohesive soils with assorted debris and waste from an old gas works. Considerable research and investigation was undertaken to determine the nature and performance of the backfill. Following a twelve month trial with an instrumented 6 m high embankment it was concluded that surcharging could be effectively employed to stabilise the fill sufficiently to support the proposed housing development and associated infrastructure. This paper describes the reclamation procedures adopted, the monitoring techniques employed and the results obtained.

RECLAMATION PROCEDURES

The surcharging was undertaken using waste sand imported from a nearby quarry. The principal adopted was to place an initial well compacted permanent 2 m thickness of sand. This layer not only provided capping to the fill which in places contained toxic contaminants, but also formed a consistent formation for the development. A further 4 m layer of sand was placed as surcharge which was subsequently removed.

The site was divided into four areas of decreasing size (see Fig 1) for progressive surcharging. Approximately 50,000 m^3 of sand was required to surcharge area 1 and area 2 was designed to accommodate the 4 m removed from area 1 and this process was then continued with areas 3 and 4.

It had been concluded from the initial investigations that the surcharging would induce about 300 mm of settlement and that the surcharging should be maintained until the rate of settlement fell below 0.5 mm per day over a two week period. Monitoring systems were therefore installed accurately to record the settlements induced across the site.

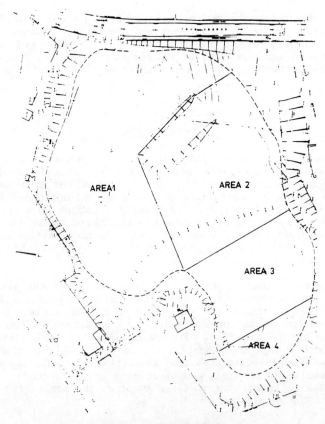

FIG.1 LOCATION OF SURCHARGE AREAS

MONITORING OF RECLAMATION

Before reclamation commenced three independent settlement monitoring systems were installed.

The first comprised survey stations located on a 25 m grid across the site. These were designed so that they could be extended upward as filling proceeded in order that access could always be obtained for levelling. The length of extensions added was accurately measured so that the readings could always be related back to the original pre-surcharge levels. The details of the survey stations are shown in Fig 2.

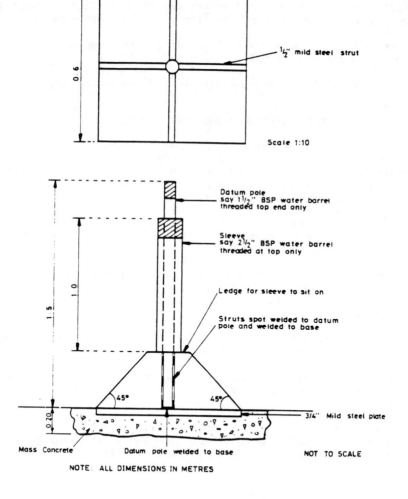

FIG.2   LEVELLING STATION ASSEMBLY

The second monitoring system comprised fourteen magnet extensometers installed at locations best suited to the surcharging areas. These instruments consisted of a series of magnets installed within boreholes at intervals for the full depth of the fill. A datum magnet was grouted into the natural sand at the base of each borehole in order to give a fixed reference to magnets above. The additional benefit of these instruments was the ability to record the settlement of the fill at various levels, thus identifying anomalies at depth which might occur. During reclamation access was also gained to the original magnetic extensometer which had been installed to monitor the performance of the trial embankment.

The final system comprised profile pipes installed in loops from manholes sunk down to the natural sand around the edge of the pit. Levels at 2 m intervals along these probes were recorded by drawing a hydraulic settlement probe through the tubes. These levels were referenced to a datum point set into the base of the manholes.

As filling proceeded the manholes were extended upwards to maintain the access to the profile tubes.

The layout of all these instruments is shown in Fig 3.

SETTLEMENTS RECORDED DURING SURCHARGING

All the measurements recorded by the three instrument systems provided consistent results and did not reveal any anomalies across the site. Readings from a typical survey station, magnetic extensometer and settlement profile are shown in Figs 4, 5 and 6. Maximum recorded surface settlements increased from about zero at the edges of the pit to 320 mm near the centre of the pit and the actual values are contoured on Fig 7. It is interesting to note that the settlements reduced to about 100 mm around the trial embankment which itself had induced about 150 mm settlement during the investigation.

The induced settlement plots, the rate of movement rapidly reduced once the surcharge embankment had reached its full height. A period of less than two months was required for the rate to drop below the specified value of 0.5 mm per day.

Once the upper 4 m of surcharge had been removed further movement was not recorded by any of the measurement systems.

Once the surcharge had been removed from Area 1 to Area 2 construction commenced on the show houses. A view of the site at this stage of development is shown in Fig 8.

The foundations adopted comprised reinforced slabs with downstand beams beneath the load bearing walls. The beam widths were designed to limit the bearing pressure to 50 kN/m^2 assuming that load was not spread into the connecting floor slab.

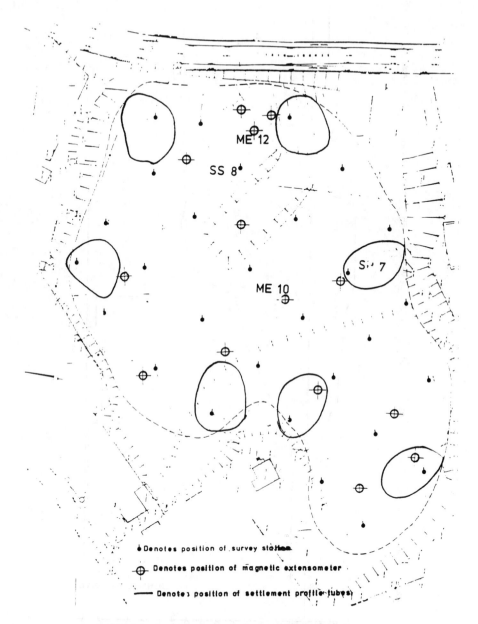

FIG.3 LAYOUT OF INSTRUMENTS

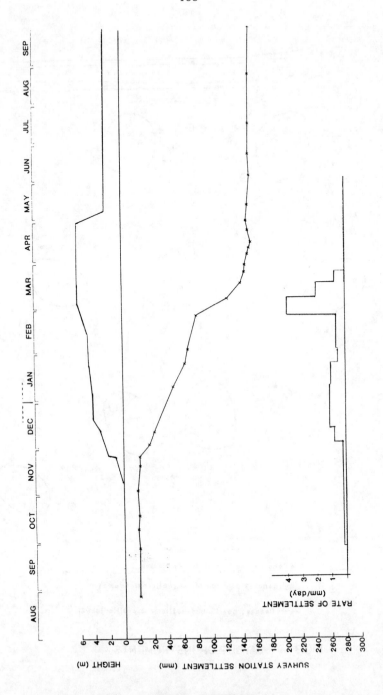

FIG.4 SURVEY STATION 8

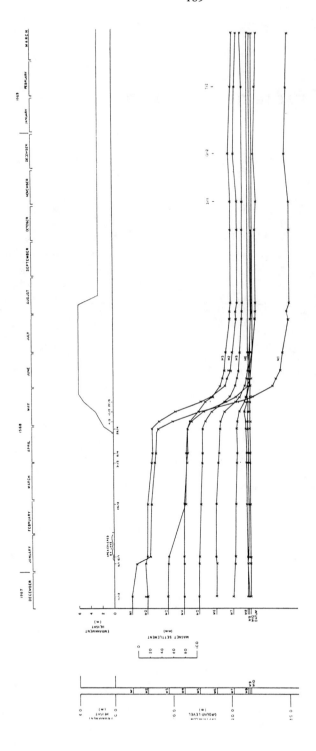

FIG.5 MAGNETIC EXTENSOMETER 10

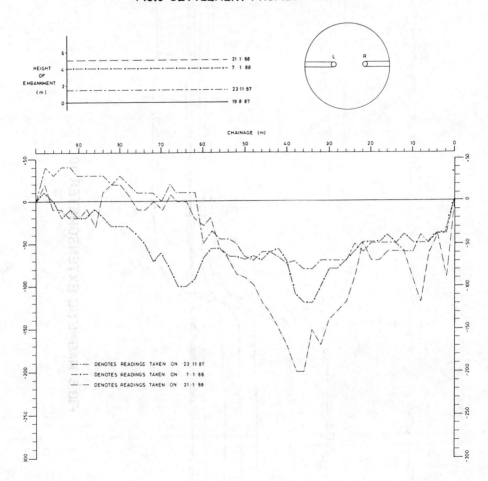

FIG.6 SETTLEMENT PROFILE PLOT 7

FIG.7 CONTOUR PLAN OF SURFACE SETTLEMENT

FIG 8 AERIAL VIEW OF SITE DURING SURCHARGING

In order to confirm the performance of the surcharged fill, two magnetic extensometers were installed through the foundations of the show block during construction. Magnets were located in the foundation concrete at the interface of the sand and the original fill. A datum was installed in the natural sand at the base of the fill. The results of measurements taken in these instruments during construction of the house and subsequently up to 1990 are plotted in Fig 9. From these it can be seen that apart from minor variations at about 1 mm to 2 mm which is within the accuracy limits of the measurement system, settlement of the foundations has not been recorded either during or after construction. The absence of settlement during construction is not surprising since the average foundation load is less than 25 per cent of the surcharge load which had been removed. The absence of subsequent movement confirms the information obtained from the other monitoring systems and clearly indicates that the surcharging successfully completed self-weight settlement of the fill in the pit.

FIG.9 MAGNETIC EXTENSOMETER 12

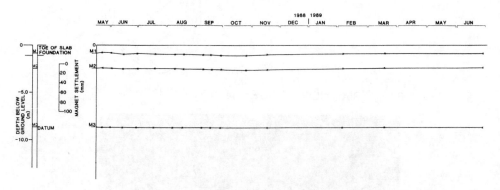

CONCLUSIONS

The development of this site was made possible by very thorough investigation, formulation of practical and economic remedial solution and very close monitoring of the reclamation works. Previously the land was valued at £300k. The expenditure of approximately £1m in investigation and reclamation released land for development valued in the order of £6m.

REFERENCE

Witherington P J. Investigation of a Backfilled Sand Pit. Proc. Land Rec 88, 1988.

CLEAN COVER RECLAMATION
THE NEED FOR A RATIONAL DESIGN METHODOLOGY

DR IRENE J ANDERS
Senior Environmental Consultant
W A Fairhurst & Partners, Environment Division
1 Arngrove Court, Barrack Road, Newcastle upon Tyne NE4 6DB

INTRODUCTION

The relative cheapness of clean cover reclamations compared to other reclamation techniques means that the method is likely to continue to be used for some time to come. However, the recent failure of a number of early clean covers and the introduction of increasingly strict environmental legislation emphasises the need for a rational design methodology which allows the design of a quantifiably safe clean cover.

A clean cover solution normally comprises:

- Topsoil
- Low permeability layer
- Main cover layer
- Capillary break/drainage layer

PERFORMANCE OF A CLEAN COVER

A rational design methodology needs to take account of the various functions which a clean cover may have to fulfil. These include:

- Provision of a quantifiably safe barrier between at-risk targets and contaminants.
- Restriction of infiltration.
- Provision of a suitable rooting medium to sustain healthy vegetation.
- Provision of a site surface suitable for the proposed development.

FAILURE OF A CLEAN COVER

Failure may result in any one or more of the following ways:

- Soluble contaminants may be brought to the surface either by rising groundwater ie flooding, or by capillary rise during hot, dry weather.

- Infiltration of surface waters through the clean cover may result in leaching of contaminants into the groundwater or onto adjacent land.

- Breach of the clean cover by deep rooting vegetation, excavation or surface desiccation.

- Absorption of contaminants by deep rooting vegetation and transfer to the food chain.

DESIGN METHODOLOGY

A rational design methodology has been developed [1] which allows a safe and economic clean cover to be designed. The model allows an assessment of three main factors to be made.

1. **Upward Migration of Contaminated Soil Water**

    Calculation of capillary rise through the soil profile is complex, a computer model based on Bloemen [2] has been developed [3], which allows the rapid calculation of capillary rise of soil water to be determined in multi-layered soil profiles. The model requires the following input:

    - Proposed clean cover layers (ie number and thickness of layers).
    - Parameters for each layer (ie suction and hydraulic conductivity).
    - The depth to groundwater from the surface of the clean cover.
    - Details of the severity of a design drought for the region ie once in 100 years, once in 50 years etc.

2. **Infiltration of Precipitation**

    It is necessary to design a cover which restricts infiltration and provides adequate drainage to remove any percolating water before it comes into contact with the contaminated materials.

    The design requires:

    - Assessment of precipitation rates and percentage infiltration to the ground.

    - Calculation of the volume of infiltration required by each layer of the clean cover to bring its moisture content to saturation.

    - Calculation of the rate of flow through the saturated layers.

    - Calculation of the rate of dissipation required by the drainage layer to prevent percolation into the contaminated ground.

    The infiltration cut off layer would also act as a groundwater control drain and a capillary break layer.

3. **Root Zone Requirements**

Where reclaimed land is to support vegetation it is essential that sufficient rooting medium is provided to sustain growth without the roots penetrating the underlying drainage layer.

Recent literature (4, 5 & 6) indicates that approximately 200-400mm of topsoil is adequate for shallow rooting vegetation. With local deepening to sustain deeper rooting trees.

4. **Development Requirements**

Having designed a quantifiably safe clean cover against contamination it is then necessary to ensure that it accommodates the development, ie the clean cover may be 500mm thick whilst foundation depths may be 1.0m.

CHECKING THE DESIGN

All empirical design methods of necessity incorporate approximations and research [7] has shown that the model actually produces a slight overestimate of capillary rise. It is therefore advisable that the theoretical design be checked by means of small diameter soil columns, which allow the soil cover profile to be monitored against infiltration and capillary rise in the laboratory environment. Fine tuning of the cover may then be made prior to installation.

References

1. Sharrock T & Cairney T - Private Communication

2. Bloemen G, 1980, - Calculation of Steady State Capillary Rise from the Groundwater Table in Multi-Layered Soil Profiles
In: Z Pflanzerernaehr. Bodenkd. 143, p701-719, 1980

3. Sharrock T - Private Communication

4. Ettala M, 1988, - Short Rotation Tree Plantations at Sanitary Landfills
In: Waste Management and Research 6, 1988

5. DTp - Department of Transport Model Specification for Landscaping Works

6. DTp - Advice Note HA/13/18

7. Anders I J, 1989, - Evaluation of the Soil Cover Reclamation Method for Chemically Contaminated Land - PhD Thesis 1989

EVALUATING THE POTENTIAL FOR SUBTERRANEAN SMOULDERING

Dr T CAIRNEY
W A Fairhurst & Partners
Newcastle-upon-Tyne

ABSTRACT

The re-use of derelict land which contains materials of high calorific value is increasing. This gives rise to the reasonable concern that subterranean smouldering/combustion could occur at some later date and cause significant infrastructure damage and cost.

Evaluating the risk of subterranean smouldering is generally conducted by calorific value testing, despite the fact that the C. V. test was never devised for such a purpose and is incapable of distinguishing more susceptible matters from others, of similar C.V., that are extremely difficult to ignite.

The method described has been successfully utilised on four colliery reclamation projects and offers quantifiable safety factors, which insurers now require as routine.

INTRODUCTION

Disused collieries, coal stock yards, and landfills with large volumes of combustible fill are increasingly being redeveloped. The concern that such sites could smoulder or burn is not unrealistic, as 84 reported cases of subterranean smouldering were noted in a single three year period [ref. 1], and some apparently non-combustible materials have exhibited large scale subterranean combustion [ref. 2].

To reduce the potential risk, various control bodies have proposed calorific value limits (e.g., the former G.L.C. limit of 7,000 KJ/kg above which sites are deemed to be at hazard. This approach assumes that a particular C.V. implies a particular susceptibility for ignition. Calorific value tests were, however, designed to measure the energy outputs of different coals, are not diagnostic indicators of combustion susceptibility and can give precisely the same value for an easily ignited bituminous coal, as for an anthracite (which is more difficult to ignite) intermixed with colliery shales. Additionally, the small size of the samples taken for C.V. testing can make replication of results much more dubious than is usually believed.

Thus a clear need exists for a simple routine test which actually measures the actual combustion susceptibility.

COMBUSTION SUSCEPTIBILITY TEST

To be appropriate, any such test has to be both cheap and easy to conduct, has to test materials in the state in which they actually occur, and has to allow the examination of a meaningfully large test sample.

Various workers [refs. 3 & 4] have researched the reactivity of potentially combustible materials. The susceptibility test now proposed is based on these basic studies. Sample material is packed into a stainless steel cylinder, which is in turn placed within a large cylinder, filled with an inert reference material. The complete assembly is then heated by heat coils, placed around the larger diameter cylinder, to 600°C over a 2 hour period. Whatever required air flow is allowed to enter the test material [fig. 1].

If the test material is not combustible, the temperature-time curves of the test sample and the reference material remain parallel [fig. 2]. If, however, the gradual heating up does cause the test material to ignite, its temperature-time graph exhibits a steeper gradient than that of the reference material. The more susceptible that a material is, the lower will be the temperature at which ignition takes place.

Normally it is best to conduct the susceptibility testing at a highish air flow (about 2.5 L.p.m.) and then retest other samples of the same material at lower air flow rates, until ignition fails to take place. This identifies the critical air flow rate which has to exist to allow ignition.

Relating the critical air flow rate to that which could occur once the site is compacted and reclaimed is simple. Site material is compacted to its design compaction and then exposed to the wind loading air pressures that will occur during extreme storms [ref. 5], and the flow of air that occurs through the test material is measured. Normally this proves that the achievable air flow is significantly less than the critical air flow for combustion and a defined factor of safety can be calculated.

APPLICATION OF THE SUSCEPTIBILITY TEST

Most sites underlain by potentially combustible material are reclaimed by either sieving out the combustible matter or by covering the compacted site with a soil cover intended to minimise the entry of atmospheric air.

The first option may well prove expensive, unless the value of the reclaimed combustible matter is high, and can actually be counter productive when the more susceptible material is very fine grained and cannot easily be removed. In this situation the overall calorific value will be reduced by the removal of large potentially combustible particles, but the combustion sensitivity may not be improved.

The second option is logical but can prove difficult to quantify to the satisfaction of insurers.

Testing on the basis described above clearly identifies the actual risk that exists, and the benefits that particular reclamation strategies will bring. To date the method has proved successful in reclamations in North-West England and the West Midlands.

REFERENCES

1. Beever, P. (1989) "Subterranean Fires In The U.K. - the problem"
   B.R.E. Information Paper T.P.3/89

2. Smith, A.J. (1989) "Investigation and Treatment of a Major Subterranean Heating Beneath an Industrial Estate at Dronfield, Derbyshire"
   B.R.E. colloquium 22/3/89. Fire Research Station.

3. Blayden, Riley & Shaw (1943) "A Study of Carbon Combustibility by a Semi-Micro Method"
   Fuel Vol. 22. No: 32-38. pp 64-71

4. Sebastian, J.J.S. & Mayers, M.A. (1937) "Coke Reactivity - Determination by a Modified Ignition Point Method"
   J. Ind. and Eng. Chemistry. Vol. 29. No: 10.

5. British Standards Institute (1975) - "B.S. 1377 - Methods of Test for Soils for Civil Engineering Purposes"

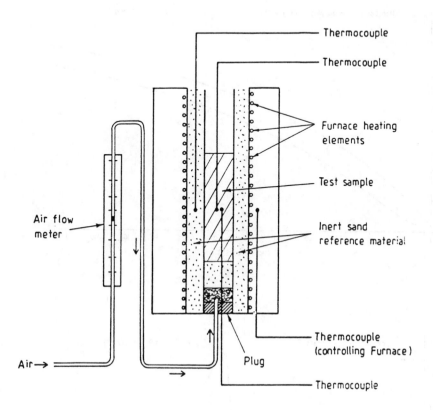

Figure 1

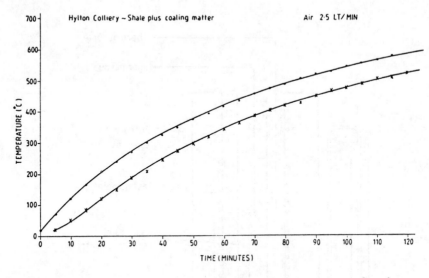

Figure 2

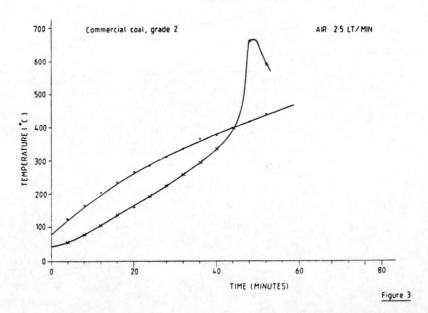

Figure 3

THE USE OF DYNAMIC COMPACTION IN LAND RECLAMATION

T. Gordon
Ove Arup and Partners
Cambrian Building, Mount Stuart Square, Cardiff, U.K.

Ground improvement by dynamic compaction offers an effective technique for reclaiming large derelict sites. This is illustrated by reference to three developments in South Wales

The site of an infilled quarry at Cwmbran was developed with industrial warehouses in 1976. Infill to the quarry consisted of some 60% industrial refuse and some 40% domestic refuse which at the time of the development varied from 5 to 15 years old. Ground improvement by dynamic compaction was carried out and floors and foundations of the warehouses were designed as flexible rafts to a settlement criterion. Monitoring has shown that settlement of the warehouses over the past 14 years has been uniform, with differential values within design limits.

Leckwith Athletics Stadium in Cardiff was built in 1987 on the old Leckwith Moors, a low lying area which was overtipped with domestic and light industrial refuse in the 1950's and 1960's. Whilst the stand and track have pilled foundations, other parts of the development were dynamically compacted. These included car parks, access roads, main drainage runs, in-field and out-field areas where it was necessary to keep settlements and consequent future maintenance to a minimum. Dynamic compaction provided a rapid method of simultaneously investigating and treating large areas, an important advantage given the inherent variability of refuse tips. It also avoided environmentally undesirable excavation of the refuse.

After the second world war, Swansea North Dock was infilled with demolition rubble from bomb clearance sites. The site then lay semi-derelict for many years. Redevelopment commenced in 1987 and Parc Tawe retail estate was constructed on a "fast track" programme. Site investigation showed that the dock infill was mainly granular but in a generally loose condition. Dynamic compaction was selected as a rapid cost-effective method of proof-testing and compacting large parts of the site at an advance stage, before building layout had been finalised. This enabled the buildings to be constructed using conventional pad footings and ground bearing slabs.

THE SUBSURFACE DRAINAGE OF DISTURBED SOILS
R A HODGKINSON
ADAS Field Drainage Experimental Unit.
Anstey Hall, Maris Lane Trumpington, Cambridge, CB2 2LF

ABSTRACT

The drainage problems of restored soils are reviewed and contrasted with those of natural soils. A new method for calculating pipe sizes for use on disturbed soils is introduced. The problems of undertaking mole drainage or deep subsoiling are discussed.

INTRODUCTION

Disturbance of soil by heavy earth moving machinery causes a loss of soil structure and reduces soil stability. Most undisturbed soils have a strongly developed structure of interconnected fissures through which excess rainfall can move or be stored in. Once soil has been moved the natural structure is destroyed and hydrologic characteristics changed. Under these circumstances precipitation is retained on the ground surface due to reduced infiltration capacity. Where cropping and cultivations have opened the topsoil sufficiently to allow infiltration, water entering the soil profile is held at the base of the cultivated layer. The net result is saturation of the upper zones of the soil profile, which are then easily damaged by traffic or livestock. Effective subsurface drainage to prevent this build up of water is an essential component of any scheme for the agricultural rehabilitation of a disturbed soil. Due to the minimal soil structure the primary drainage function is to intercept surface and shallow moving water. Mole drainage, and or deep subsoiling, help to improve soil structure and improve drainage efficiency. Due to the lack of natural structure to provide temporary water storage in the soil peak drainflow regularly exceeds drainage designs based upon the behaviour of undisturbed soils.

When disturbed soil has been stored in mounds during mineral abstraction or other activities a greater reduction in stability will occur due to changes in the amount and types of organic matter present. Re-establishing soil structure is therefore not easy on most disturbed soils.

REVIEW OF RELEVANT RESEARCH

Current UK design for subsurface drainage of undisturbed soils uses maximum rainfall occurring within a 24 hour period [1] for the selected return period, to calculate the required pipe discharge capacity. Rainfall mm/24hr is converted to flow by use of a drainflow coefficient which takes account of slope, soil type and landuse. Research has shown that peak flows on an undisturbed soil frequently exceed pipe capacity calculated from a 24hr rainfall value [2] [3]. When the pipe capacity is exceeded, water is stored in the drain trench, surcharging the pipe and can submerge mole drains causing early channel collapse. Rainfall/drainflow data, from several restored sites, has been used to develop an equation to predict peak drainflow for a given return period from rainfall statistics. This peak flow is used, in conjunction with information on the characteristic shape of hydrographs from these sites, to derive triangular synthetic hydrographs. These are used as an input to a computer model [4] which routes the flow through a given combination of drainage parameters to determine the probable surcharge. Drainage parameters can then be manipulated until an appropriate level of protection has been achieved. This computer model is now used by ADAS as the basis for drainage design and can be applied to any disturbed soil. Mole drainage and deep subsoiling increase the interception of water by drainage systems. Subsoiling is widely used on restored opencast coal sites both to improve drainage efficiency and to assist the development of soil structure. Due to the conditions found on these sites the tendency has been to use the largest winged subsoilers available and to bring the permeable backfill over pipe drains to within 150mm of ground level or closer. It has been shown that the passage of a winged subsoiler through the drain trench tends to move the permeable backfill need only forwards and upwards out of the trench [5]. If large winged subsoilers are to be used the amount of permeable backfill used can be reduced to only overlap the subsoiling depth by 150mm. This will maximise the cost effectiveness of the system. Regular usage of these large implements may disperse the permeable backfill to such an extent that subsequent soil loosening operations using ordinary farm equipment may fail to connect with it.

ACKNOWLEDGEMENTS

The support of British Coal Corporation Opencast Executive is gratefully aknowledged.

REFERENCES

[1] Ministry of Agriculture, Fisheries and Food (1982) "The design of field drainage pipe systems" Reference Book 345, HMSO, London.

[2] Hodgkinson R A (1988) Evaluation of drainage design criteria on a restored opencast coal site. In "Ten Years of Research - What Next", British Coal Opencast Executive, Mansfield.

[3] Hodgkinson R A (1989) The drainage of restored opencast coal sites, Soil Use and Management Volume 5, Number 4, 145-150.

[4] Harris G L, Pepper T J and Goss M J (1988) The effect of different tillage systems on soil water movement in an artificially drained clay soil. International Soil Tillage Research Association 11th National Conference, Edinburgh, Scotland.

[5] Hodgkinson R A (1988) The displacement of permeable backfill from the drain trench by winged subsoilers. In "Ten Years of Research - What Next", British Coal Opencast Executive, Mansfield.

ENGINEERING TECHNIQUES FOR ENVIRONMENTAL CONTROL OF THE COASTAL ZONE

R.A.PEREVERSEV and J.F.FARBEROV
Leningrad Shipbuilding Institute,
Lotsmanskaya 3, Leningrad, 190008 USSR

ABSTRACT

Stationary and special floating structures protecting coastal zone from pollution effects are considered. The matters of land reclamation and shallow water areas treatment are discussed in relation to possible oil and chemical penetration to bedded deposits. To calculate and design hydraulic engineering systems and structures the ADHESIVE finite-boundary element methods computer program is used.

To prevent erosion and avoid property destruction in coastal zone, various engineering structures are erected, such as embankments, breakwaters, and dams. Special floating structures are also used to protect a coastal zone.

Actions and technologies preventing penetration of harmful substances into water/environment near shiprepair and marine structures, service stations, berths of small vessels etc. are considered.

The existing enclosures of water basins in locations of expected pollution have a serious disadvantage so long as their efficiency is reduced by waves. A high-efficiency device which based on the analysis of the phenomena's physical nature is offered. It is an echeloned system of active vibrotraps. The system comprises special means of surface water analyses and automatic control of vibrotraps.

The arrangement of a shore facility whose activity is related to environmental risk, such as a technical service station for small vessels is reviewed. The facility comprises a platform provided systems of oil-contained substance collection and disposal. The facility precludes penetration of harmful substances into soil, atmosphere, and water.

Land reclamation and shallow water areas treatment are considered in relation to possible oil and chemical penetration to bedded deposits. Efficiency of such efforts depends on the degree of information on the state of the area to be reclaimed. After the processing of the information on environment parameters, analyses of hydrology and chemical hydrology phenomena, and pollution processes, - complex models are formed and on their basis, environment condition are predicted and optimum procedure for reclaiming work is developed.

The computer programs based on the filtration theory have been developed to calculate the predicted volume to be cleaned and to select the required power of technical means.

Special attention is paid to calculation and design of various hydraulic engineering structures for irrigation. For certain problems, such as the analyses of dams, earth embankment, etc., the finite element method can be successfully employed.

The ADHESIVE computer program is used for the calculation and design of hydraulic engineering structures. ADHESIVE is an integrated system for strength analyses and it realizes finite and boundary element methods. Certain program modules of the system allow to consider material and geometric nonlinearities, static and dynamic loads. It is also possible to consider a wide spectrum of external effects (including thermal and hydrothermal loads). The ADHESIVE program includes the following principal components: preprocessor, processor, postprocessor, database, monitor. It is implemented on ES 1060 (IBM 370) and IBM PC/AT computers. The use of color computer graphics allow to effectively display object calculation results on the visual display unit screen.

SUBSIDENCE AND DRAINAGE ON ABANDONED SOUTH WALES
MINES SITES - HAZARDS AND ENGINEERING SOLUTIONS

M.J. SCOTT and I. STATHAM*

Introduction

Following the decline of South Wales coalmining in recent decades, many colliery sites were cleared of buildings and quickly redeveloped. Shafts were often backfilled or capped and adit entrances overtipped, with little consideration of long term stability. In many cases, those remedial measures which were carried out were inadequately recorded or not recorded at all. Associated subterranean structures such as Fan drifts, culverts, basements and access tunnels were buried and are now forgotten, yet remain as substantial voids at shallow depth.

Mineworkings under the valleysides were often drained by well-constructed drainage adits, which were well maintained over a long period of time. Many of these still remain and now form an integral part of the groundwater regime, although are now in a badly deteriorated state. The consequences of them failing and becoming blocked are in many cases unpredictable and could result in severe groundwater or flooding problems in areas which hitherto had been dry for a long period of time.

There is a risk that, as detailed knowledge of the old mine sites passes out of local memory, redevelopments may fail to take full account of the hazards that remain. Two case histories are described briefly below which illustrate the risks at old mine site and the engineering consequences of redevelopment.

Pandy Shaft, Naval Colliery

The Pandy Shaft of Naval Colliery, Penygraig was a typical Victorian Colliery Shaft, some 480m deep and 5m in diameter. National Coal Board records indicated that it was backfilled to 89m depth in 1959 and finally filled to surface in 1969. However, it collapsed without prior warning in 1989, to reveal a hole 60m deep, filled with water to 20m below ground level. The collapse revealed that steelwork forming the cage landings, cage guides and other girders had been left in place and the pithead simply had been covered with loosely bulldozed demolition debris from the old colliery buildings. Colliery surface plans indicated that the fan drift, formerly associated with the shaft, led to a fan house about 45m away. This was proved by probeholing to be open to its full diameter of 5m; it also passed beneath a later factory building with its crown about 1m below floor level.

Remedial measures involved backfilling the shaft with site won, suitable material to just below rockhead, at a depth of some 19m. A grouted hardcore plug was then placed and finally, a reinforced concrete cap was cast within the superficial glacial deposits. The fan drift was broken out and backfilled with suitable compacted fill, except beneath the factory, where it was stowed using uniformly graded, granular material. During the excavations an extensive culvert system was discovered, which formerly drained the colliery surface. However, much of the system remained "live", taking stormwater from the site and surrounding street drains to the River Rhondda, despite being unknown to the drainage authority. The system was surveyed and repair works carried out where collapse was threatened.

<u>Yard Adit, Tredegar</u>

Yard Adit was formerly a major drainage level on the western side of Tredegar. Its entrance was close to the Tredegar Ironworks, from where it trended to the south-west for many kilometres, draining a large area of coal and ironstone workings.

This drainage level passes at shallow depth beneath a large redevelopment area in the centre of Tredegar. It was found to be open and accessible for a considerable distance and still carried a substantial volume of mine drainage. It also passes beneath a main road with less than 1m of cover to the crown of the arch. A condition survey of the level revealed that it had been maintained over a very long period of time, with various types of liner including masonry, brick, cast iron, mild steel and corrugated sheets. Several collapses were present and substantial voids were visible above the liner. Boreholes showed that the remaining rock cover at some collapses was negligible and that ground surface instability to form crownholes was imminent.

Stabilisation of the adit was needed over a length of more than 100m, whilst maintaining the flow of drainage water to avoid unpredictable changes in local hydrology. This was achieved by laying a 900mm diameter culvert within the adit from two purpose-built bolted segmental shafts. The annulus was then backfilled with groundwater material. Boreholes drilled from surface were used to backfill voids above the level; again granular material was used, topped up to the roof with stiff grout. The maximum height of void migration discovered was 4m above the original adit roof, close to foundation level of nearby buildings.

*   ARUP GEOTECHNICS
    CAMBRIAN BUILDINGS
    MOUNT STUART SQUARE
    CARDIFF
    CF1 6QP

RECLAIMING THE FERTILE LANDS OF THE LAKE FAGUIBINE SYSTEM

M.M. VIERHOUT, Land and Water Use Engineer,
HASKONING, Royal Dutch Consulting Engineers and Architects,
Nijmegen, The Netherlands

THE HYDRAULIC SYSTEM

The lake Faguibine System is located on the leftbank of the river Niger, in the Tombouctou Region of the Republic of Mali. The System consists of a series of large and interconnected depressions along the outer edge of the floodplain which fill up with spill water from the Niger, through a number of spill channels (see situation map below). The annual spill volume is a function of the Niger flood levels and the hydraulic conditions of the spill channels. Water levels in the lake are subject to inter annual fluctuations, depending on the difference between annual inflow volumes and losses from the lakes. The annual drawdown of the water level in the Lakes is in the order of 1.5 meter. The maximum water depth in Lake Faguibine, the biggest of the 4 lakes, is some 8 to 10 meters, corresponding to an inundated area of 55.000 ha. The maximum submerged area of the 4 lakes is about 90.000 ha.

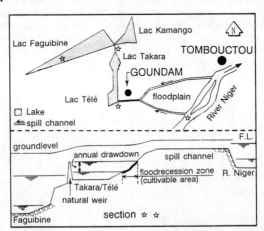

Schematic lay-out and cross-section of Faguibine System.

THE AGRICULTURAL SYSTEM

At the start of the flood recession in the lakes farmers plant crops (mainly sorghum) on the wet fringes of land just above the falling water table. The residual soil moisture in this recession zone, supplemented by rainfall at the end of the growth season, is usually sufficient to grow one crop. The total cultivable area is dependent on the slope of the recession zone, the soil characteristics and the duration of the recession period. The flatter the slope of the recession zone the larger the cultivable area will be. The cultivable area in a given year, therefore, is a function of the maximum flood level in the lakes, which in turn is a function of the spill volume or the Niger flood level. As the slopes of the lake bottoms increase with height, cultivable areas decrease when lake water levels increase. The potential cultivable area in Lake Faguibine attains a maximum of 20.000 ha when the inundation depth is 1.5 to 2.0 meter.

The maximum cultivable area of all lakes under natural conditions (no flow regulation or land improvement) is estimated at 36.000 ha, which offers a potential yield of about 50.000 tons of cereals. This traditional low input farming system has been practiced for several ages, producing enough grain for the well-known city of Tombouctou and its surroundings. In addition to crop production, the lakes are used for fishing and in the dry season for livestock-watering and cattle grazing.

THE CURRENT PROBLEM

Since the early seventies flood levels in the Niger river have been below the long term average, resulting in low spill volumes and reduced inflow into the lakes. Consequently the inundation levels in the lakes decreased as the drought period persisted, causing the Faguibine Lake to completely dry up in 1983 and subsequent years. This situation, which was aggravated by the extremely low rainfall in the sahelian zone, forced a great deal of the local population to abandon their agricultural plots in the lakes. In 1986 the Government of Mali, with assistance from the United Nations Sudano-Sahelian Office (UNSO) launched a project aimed at rehabilitation and improvement of the Faguibine system and re-instating the pre-drought situation. The project is an attempt to reclaim back the drought affected lands and to stop further desert encroachment.

THE PROJECT

Preliminary studies conducted by HASKONING concluded that in order to re-instate the pre-drought situation two priority measures are required:
1. Increasing the hydraulic capacity of the spill channels with the aim to inundate the lakes during extreme low Niger floods as well; and
2. Protecting the spill channels against encroaching sand dunes (dune fixation) and preventing siltation int he channel beds.

The required increases of the channel capacities have been determined by means of a mathematical flow model of the conveyance and lake system. The immediate result of the capacity increase would be a cultivable area of some 10.000 ha. The priority measures are currently implemented. Further studies are being conducted to define permanent improvements which would allow an optimal use of land and water resources in the Faguibine system and to define a long term strategy for rural development of the area, while conserving the productive capacity of the natural sahelian environment.

A CASE HISTORY - BRETT GRAVEL'S FAVERSHAM QUARRY

L.M.S. Williams BSc, MSc, PhD
Planning, Restoration and Aftercare Manager
Brett Gravel Ltd
Brett House, Wincheap, Canterbury, Kent CT1 3TZ, UK

INTRODUCTION

The following notes are intended to give one example of the sand and gravel industry in Kent, historically making good the environmental damage caused by gravel extraction.

Extraction of sand and gravel at the 250 acre Marsh Works Quarry, Faversham, Kent has continued uninterrupted since 1934. Before this time the site was, for 300 years, the location of Favershams' Gun Power Works and many of the quarry buildings date back to the 1600's.

The gravel is a low-level alluvial deposit, varying in depth from 1-7m, overlain by soils and in places, peat and alluvial clay. The site has a high water table and is currently worked dry by pumping; in the past the deposit was worked wet. Underneath the gravel lies Thanet Sand.

PLANNING HISTORY

Much of the land was quarried before current planning legislation came into force and consequently there was no obligation to restore the land to a beneficial afteruse. Indeed the 1948 Interim Development Order covering part of the site states:-

"That the applicants shall .... fill the excavated areas and respread the topsoil to the levels shown marked (ii) on the deposited plan lettered 'B' - the top surface of the land shall be left reasonably level and suitable for cultivation."

AREAS OF NATURE CONSERVATION

Despite the long history of quarrying at Faversham Marsh Works much of the area has developed into a wildlife paradise with a variety of habitats ranging from open water, marginal vegetation, small ponds, gravel beach areas, reed beds, willow/alder carr and woodlands. In view of the quarry's close proximity to the Swale Site of Special Scientific Interest on the North Kent Coast the area is much favoured by local birdwatchers.

The reed beds (formed on silt lakes) in particular form a haven for reed warblers and bearded tits. However, the reed habitat is of a transient nature; as soon as the silt from the washing plant ceases to flow, the reed bed dried out. Common reed (Phragmites australis) becomes out-completed by horsetails (Epilobium spp) and willow herb (Epilobium hirsutum). Once this drying out process has begun the reed bed is quickly inundated by alder (Alnus glutinosa) and willows (Salix spp). Whilst the area is dominated by Phragmites it is a feeding ground for yellow wagtails, meadow pipits, linnets and goldfinches.

MORE RECENT RESTORATION

Since the '70's as gravel extraction has progressed, so restoration has been more in keeping with current good practice.

Lakeside margins are graded, overburden and soils are spread over restoration areas and these areas are rotovated and grass seeded. Grassed lakeside banks are mown regularly for good-housekeeping. Trees have been planted in blocks bordering the lakes and hedgerows have been thickened. Other areas have been infilled and restored to grazing pasture.

DISCUSSION

Had the site been the subject of a modern planning application and Environmental Statement, both the method of working and final restoration and aftercare proposals would have been significantly different. The area open for quarrying at any one time would have been minimal with backfilling rapidly following extraction. Infilling would have been a priority with final restoration consisting of larger areas of grazing pastures with less opportunity for habitat creation and species diversity.

SECTION 3 : CONTAMINATED LAND

BEHAVIOUR OF POLLUTANTS IN SOILS

R A FAILEY
DR R M BELL
The Environmental Advisory Unit of Liverpool University Ltd
Merseyside Innovation Centre, 131 Mt. Pleasant, Liverpool L3 5TF

ABSTRACT

Contaminated land cannot be successfully reclaimed unless the behaviour of the chemicals comprising the contamination is fully understood. Different chemical contaminants partition to varying degrees between the vapour, solid, and liquid phases of the soil and thus behave in different manners. This behaviour is dependent upon chemical and physical characteristics of such contaminants which will in turn govern the incidence of bioaccumulation, leaching and migration. Contaminant interaction with soils will therefore influence the integrity and life span of a reclamation scheme.

INTRODUCTION

The behaviour of any compound, including those compounds generally known as pollutants in the soil, is dependent upon both the physical and chemical properties of the compound and those of the soil.

Many processes act to transport the pollutant from its point of origin or disposal and govern its spread throughout the soil. They include molecular diffusion and both liquid and gaseous flow. Many other processes are known to determine the fate of a pollutant once it is within a soil. They include physical, chemical and biological processes such as leaching, adsorption, desorption, photo-decomposition, oxidation, hydrolysis, and metabolism. These are shown in Figure 1.

As any number of the above processes can be acting on any given pollutant at any time, an assessment of the environmental fate of the pollutant appears extremely complicated. This is not

FIGURE 1: TRANSFORMATION OF POLLUTANTS IN SOIL

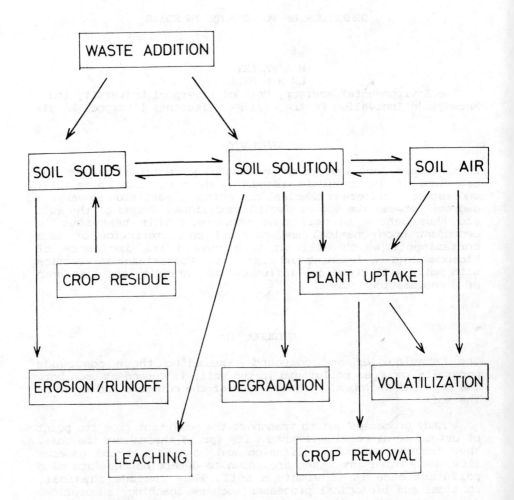

the case. Different processes may be dominant for different pollutants and much scientific research has been undertaken to accurately understand and describe pollutant pathways such that it is now possible to predict behaviour.

Pathways, soil and chemical reactions, and their interactions, need to be understood and clearly identified before any reclamation of a contaminated site can be described as long term and safe. They are however frequently forgotten or ignored.

In the soil, a pollutant will distribute between the solid phase (either undissolved particles or adsorbed onto receiving surfaces), the liquid phase (normally dissolved in the soil water), and the vapour phase. In the solid phase the pollutant can be regarded as inert, as it cannot migrate throughout the soil mass and is largely unavailable to biological organisms. Once it is in the liquid or vapour phase however, the pollutant can become active and start to affect its immediate surroundings, through biological or chemical reactions. It is therefore the properties of the pollutant and in particular its partition between the solid, liquid and gaseous phase, which governs its impact within the soil environment.

This paper addresses some of the simple chemical and physical parameters that determine the distribution and fate of pollutants in the soil with an aim of assisting with the scientific basis of land reclamation.

Soil/Pollutant Interactions

Soil is normally heterogenous and is formed from weathered bedrock. There are therefore a great variety of soils and a great variation in soil quality. A silt loam topsoil in good condition for plant growth, for example, typically consists of 20-30% by volume of air, 20-30% water, 45% minerals and 1-3% organic matter.

When soil is in-situ and undisturbed it is characterised by three layers referred to, from top to bottom, as the A, B, and C horizons. The A horizon, usually called the topsoil and which supports vegetation, is normally higher in organic matter than the other horizons, it contains the majority of the nutrients within the soil, and it is subjected to leaching of soluble materials. The B horizon, or subsoil, is usually of greater depth than A, has a greater clay content, and is a zone of accumulated leached material. The C horizon is usually called the substratum, and consists of weathered bedrock and other parent materials.

When pollutants are applied to the soil surface, they are in fact being applied to the A horizon. If pollutants are disposed of down a trial pit, they could be within the B or C horizon. Pollutants behave differently in different soil horizons.

The major physical soil parameters affecting pollutant partitioning between the solid, liquid and gaseous phase within the soil are ;
   soil texture and soil compaction,
   clay content and,
   soil organic matter - this is found primarily in the A soil horizon, and consists of a mixture of plant and animal residues in various stages of decomposition, of chemicals synthesized chemically and biologically from their breakdown products, and of microorganisms and small animals and their decomposing remains (1). It is the most active area of the soil.

The major chemical constituents affecting this same partitioning are;
   soil acidity - which is a measure of the hydrogen ion concentration in the soil and influences all soil reactions and biological activity, and,
   cation exchange capacity - which refers to the capacity of the soil to hold or sorb exchangeable cations. This property arises in the soil organic matter and clay minerals and again influences biological activity and the potential of the soil to be leached.

When a pollutant is within the soil, diffusion by the molecules of the pollutant occurs continuously and spontaneously in an attempt to even out any concentration gradient and thereby achieve a uniform distribution of the pollutant throughout the soil. Thus, molecular diffusion is a natural process which attempts to remove concentration hotspots. It rarely occurs to completion partly because of the time involved for molecules to move or migrate any distance from their source.

Mass flow is a further and more significant process which transports pollutants within the soil. It occurs by water infiltration of the soil mass causing leaching of chemicals to the underlying groundwater. In temperate regions, the movement of pollutants through soils via leaching shows a seasonal distribution (2). In the spring much rainfall evaporates from the soil surface or is taken up from the upper layers of the soil by plant roots. In the winter, both rainfall is larger and plant usage less, so that leaching becomes more significant.

Movement of water in the soil is both upward (3) and downwards, and thus transport of any pollutant within the soil water follows the same route. The presence of vegetation on the soil surface with its associated water requirement for transpiration can result in considerable quantities of water being drawn to the soil surface, especially in those areas where high evapo-transpiration ratios are prevalent. This influence of a crop on the movement of a compound in the soil is estimated to increase as the mass of the crop increases (2).

The upward transport of heavy metal and various anions, including sulphates and complex cyanides, have been reported from polluted soils that were overlaid by 600 mm of clean soils, primarily through the translocation pull of overlying vegetation (4).

Vegetation growing upon a soil can significantly affect many of these characteristics and reponses. Depending upon the nutrient source, for example, plant roots can make the soil near them either more acidic or more alkaline than the soil at distance from the root. Generally however plant roots can be regarded as only penetrating the upper soil horizons and are not of significance when regarding the soil as a whole.

Sorption and desorption

Once a pollutant is within the soil, a dynamic equilibrium is established between its solid and its liquid or solution phase, which is known as the partitioning effect. The driving force for this partitioning is sorption.

Sorption is an attraction causing the accumulation of the pollutant at an interface, for example an accumulation of a pollutant on the surface of a solid at the clay colloid-water interface. Adsorption is where the partition moves from the liquid to the solid, while desorption is where the partition is affecting behaviour of all chemicals in the soil and can occur through either chemical or physical attractions. Anionic or cationic compounds are chemically sorbed, whereas nonionic compounds are physically sorbed (5).

Several workers have reported that adsorption is highly correlated with the organic matter content and cation exchange capacity of the soil (6,7,8).

A study of the behaviour of 30 pollutants, within four different agricultural soils, allowed a relationship between sorption defined as the pollutant partition between organic matter and water and the octanol water partition coefficient (K_{ow}) of the pollutant to be obtained (9). The K_{ow} is defined as the ratio of the chemical concentration in n-octanol to that in water when an aqueous solution is intimately mixed with n-octanol and then allowed to separate. This equation:

$$\log K_{om} = 0.521 \log K_{ow} + 0.62$$

followed a linear regression equation where 0.521 and 0.62 are data-fitted coefficients. In essence the relationship means that one term, the K_{ow} of a pollutant can be used to describe its sorptive capacity in the soil.

Many investigators have since assessed this equation with different groups of environmentally active chemicals and varying soil types (10,11). It has been concluded that the practical utility of this form of relationship is maximised when a broad range of chemicals is being investigated (12).

To obtain the above relationship it is necessary to assume that pollutant adsorption to soil is exactly reversible, that is, the pollutant will be released from the organic fraction of the soil into the soil solution in the direction of the concentration gradient. This is not quite the case. The ease of desorption appears to depend upon the actual strengh of the adsorption process, so that herbicides are recovered from soils and sediments with low organic fractions more readily than they are from organic rich soils (13).

Desorption was found to be fully reversible if desorption took place immediately after uptake when the sorbing soils were still wet (14) but was modified if the soils were allowed to dry thoroughly (15).

In most instances, sorption can be regarded as advantageous, as sorbed compounds in the soil are environmentally less active. The addition of sludge containing PCB's to soil, for example, did not increase the environmental hazard of the PCB's because of the additional sorptive capacity of the added sludge (16).

It has also been reported that the K_{ow} reflects the bioaccumulation potential of a pollutant (17); which in turn is defined as the concentration of the pollutant in an aquatic organism compared to the concentration of the pollutant in the water to which the organism is exposed, (18). The bioaccumulation factor is, in fact, a partition of the pollutant between water and the lipid and protein phases of the organism (19). It suggests that in terms of chemical attractions, the soil organic matter and lipid and protein phases of an organisms behave similarly.

Considerable effort has been expended to determine the relative importance of clay and organic matter in pollutant sorption. Both materials sorb chemically and physically and both can exhibit a wide variation in their properties. It is now however generally accepted that organic matter is more important than clays except in dry soils or where the organic carbon content of the soil is low (5, 20, 21).

As mentioned above, soil pH and moisture are further critical factors affecting pollutant behaviour in soils. The pH affects the dissociation constant of the pollutant which in turn affects its chemical bonding potential and thus the strength of sorption.

Sorption of pollutants is more complete in dry soils than in wet soils as surfaces become available to the chemical which would normally preferentially adsorb water (22).

Temperature may also influence sorption both through its effects on the solubility and vapour pressure of the pollutant.

Generally, an increase in soil temperature leads to decreased adsorption, however there are exceptions, where higher temperatures result in lower soil water conditions (13).

Volatilization

The partition of a compound from its liquid or soluble phase to its gaseous phase is decribed as volatilization. This is a further process by which pollutants can be spread throughout the soil and is particulary important for some pollutants, as transport through the soil in the gaseous phase is many times greater than that in a liquid phase (23).

Ficks first law of diffusion is applicable to all diffusion processes in soil, including movement of pollutant through air spaces. This means that the rate of movement of a chemical is directly proportional to the concentration of the chemical and its diffusion coefficient, which is a quantitative expression of the diffusion rate through different media (24).

The division of a compound between the soil solution and the air spaces in the soil is often described by Henry's Law and the extent of partitioning described by Henry's Constant (10). Henry's Law is expressed:

$$C_g = K_h C_l$$

where C_g is the concentration of chemical in the vapour phase (g/m^3 soil air),

K_h is Henry's Law Constant which has no dimensions, and, C_l is the solution concentration (g/m^3 soil solution).

Those pollutants with a high vapour pressure, which would be reflected in a high Henry's Constant, will easily move from the soil solution into the soil air, and will be quickly spread throughout the soil. There have, unfortunately, been no all embracing studies to determine the level at which Henry's Constant describes the predominant transport route although it has been suggested that a partition between vapour and aqueous phase of 10^{-4} is normally sufficient to ensure vapour effects when working with pesticides (25).

There have been investigations aimed at assessing the rate of volatilization, and thereby loss of the chemical from the soil. These have determined that the vapour density of a chemical in soil is the main factor controlling volatilization (26). Much of this work has been carried out with soil fumigants, or herbicides that act through their vapour phase. These compounds normally have high vapour pressures with low water solubilty (27, 28).

Volatilization itself is influenced by many soil and compound parameters. The position of the compound in the soil will, for example, affect its chance of being volatilized and thereby leaving the soil. For example, only 3% of applied heptachlor was recovered from surface treated soil, 3-4 months after application compared to 15% of the insecticide incorporated into the soil after 5 months (29).

Soil type also affects the degree of potential volatilization of a chemical from the soil. Volatilization rates for DDT and lindane were dependent on adsorptive characteristics of the soil and concentration of the chemical (30).

Degradation

While sorption and volatilisation describe the behaviour of pollutants in soil, they are only of significance if the pollutant actually reamins within the soil long enough for partitioning to occur. The concentration of most pollutants in the soil decreases with time providing that no further additions are made or that the compound is not being synethesised via the degradation of other compounds.

Degradation of a pollutant in the soil can occur by biological or non biological processes. Non biological degradation is more numerous than biological degradation. It includes degradation by light as well as hydrolysis, oxidation, or reduction, which may be cataylsed by soil colloids.

Biodegradation is generally defined as the molecular degradation of a substance resulting from the complex action of living organisms. A substance is said to be biodegraded to an environmentally acceptable extent when its undesirable environmental properties are lost (31). While it is primarily organic pollutants that can be biodegraded, the chemical form of a metal pollutant may be changed by biological activity such that the metal becomes more or less soluble in water. This may, in due course, affect the impact and ultimate fate of the pollutant.

Within the soil there are many biological agents with the capacity to directly degrade, or otherwise change, pollutants. These include bacteria, fungi, algae, and higher plants. Generally, biological degradation seems related to the overall level of microbiological activity rather than the presence of any one particular agent. Biological degradation tends to occur more in soils of relatively high organic matter contents, providing soil sorption does not protect the target compound.

The rate at which degradation occurs is described as

$$M(t) = M(0) \exp(-ut)$$

where $M(t)$ is the quantity of the compound remaining in the soil at time t, (10). The rate of degradation of individual pollutants is normally described by the pollutant's half life, which is defined as the time taken for one half of the initial concentration of the compound to be degraded, and is expressed as

$$T_{1/2} = 0.693/u$$

Table 1. The Half life ($T_{1/2}$) for some compounds

	log K_{ow}	$T_{1/2}$	Hc
Acrolein	-0.09	b	3.3720
Aldrin	-0.14	c	0.0022
Chlordane	2.78	c	0.0001
DDD	5.99-6.08	c	0.0000
DDE	5.69	b	0.0009
DDT	4.89-6.19	c	0.0190
Dieldrin	2.9	c	0.0003
TCDD	6.14	c	nd
Phenol	1.46	a	0.0000
Nitrobenzene	1.85	c	0.0005
Benzene	2.13	a	nd
Toluene	2.69	a	0.2703
2,4 - dichlorophenol	2.75	a	0.0002
2,4,6 - trichlorophenol	3.38	nd	0.0135
1,2,4 - trichlorobenzene	4.26	a	0.1387
Hexachlorobenzene	6.18	c	0.0251
Pentachlorophenol	5.01	a	0.0001
2 - nitrophenol	1.76	c	0.0036
Acenaphthene	4.33	c	0.0025
Acenaphthylene	4.07	c	0.0021
Fluorene	4.18	c	0.0032
Naphthalene	3.37	c	0.0100
Anthracene	4.45	c	0.0434
Fluoranthene	5.33	c	0.0004

where log K_{ow} is taken from USEPA, 1979

where half lives are assessed as
    a = less than 10 days
    b = 10-50 days
    c = greater than 50 days
    nd = not determined
based upon principal fate in the environment from USEPA, 1979

where Hc has been calculated according to Thibodeaux, 1979 from data supplied by USEPA, 1979.

The half lives of many compounds have been published (32, 33), and some are shown in Table 1. Actual half lives are however very dependent upon local environmental conditions and published material can only be taken as a guide.

Conclusions

The behaviour of a pollutant within the soil can be simply described by its K_{ow}, its Henry's Constant and its half life.
Each of these parameters has been well described. The organic matter content of the soil best describes its ability to sorb compounds, while the K_{ow} best describes the ability of the compound to be sorbed.

Gaseous transport in soils is affected by both physical and chemical aspects of the soil as well as the vapour pressure of the pollutant.

Biological, physical and chemical degradation in soil continually act to reduce the concentration of a pollutant with time.

Understanding the behaviour of pollutants in soils, the processes which govern that behaviour, and the environmental fate of the pollutants, is essential in determining safe reclamation practises which aim to isolate pollutants from their targets.

REFERENCES

1. Lee,C.R.,J.G.Skogerboe,K.Eskew, R.A.Price, N.R.Page 1985 Restoration of problem soil materials at Corp Of Engineers construction sites. US Army Corp. of Engineers. Waterways Experimental Station. Vicksburg, Mississippi.

2. Leistra,M., 1980. Transport in solution. <u>Interactions between herbicides and the soil.</u> Academic Press. Ch. 2. 31-57.

3. Bailey,G.W., J.L.White 1968 Review of adsorption and desorption of organic pesticides by soil colloids with implications concerning pesticide bioactivity. Agric. Food Chem. 12, no 4, 324-332.

4. Bell,R.M.,G.D.R.Parry, 1984. Upward migration of contaminants through covering systems. In Proc. Conf. Management of Unconrolled Hazardous Waste Sites. HMCRI Washington DC.

5. Kaufman, D.D. 1983 Fate of toxic organic compounds in land applied wastes. in <u>Land treatment of hazardous wastes.</u> Eds. Parr, J.F., P.B. Marsh, J.M. Kla Noyes Data Corp., Park Ridge, N.J.

6. Rhoades, R.C., I.J. Belasco, H.L. Pease 1970 Determination of mobility and adsorption of agrochemicals on soils. J. Agric. Food Chem. 18,no 3, 524-528.

7. Bailey,G.W, J.L.White, 1970. Factors affecting the adsorption, desorption, and movement of pesticides in soil. Res. Revs. 32,29-92.

8. Means, J.C., S.G. Wood, J.J. Hassett, W.L. Banwart 1982 Sorption of amino and carboxy-substituted polynuclear aromatic hydrocarbons by sediments and soils. Environ. Sci. Technol. 16, 93-98.

9. Briggs, G.G. 1973 A simple relationship between soil adsorption of organic chemicals and their octanol water partition coefficients. Proc. 7th Br. Insect. Fung. Conf., Nottingham, U.K. 83-86.

10. Jury, W.A., W.F. Spencer, W.J. Farmer 1983 Behaviour assessment model for trace organics in soil; 1. Model description. J. Environ. Qual. 12, no 4, 558-564.

11. Karickhoff, S.W. 1984 Organic pollutant sorption in aquatic systems. J. Hydraulic Eng. 110, no6,707-735.

12. Brown, D.S., E.W. Flagg 1981 Empirical prediction of organic pollutant sorption in natural sediments. J. Environ. Qual. 10, no3,382-386.

13. Harris, C.I., G.F. Warren 1964 Adsorption and desorption of herbicides by soil. Weeds 12,120-126.

14. Graham-Bryce, I.J. 1967 Adsorption of disulfoton by soil. J. Sci. Food Agric. 18,73-77.

15. Sims, R.C.,D.L. Sorensen, J.L. Sims, J.E. McLean, R. Mahmood, R.R. Dupont, K. Wagner 1984 Review of in place treatment techniques for contaminated surface soils. EPA-540/2-84-003b November, 1984.

16. Fairbanks, B.C., G.A. O'Connor 1984. Toxic organic behaviour in sludge amended soils. Proc. Conf. &th Madison Waste Conf. 161-165.

17. Dawson, G.W., C.J. English, S.E. Petty 1980 Physical and chemical properties of hazardous waste constituents. Attachment 1, Appendix B in background document (RCRA subtitle C) identification and listing of hazardous waste. Office of Solid Waste, EPA May 2, 1980.

18. Brown, K.W., G.B. Evans, B.E. Frentrup 1983 <u>Hazardous waste land treatment</u> Ann Arbor Sci., Ann Arbor, MI.

19. Briggs, G.G. 1981 Theoretical and experimental relationship between soil adsorption, octanol water partition coefficients, water solubilities, bioconcentration factors, and the parachor. J. Agic. Food Chem. 29,1050-1059.

20. Wahid, P.A., N. Sethunathan 1978 Sorption-desorption of parathion in soils. J. Agric. Food Chem. 26, no1,101-105.

21. Wahid, P.A., N. Sethunathan 1979 Sorption-desorption of a,b and g isomers of hexachlorocyclohexane in soils. J. Agric. Food Chem. 27, no5, 1050-1053.

22. Yaron, B., S. Saltzman 1972 Influence of water and temperature on adsorption of parathion by soils. Soil Sci. Soc. Amer. Proc. 36, 583-586.

23. Mayer, R., J.Letey, W.J.Farmer 1974 Models for predicting volatilization of soil incorporated herbicides. Soil Sci. Soc. Amer. Proc. 38, 563-568.

24. Goring, C.A.I. 1962. Theory and principles of soil fumigations. In R.I. Metcalf (ed) Advanced Pest Control Research. 5,47-84. Interscience Publishers New York.

25. Graham-Bryce, I.J. 1984 Optimization of physicochemical and biophysical properties of pesticides. In Pesticide synethesis through rational approaches. Eds. Magee,P.S., G.K.Kohn, J.J.Menn. Amer. Chem. Soc. Washington DC.

26. Farmer, W.J., K. Igue, W.F. Spencer, J.P. Martin 1972 Volatility of organochlorine residues from soil: Effect of concentration, temperature, air flow and vapor pressure. Soil Sci. Soc. Amer. Proc. 36, 443-447.

27. Harvey, R.G. 1974 Soil adsorption and volatility of dinitroaniline herbicides. Weed Sci. 22, no 2, 120-124.

28. Goring, C.A.I. 1967 Physical aspects of siol in relation to the action of soil fungicides. Ann. Rev. Phytopath. 5, 285-318.

29. Saha, J.G., W.W.A. Stewart 1967 Heptachlor, heptachlor epoxide, and gamma-chlordane residues in soil and rutabaga after soil and surface treatments with heptachlor. Can. J. Plant Sci. 47,79-88.

30. Guenzi, W.D., W.E. Beard 1970 Volatilization of lindane and DDT from soils. Soil Sci. Soc. Amer. Proc. 34,443-447.

31. Rochkind,M.L.,J.W.Blackburn, G.S.Sayler.1986 Microbial decomposition of chlorinated aromatic hydrocarbons. EPA/600/2-86/090.

32. Ryan, J.A. 1976 Factors affecting plant uptake of heavy metals from land application of residuals. Proc. Conf. Disposal of residues on land. 98-105. St Louis.

33. Smith, L.R., J. Dragun 1984 Degradation of volatile chlorinated aliphatic priority pollutants in groundwater. Environ. Intern. **10**, 291-298.

REDEVELOPMENT OF A FORMER STEELWORKS

DAVID L. JONES
ASSOCIATE
L.G. MOUCHEL AND PARTNERS LIMITED WEST HALL, PARVIS ROAD,
WEST BYFLEET, WEYBRIDGE, SURREY KT14 6EZ

ABSTRACT

The James Bridges Steelworks at Park Lane, Darlaston was once a major producer of steel within Europe. Changing fortunes has now led to its substantial closure, the site being redeveloped for light commercial and retail premises and housing. The former steelworks was constructed over a mine shaft for which treatment was required to enable the new construction to proceed. The by-products of the works, namely slag and foundry sand, form an extensive area rich in contaminants. These were isolated and removed with a barrier capping being incorporated in the final design. Treatment of the regraded foundry sand was by dynamic compaction and vibrocompaction the former producing unacceptable ground accelerations leading to a reduction in the DC area. The numerous geotechnical difficulties presented by the site were overcome within the allocated time scale.

SITE HISTORY

The 100 acre site of the James Bridges Steelwork at Darlaston in the West Midlands has seen a change of fortune during its lifetime. The site has reportedly been producing steel on the present site continuously since 1879, and at the peak of its production this plant was the largest of its kind in Western Europe with 5000 employees. A changing economic climate led to a drastic reduction in steel production between 1980 and 1985. Now the majority of the site has been redeveloped to accommodate the light commercial units much favoured these days.

During steel production, major by-products were foundry sand and slag. These materials were kept on site, either forming infill for the expansion of the works or for general land raising. A significant depth of foundry sand thus existed prior to site redevelopment.

Being located in the West Midlands it is not surprising that mining is also associated with the site. A search of NCB records revealed three disused mine shafts in the area of the works, but do not reveal any evidence of working. Any such workings would have taken place in the latter half of the last century prior to records being kept. There is a possibility that one or all of these shafts serviced mine workings extracting the Blue Flatts ironstone seam, for which no records are available. The iron ore from this seam may have been used in the foundry for pig iron production. The seam, located between 30m to 60m below the site, was extensively worked during the middle of the last century.

Two of the shafts had been built over by the steel works, one in the light fettling shop and one in the mini-mill.

Over the remainder of the site the course of the River Tame was frequently changed as a result of infilling with foundry sand and slag.

GEOTECHNICAL AND MINING CONSIDERATIONS

Several investigations had been undertaken over the years. These related to an assessment of the ground contamination due to the industrial processes as well as the underlying geology and geotechnical properties. A specific investigation was also undertaken to detect a mineshaft. These investigations were carried out over the period 1984 to 1988.

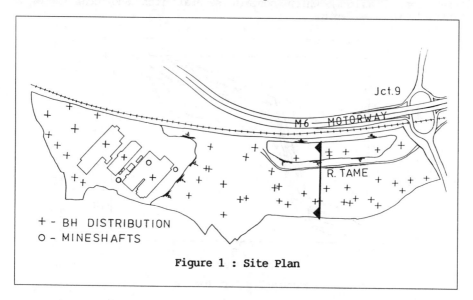

Figure 1 : Site Plan

The distribution of boreholes and trial pits are shown in Fig. 1. The investigations within the steelworks area was more limited due to the presence of existing buildings. A typical cross section across the site, indicating a simplified geological profile is shown in Fig. 2. As can be seen, made ground occurs extensively over the site, the depth varying depending upon locality. The greatest depths are in the region of the steelworks and between the River Tame and the railway.

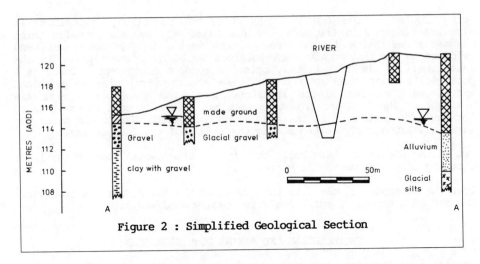

Figure 2 : Simplified Geological Section

The made ground is predominantly foundry sand of a well defined grading (See Fig. 3), overlying glacial gravels and tills. Locally, alluvial clays and sands, associated with the River Tame are also encountered. Beneath the generally stiff glacial tills lie rocks of the Westphalian A (Lower Coal Measures) which contain several iron and coal seams at relatively shallow depth.

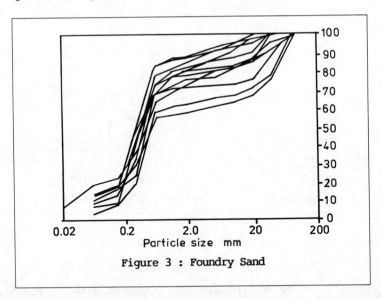

Figure 3 : Foundry Sand

The geotechnical investigations, comprising shell and auger techniques, rotary holes and trial pits, were the standard geotechnical approach to ground investigations. Of particular interest, however, was the identification of the mine shafts. This commenced with a study of available records, including Ordnance Survey maps and British Coal records. These identified the presence of three shafts, the position of one varying depending upon the record considered.

As the form of the development was known, only one of the shafts was of specific interest as it was shown to be immediately adjacent to a proposed retail store. It was therefore necessary to positively identify the location of the shaft and establish the nature of the infill, followed by grouting up of the shaft (1).

From the conflicting records, a best estimate was taken of the shaft location and a borehole grid pattern set out at 1.25m centres. Probe drilling took place using rotary open hole percussive techniques drilled with a 75mm cross bit to depths varying from 13.0m to 30.0m using air flush. The nine probe holes readily identified the shaft location and suggested an internal diameter of 3.6m.

Having identified the location, the shaft infill and depth was proved by rotary coring. The base of the shaft terminated in sandstone at a depth of 62m below ground level. The shaft infill comprised relatively loosely packed mudstone to a depth of 30m as well as considerable amounts of timber, iron and steelwork. At a depth of 36m there was evidence of a timber staging beneath which was a very loose fill comprising weathered mudstone, clay and gravel of foundry ash clinker glass.

The shaft was grouted from the base via the drilling rods. The grout comprised 8.5 parts PFA to 1 part OPC with water added to give a flow rate of 300-600mm. The grout was injected until the pressure was maintained at 200 kN or a maximum pressure of 10kN per metre of overburden was obtained. The grout mix had a compressive strength between 2-5 N/mm^2.

A total of 97 tonnes of grout was injected from the shaft base of 62m up to 12m below ground level via five drill holes.

Various options were considered for capping of the shaft, v.i.z.:-

a) Installation of a secant piled wall around the shaft with the wall being capped at ground level.

b) Formation of a larger diameter cap on a ring of eight cast in situ piles.

c) Excavation through the fill and forming the cap directly on top of the shaft.

The latter option was adopted. The cap was 7m x 7m and 450mm thick. Between the cap and the grouted column of the shaft a 300m granular layer was placed containing a 100mm pipe to act as a gas vent. Overlying the shaft a HDPE liner was placed to act as a gas seal. Gas had previously been detected in the ungrouted shaft, the venting being installed as a precautionary measure even though the shaft was grouted as part of the shaft treatment.

GROUND CONTAMINATION

Ground contamination posed a further difficulty. The on going industrial activities suggested that lead, copper, zinc, arsenic and cadmium would be expected on the site along with foundry slag. Records also showed that leaking of phenols from foundry sand had at some time caused contamination of the River Tame.

DD175 (2) gives advice on sampling density, but this was not wholly applicable as the development plan was to retain as much material on site as possible and produce a level site by regrading of the foundry sand mounds present. The chemical investigation was therefore geared to identifying overall levels of contamination within the foundry sand; identifying and classifying visually different material from the slag; identifying the type and expansive nature of slag and whether any obvious zoning of contaminants existed.

The investigation was based mainly on trial pits excavated to between 2.5m and 4.0m deep. Identifiable waste dumps on the site were also sampled.

The waste dump were clearly related to fly tipping and it was noticeable that these contained high levels of contaminants, notably cadmium, lead and arsenic. Being readily identifiable these were removed from site.

Phenols were quite widely detected. These were not all anthropogenic, with naturally occurring phenols existing in an area which had been used as a sports field. Closer examination of the sports field revealed this to be clean, thus allowing this area to be used for domestic dwellings.

Over the main development area, containing foundry sand, 70% of the pH values recorded were above 9.5 and 37% greater than 10. It was concluded that the lack of buffering capacity in the sandy profiles must to some extent be responsible for the high pH values.

The main concern with the foundry sand was the presence of slag. Large, boulder size slag had mainly been stockpiled and not intermixed with the sand, but the concern was of finer slag particles within the sand. The steel making process had produced a blast furnace slag and latterly an electric arc furnace slag, the latter being highly expansive. Fortunately the electric arc furnace slag appeared to be concentrated in one area and could be readily isolated and removed.

To identify the expansive properties of the foundry sand and slag special expansion tests were carried out on the sand and both types of slag to establish a benchmark. This involved establishing the amount of free calcium oxide and relating this to an accelerated expansion test. The electric arc furnace slag showed up to 8% of free lime and a measured expansion of 19%. The blast furnace slag indicated free lime less than 2% with expansion up to 2%. The tests confirmed the differences in the slag and that it should be removed to avoid future swelling problems. As the slag was generally of boulder size, identification and removal was not difficult. The sand posed a greater concern as it could contain fine slag particles.

Free lime within the slag ranged from a maximum of 0.5% to being non-detectable the average being 0.05%. The maximum measured expansion was less than 4% and typically less than 0.2% to non-recordable. The sand was therefore not considered to be a problem with respect to expansion.

GROUND TREATMENT

The nature of the ground readily lends itself to ground treatment by dynamic compaction and vibrocompaction. This was dealt within two stages. The first stage comprised the area containing the former steelworks, at a higher elevation than the rest of the site. This was to be the site of an international furniture store, the elevated profile giving it prominence to the nearby M6 motorway.

Treatment consisted of dynamic compaction using a 10 tonne pounder dropped from a height of 15m, the aim being to achieve a bearing pressure of 200 kN/m^2. Due to the development strategy of the site, this element of the work was contractor designed and is not discussed further.

For the remainder of the site, which was Engineer designed, the aim was to uniformly compact the foundry sand which had been reworked over the site. This would thus not put any restriction on the siting of structures within the development.

To establish the optimum energy level various trial areas were designated, based upon a known knowledge of site conditions. At each location both 8 tonne and 15 tonne pounders were dropped from varying heights with varying number of drops. By plotting the net volume of imprint formed and cumulative energy, the optimum energy level was achieved. It was noticeable that the maximum drop height did not necessarily correspond with optimum compaction, as measured by the heave tests. The energy levels adopted varied between 160 Tm/m^2 to 200 Tm/m^2 generally using a 15 tonne weight dropping from a height of 11m. Tamping infill passes were achieved using an 8 tonne weight.

The development is adjacent to a residential area and school. There was therefore concern that the vibrations produced by the DC operations would cause unacceptable horizontal ground accelerations. A buffer zone of 30m was therefore imposed between the site boundary and DC works. Compaction within this area was by vibrocompaction.

A series of trials was undertaken varying energy levels and distance, and measuring the resultant peak particle velocity and vibration frequency. The guidelines for assessment were BS 6472 : 1984 (3). The difficulty comes in interpreting the standard, i.e. is a DC operation a continuous or intermittent vibration?

Depending upon the energy level adopted results showed velocities of around 2mm/s at 150m and 5-6mm/s at 45m. At 20m values of 30mm/s were recorded. The levels recorded were unacceptable to local residents. Accordingly, after discussions with the Environmental Health Officer, an upper limit of 4mm/s was adopted combined with an increased buffer zone of 65 metres. Restrictions were also put on the hours of working to further minimize the nuisance to residents.

The above meant that the vibrocompaction area had to be increased to compensate for the reduction in DC. This strategy change initially created difficulties with the apparent inability of the vibrocompactors to readily penetrate the fill, but these were eventually overcome.

OTHER WORKS

As part of the development the existing course of the River Tame was altered and a new channel formed. In soft ground areas the new channel was lined with limestone armour, some of which had to be rejected as it contained an unacceptably high cadmium level. Although all pollutants could not be removed from the site, all import had to comply with either the ICRCL (4) acceptance levels for contamination or those produced by the Black Country Development Corporation, which were generally tighter.

Upon completion of the groundworks the site was capped with a 500mm granular fill to act as a capillary break.

Special details were required for landscaping to ensure that trees did not penetrate either the clay or underlying granular fill. As only the infrastructure has so far been developed details of foundations can not yet be given.

ACKNOWLEDGEMENTS

The author wishes to thank Triplex Gallagher for permission to publish this paper.

REFERENCES

1. Healy, P.R. and Head, J.M. Construction over abandoned mine workings, CIRIA Report No. SP32, London 1984.

2. BS DD 175 : 1988, Draft for Development, Code of practice for the identification of potentially contaminated land and its investigation, British Standards Institution, London 1988.

3. BS 6472 : 1984, British Standard Guide to Evaluation of Human Exposure to Vibration in Buildings (1 Hz to 80 Hz), British Standards Institution, London 1984.

4. ICRCL 59/83 (2nd Edition), Guidance on the assessment and redevelopment of contaminated land, Department of the Environment, London 1987.

ENVIRONMENTAL ASSESSMENTS OF RECLAIMED LAND IN THE UNITED STATES OF AMERICA

Matthew Leonard - Golder Associates Inc. 20000 Horizon Way, Suite 500, Mt. Laurel, New Jersey, USA 08054

Kevin Privett - Golder Associates (UK) Ltd. 5 New Mill Court, Phoenix Way, Enterprise Park, Llansamlet, Swansea, SA7 9EH, Wales.

ABSTRACT

United States law, both federal and state, requires environmental assessments of certain industrial establishments prior to closure or transfer to a new owner. If a site is to be reclaimed it is the current owners responsibility to prove to the satisfaction of the authorities that it is clean. One of four major phases of study that must be undertaken is the Remedial Investigation Feasibility Study which includes gathering site data and assessing the feasibility of different remedial actions. For sites that are on the Superfund list there are rigorous procedures defined by the United States EPA. The investigation work for a large site may take years to complete and cost several million dollars. Site assessments for property transfers are on a much lower scale (but can lead to the discovery of contamination sufficient to place the site on the Superfund list). There are no detailed guidelines and current practice has evolved a three tier level of assessments necessary to address the issue of how much investigation is enough. These three levels are the Screening Phase, Environmental Assessment and In Depth Environmental Assessment. Given the ever increasing environmental liabilities, it is vital that

Environmental Assessments are carried out by suitability qualified professionals to very high standards of care, compared to more traditional forms of site investigation.

INTRODUCTION

The term "Environmental Assessment" in the USA has several meanings, from a broad Environmental Impact Statement encompassing issues such as wetlands, noise, dust, traffic, pollution, site history, etc., to narrower studies focusing on precise fields such as biology or groundwater chemistry. Environmental assessments in this paper refer to site characterization in terms of soil and groundwater composition with respect to existing laws, standards, or practices on acceptable chemical concentrations in soil, surface or groundwater.

The cycle of demolishing existing industrial facilities to construct new ones, or simply to reclaim the land, is frequently accompanied by an environmental assessment. Although the prime objective of the assessment is to make the seller perform any necessary clean up or remediation on a site by site basis, the overall result is intended to gradually restore the environment in areas suffering from industrial or agricultural contamination. The framework for these assessments in the USA, as well as some current practices in their performance, are discussed in this paper.

REGULATORY FRAMEWORK

The federal Superfund program was designed to provide financial and technical guidance to clean up seriously contaminated sites in the United States. The law, known as the Comprehensive Environmental Response Comprehensive Liability Act (CERCLA), is administered by the Environmental Protection Agency (EPA). The law does not provide specific action or clean up levels but does provide administrative procedures and a framework for determining if a site is to be placed on the National Priorities List (NPL). Sites are evaluated with a Hazard Ranking System (HRS), which adopts a system of points to evaluate the overall threat to human health.

In parallel with the federal program, most stages have their own environmental laws which often mirror federal regulations. The State of New Jersey has a very comprehensive set of environmental regulations, and in particular, the Environmental

Clean-up and Responsibility Act (ECRA). The law addresses the closing, terminating, or transferring of certain industrial establishments. The law places a burden on the owner and the operator to make the necessary filing of ECRA documents to the State. A Sampling Plan to evaluate the site must be included in these documents. The Sampling Plan shall be designed to determine the presence of and delineate any contamination, including any off site contamination which is emanating or has emanated from the industrial establishment.

The Sampling Plan has to be approved by the State, and then be implemented by the owner. The industrial establishment can only be transferred, sold, or terminated if the report issued after site sampling states there has been no discharge of hazardous waste, or the discharges have been cleaned up to the current satisfaction of the State. The State does have "informal" guidelines for chemicals in soil and groundwater that are used to evaluate the need for environmental clean-up. If hazardous materials remain, the new owner must provide a letter accepting ownership of the materials. Also the parties can proceed with transfer if the current owner enters a Consent Order for clean-up, with a bond.

The above concerns on the presence of hazardous materials have led to a frequent need for Environmental Assessments, particularly when industrial properties are transferred. The same concerns can apply for abandoned sites that are to be reclaimed, as there is frequently an accompanying change in ownership.

SITE EVALUATION PROCEDURES

Sites that meet the NPL criteria for inclusion on the Superfund list undergo four major phases of study or actions. These are Remedial Investigation Feasibility Study, Remedial Design, Remedial Action, and Operation and Maintenance. This paper is concerned with the Remedial Investigation Feasibility Study (RI/FS) which includes gathering physical data related to on-site chemistry, and studying the feasibility of different remedial action alternatives. The overall procedures to be followed are very specifically defined in EPA guidance documents. The study will include a risk analysis to evaluate the risk to human health and the environment. The RI/FS process can take several years, usually involves large, costly field sampling and analytical programmes, and can cost several millions of dollars on large sites. Several consultants may be employed together to perform the RI/FS, which is very site specific.

Site Assessments for property transfers are on a much lower scale from all aspects when compared to the RI/FS procedures, but can be the initial stage for discovery of the potential contamination to place a site on the NPL. There are no detailed guidelines for the performance of Site Assessments, which have and will lead to debate on the required level of care and diligence needed to make them. A site of a well run modern facility would not be expected to be subjected to such a rigorous investigation as an old derelict site with a history of environmental problems. However, a new facility may have been constructed on a contaminated site, on which all visible signs of past activities have been landscaped out or built upon.

Current practice has established a three tier level of assessments to address the difficult issue of how much investigation is enough? The three levels of investigation are summarized below.

Level 1 - Screening Phase

The screening phase is essentially a paper study to determine the likelihood of hazardous materials being present on the site. The study would review any available borehole and groundwater data, site and aerial photos, and historical well information. A procedure known as a Title Search would be performed to establish all past site owners, and where possible, neighbours and site workers would be interviewed on past site activities.

Level 2 - Environmental Assessment

The assessment will include field sampling, the installation of monitoring wells, and analytical chemical testing of soil and/or water samples. In general, there is more concern over the choice of chemical tests than the locations of the sampling points.

Monitoring wells are generally installed at all four sides of the site and certainly on the inferred up and down groundwater gradient sides when this can be readily identified. Monitoring wells or soil sampling boreholes are also placed around structures, particularly near drains, sumps, tanks and storage areas. The general philosophy is that chemicals of concern are likely to be detectable in the groundwater, which acts as a means to spread chemicals downgradient. However, certain hazardous materials such as dioxin, PCB's and some heavy metals are relatively insoluble. In this case suspect areas such as lagoons would be sampled.

The nature of the chemical tests to be performed on soil and/or water samples depend very much on previous activities at the site. The choice of test must be made by an experienced team of professionals, such as a geochemist or chemical engineer and a hydrogeologist. The type of test is well defined if the site was a location for petroleum refining or storage, industries using heavy metals, PCB's or pesticides. Sites covered by the theme of this conference should fall into these test categories. The major difficulty is when there is no previous indication of industrial activity.

The field investigations must include enough QA/QC samples to ensure that samples are not contaminated by improper washing of sampling equipment and containers, poor sample storage, or cross contamination between samples. To meet these requirements, drilling equipment has to be steam cleaned, sampling equipment has to be washed with several chemicals and then with distilled water, and the samples have to be sealed such that they cannot be tampered with. Careful field procedures have to be followed, and the average geotechnical driller will not be able to understand the level of care required unless he also offers environmental drilling services.

It should also be mentioned that the analytical laboratory must be sufficiently qualified to know how to perform the tests. In order to monitor the level of decontamination used in the field and laboratory, additional samples are tested which have either been filled by passing distilled water over sampling equipment (samples called equipment rinsate blanks), or are made up to have known concentrations of specific chemicals (called spike blanks). Field duplicate samples are often analysed to assess the precision of the sample collection and analysis procedures. In this way, the performance of the field and laboratory staff can be monitored.

The results of the tests are interpreted to assess if the site shows sings of contamination.

Level 3 - In Depth Environmental Assessment
The Level 3 assessment is a more in depth, or third phase investigation. Such an assessment is generally only performed if the Level 2 assessment reveals contamination,

or if the site has a high probability of serious environmental problems. The distinction between Levels 2 and 3 investigations is made easier by the recent development of soil gas surveys, which "sniff" air in the unsaturated soils for traces of volatile compounds. This enables certain sites, for example filling stations, to be rapidly investigated to locate areas of potential contamination. Once the potential problem areas have been found using soil gas as the Level 2 investigation, Level 3 investigations with much more costly wells and boreholes can be performed in suspect areas.

The performance of Environmental Assessments is most often made by consulting engineers or geologists with a hydrogeological or geotechnical background. The liabilities associated with the assessments are as great or greater than with traditional studies. A fine balance between project costs and the conclusions which can be reached must be made. It is important to realize that it is only possible to make representations on the basis of the results and observations at the locations sampled. Stronger representations, in the form of certifications and warranties are misleading, and in reality serve no useful long term purpose.

CONCLUSIONS

There is no doubt that Environmental Assessments in the USA have assisted in the identification of contaminated sites. Unfortunately their cost, and that of the clean up, has dissuaded or prevented a number of owners from selling, leading to an ever increasing number of derelict facilities owned by bankrupt entities. An Environmental Assessment is only meaningful if it has been performed by qualified professionals to very high standards of care, compared to more traditional forms of site investigation. Given the ever increasing environmental liabilities, both in terms of scope and cost, it is advisable that reclaimed properties be appropriately examined for signs of environmental contamination. An equal level of examination should also be applied to issues such as the quality of fill materials brought on site to make up levels, or the liabilities associated with materials taken off site.

RISK ANALYSIS AS A GUIDE TO OPTIMAL RECLAMATION STRATEGIES FOR CONTAMINATED LAND.

BY DR M LOXHAM
Delft Geotechnics Ltd. Black Lion Court, Congleton Cheshire.

Abstract.

Given the costly and disruptive nature of remedial actions on contaminated sites, there is an urgent need for a rational design methodology to avoid wasted effort and resources. Such a technique is the Source-Path-Target methodology described and illustrated here.
The methodology can be used to analyze the impacts of a contamination, to rank the seriousness and urgency of possible problems, to classify the responses available and to establish the environmental engineering design points for the remedial measures chosen.

Introduction.

Soil pollution issues have started to move to the centre of attention in environmental engineering and legislation. There are several driving forces behind this movement. Firstly it has become recognised that polluted soils can present a real health hazard to people coming in direct contact with them, or more disquieting, remaining for longer periods in their vicinity. Secondly, soils as an essential link in the natural ecological cycle, can be damaged by pollution. Thirdly the value of soils as a raw material and as a source of groundwater can be severely reduced by contaminants. Finally the use of soils as an agricultural substrate is significantly restricted by pollutants that can be taken up into the growing plants.

In most modern countries there is a body of legislation in place to prevent new pollution and to ensure the clean-up of, or use restriction on, polluted soils and this is usually coupled to legislation ensuring that polluters and\or owners of polluted soils are penalised and forced to bare much of the cost of clean-up.

The public perception of the risks presented by polluted soils is highly developed, perhaps less so in the UK, and the Bruntland doctrine of "clean up your own messes" has started to win ground as a basic ethical underpinning of soil protection legislation.

All these factors are contributing to a significant coupling between the value of land and its quality. Land values, property values and land transfer have become dominated by the soil pollution aspect and many companies have had to make major financial reservations to meet future contingencies in this respect.

On the other hand, the technical reposes available to clean-up or rehabilitate polluted soils are time consuming, imperfect, often temporary and in all cases very costly. Typical clean-up actions cost in the range of 100000 pounds to tens of millions of ponds with exceptional cases being documented to above hundreds of millions.

The nature, and thus the cost of a remedial action is determined by three factors ::-
* How rigorous has the clean-up to be ?
* How quickly has the response to be implemented ?
* What technologies are available the effect the response ?

With so much at stake, it is necessary to develop a rational basis for deciding upon the answers to the above questions and to generate the design points and constraints for the environmental engineering required. One such basis will be described in what follows.

Risk analysis and the Source-Path-Target methodology.

An objective motivation for contaminated land remediation is to prevent danger to people using it or living on or near to it.

This danger can in principle be defined in toxicological terms. Unfortunately however the necessary data is not available for most of the wide range of chemicals found in contaminated sites and is unlikely to be available in the near future. In particular the long term effects of continuous exposure to low intakes of chemicals is a cause for serious concern.

A second factor that has to be taken into consideration is that of risk perception rather than risk. This can lead to action levels much lower than otherwise justified and form an intractable element in the problem.

This has lead legislators and environmental engineers to seek for substitute criteria in terms of action and monitoring levels that are simple to use and easily grasped by most people. The most common is a Maximum Allowable Concentration

Targets are usually seen in terms of people or food chains but this is not necessary, for example the underground services of a factory or foundations form a major category.

The source and target are connected by a pathway along which harmful pollutants can migrate. The characteristics of the pathway are those of a vector transfer function :-,

* direction
* velocity
* transfer time

In addition the pathway is usually physically and chemically active leading to three other characteristics :-

* attenuation.
* dilution.
* dispersion.

The migration of chemicals along various pathways and chains has been the subject of intensive study in the last 50 years and much progress has been made. Several excellent modelling techniques to define the path transfer functions are available. The use of the SPT methodology is in principle at least, simple. The parameters of source path and target as given above are established. For each of the target-pathway combinations identified and modelling calculations are performed to predict the future impact at target expressed in terms of concentrations. These are compared to the MAC values and a ranking list is established of the pathways associated with the site.

The results can often be expressed in terms of the toxicity ratios :-

Toxicity Ratio = (predicted concentration)/(MAC)

If a safety factor is included to reflect the value of the data base or the trust in the modelling results, action is then required if and when :-

(toxicity ratio) > 1/(safety factor)

At this point the remedial action design criteria and time scales are known as are the factors that contribute to the unacceptable impacts and the remedial action design exercise can begin.

The Source-Path-Target methodology and Site Investigation.

In order to apply the methodology effectively, information on the site and its surroundings is required. This information is costly and the process of gathering it, time consuming. Furthermore all information obtained by measuring in the soil has a wider margin of uncertainty than in other fields. This has led to the development of the "need to know" principle for site investigation.

The methodology is iterative. In the first cycle the SPT model is worked through with available or best guess data usually based on historical or walk-about information. The results of the model are subjected to a sensitivity analysis of the data base and key parameters that have a significant impact on the results are identified and where necessary refined by site investigations. This cycle is often run in conjunction with a similar exercise on the choice of potential remedial actions.

The second cycle takes the new field information and repeats the process until the desired degree of discrimination between real alternative scenarios has been achieved. This iterative process can never eliminate collecting data which after the event turns out to be redundant, but it can improve significantly the scoring chance above the 5% which many investigations hardly achieve.

It is usual to examine five categories of pathways from a site. These are :-

* vapour phase diffusion and convection.
* surface water and airborne erosion.
* uptake into the food chain.
* direct site intrusion.
* leaching to surface and groundwater.

Each of these pathways becomes of greater or lesser importance as times and circumstances change. In some special cases such as mine tailings depots and sea dumping, the large scale physical displacement of the source has to be considered.

Source-Path-Target methodology and Remedial Actions.

The SPT methodology can be also be used to classify the various options for remedial action where necessary and to calculate their value in terms of reduction of impact at target. It should however be noted that other, less rational and risk perception driven considerations can determine the remedial action chosen.

Remedial actions can be directed to addressing the source, the path or the target. Often a combination is the most cost effective strategy.

The basis of source strategies is to reduce or eliminate the emission along the pathway. There are several methods of achieving this end :-

* removal by excavation.
* removal by (forced) elutriation or leaching.
* destruction by chemical techniques (detoxification).
* destruction by biological techniques.
* destruction by physical techniques. (heating, consolidation etc).
* emission reduction by desaturation (of oils usually).
* emission reduction by bulk stabilisation (fixation).
* emission reduction by capping.

It is important to emphasis that the basis of source related techniques is emission reduction and this does not necessarily demand source clean-up to high levels. Indeed in many cases such as an operating chemical factory or a waste disposal site, this is impossible.

Target related techniques involve two main strategies :-

* Target removal.
* Target hardening.

Both these might seem a little bizarre when human targets are considered but are quite rational if the target is for example the underground infrastructure of a city or a factory. Steel water mains can be substituted for pvc ones and cabling can be sheathed or relayed above ground.

The manipulation of the pathway forms a major category of remedial action options. The pathway can be :-

* redirected.
* activated to enhance attenuation, dilution or dispersion.
* retarded.
* interdicted.

Examples are known of all these techniques. Redirection usually involves manipulating the geohydrology as in the conventional pump and treat methods, or by giving escaping vapours an alternative route away from the target. Enhanced dilution and dispersion can be realised by massive infiltration regimes and biochemical activation can be achieved by changing the RedOx status on the path. Containment, barrier or other interdiction techniques are well known and many cut-off walls or guard wells have been installed over the past years.

In many situations where there is a concentration of polluted sites, such as industrial estates, petrochemical complexes or port areas, it is only practicable to address all the sites together as a cluster. This cluster approach is being used with success in the Rotterdam Europort and Botlek areas.

of contaminants at some point in the environment such as ground or groundwater.

There is much debate on the validity or relevance of the often very low MAC values chosen but in fact they represent a socio-political compromise between risk and risk perception in the educated community at large and are often used with their admitted limitations for want of better.

Risk is thus defined in terms of (the chance of) exceeding the appropriate MAC value at some point surrounding the site or in the associated ecological chain.

This restriction on the classical risk analysis methodology has lead to the use of a simpler method of arriving at an evaluation of the nature and extent of the problem and in specifying the design points for remedial action. This is the so-called Source-Path-Target methodology.

In the SPT methodology the source is identified by the following parameters :-

* position in the environment.
  this can be a geographical point but also a conceptual position in say, a food chain.
* composition.
* extent.
* emission strength.

These parameters are usually a function of time and vary in a strong and complicated manner.

The effects of a polluted source often manifest themselves at the target in the distant future whilst action has to be considered now in order to be effective. This requires that the effects be predicted with modelling techniques, usually requiring complicated computer codes. The ability to make the predictions and their certainty, is at the heart of modern environmental engineering.

The assessment and modelling techniques to predict long term source behaviour are in their infancy and the subject of active research. Many investigations have ignored this aspect entirely.

The target is characterised by its :-

* position in the environment.
  again this can be a conceptual or a physical position. All sources are surrounded by three dimensional potential target envelopes, although this is often ignored.
* Its sensitivity, expressed as one or more MAC criteria.

Engineered remedial action techniques can fail at some point in time and pathway interdiction and redirection strategies are especially sensitive to this problem. This has led to the development of two design methods for sensitive sites :-

* design on failure methods.
* fail safe techniques.

The former involves considering the various failure modes of a strategy and calculating through the consequences of such a failure. This has been applied in the Netherlands for isolation and interdiction techniques and has lead to a formal design practice based on probabilistic design and risk analysis.

The fail safe design methodology recognises two facts, firstly that sooner or later all engineering will fail and that it is only in the small time window of the construction phase that processes can be initiated that will prevent problems at failure which will occur possibly in times beyond the maintenance of institutional control. Typically long term processes in the source are initiated to prevent significant emission in the event of distant containment failure.

One very fruitful combination of strategies to achieve failsafe behaviour is to contain a site by capping or curtain walls to eliminate the immediate risk and to initiate in-situ clean-up or detoxification processes such as bio-remediation or enhanced leaching. The time scales for the life of the wall are then determined by the effectiveness of the in-situ clean-up method, however they are significantly longer that those usually available and can be as long as 50 years. This seems to be the best way of moving ahead on the problem of polluted, operating, chemical and refinery sites.

Conclusion.

The source -path-target methodology is a flexible and useful tool for analyzing the impact of polluted sites on their surroundings and in helping to choose the appropriate remedial actions required. It deserves further study and implementation in this field.

AN ASSESSMENT OF THE EFFICIENCY OF REMEDIAL TREATMENT FOR METAL POLLUTED SOIL

STUART MUSGROVE
Department of Biotechnology, South Bank Polytechnic, 103 Borough Road, London SE1 OAA

ABSTRACT

Beaumont Leys is the site of a former sewage farm which is the subject of comprehensive redevelopment by the City of Leicester council. Remedial measures agreed between the City Council and the Interdepartmental Committee for the Redevelopment of Contaminated Land (ICRCL) were designed to minimise any risk to human health from metallic contamination in the soil.

This study has assessed the effectiveness of soil stripping and respreading operations carried out on site.

Detailed soil analyses before and after redevelopment have demonstrated the success of this strategy; significant lowering of average contamination levels and reduction of maximum metal levels has been achieved.

INTRODUCTION

Beaumont Leys is an area to the North of Leicester which was occupied from the 1890's to 1966 by the City of Leicester sewage works. Extensive site redevelopment is now underway. The design of the development took credence of the results of soil analyses carried out in 1971-2 [1,2], with measures agreed between the City of Leicester Council and the ICRCL designed to minimise any risk to human health, during the projected lifetime of the development, from metallic contamination present on the site.

Briefly, topsoil from the sludge spreading area was removed and the land used for industrial development. This highly contaminated topsoil, e.g. with levels in excess of 20 mg/kg cadmium dry soil (D.S.), was used as a base material for landscaping projects where it was covered with other less contaminated topsoil, (cadmium 10-15 mg/kg D.S.) before being seeded to form open grassy areas of amenity "parkland". Local authority and private housing schemes were provided with limited garden areas and considerable care was taken to ensure that the topsoil used for the gardens contained no more than 5 mg cadmium/kg D.S.

The cadmium content of the soil was most concern on the Beaumont Leys development on the grounds of toxicity or absorption by crops; therefore the concentration of cadmium was used as the indicator of acceptability of the soil for a particular end use. Before a new area was developed for housing the results of the 1972 survey were examined and any sample point (on the intersection of a 200 ft. grid) which had a cadmium content in excess of 5 mg/kg D.S. was identified. This point was then judged to be the centre of a square with 50ft sides and the topsoil contained within it was removed to be used elsewhere on the development in a less critical area. This approach did not result in a shortage of topsoil because only 60% of the original topsoil was required for the final development due to the construction of roads, paths and buildings.

Where an area of land was identified with an acceptable soil cadmium concentration of less than 5 mg/kg the remaining topsoil was stripped off by a mechanical scraper and deposited in horizontal layers to form a composite heap of stripped soil.

When the housing development was completed this topsoil was recovered by digging vertically into the heaps and respreading this 'mixed' soil back into gardens and other areas of human contact. It was assumed that this mixing produced a more homogeneous concentration of metals in the soil over the whole site and reduced any hot spots, i.e. small areas with high metal levels not discovered in the original soil survey.

In 1979 the Department of Environment's Central Directorate of Environmental Protection awarded a contract to investigate the efficiency of the stripping/respreading operations, and to determine in practice, using normal building site activities, whether the presumed reduction of maximum metal contents and lowering of average contamination levels were achieved.

The investigation described below covered an area of approximately 4 hectares (10 acres) on a local authority housing development in an area known as Heatherbrook. The design essentially followed that of the 1972 survey (1, 2); 20 grid points were taken at the intersections of the 200ft grid used in the original survey. (See Figure 1).

Figure 1

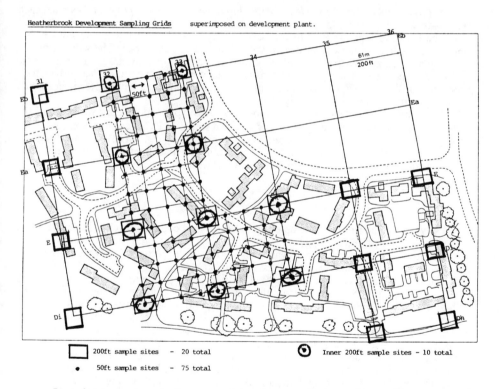

In order to assess whether more intensive sampling would provide more precise information on the distribution of metallic contaminants a further 75 samples were taken on a 50 ft grid within the central area.

All 95 topsoil samples were collected from the greenfield site in 1979, and after the topsoil was stripped, 18 subsoil samples on the 200ft. grid were obtained.
Subsoil samples were analyzed because results in the 1972 survey suggested that the metallic contamination was not confined to the surface layers but had penetrated into the subsoil (1.2.). However, recent work (3) has shown that the metal contamination was confined to the topsoil at Beaumont Leys.

By 1981 the housing development was complete and further soil samples were taken.

Where it was not possible to resample at the precise location of the original samples, the closest suitable place was chosen. In total, 17 of the original 20 major 200 ft grid points were resampled for both top and subsoil and all 75 topsoil plus 73 subsoil samples from the 50 ft grid were collected. The results of the analytical studies are presented in tables 1-2.

MATERIALS AND METHODS

Materials

Glen Creston Cross Beater Mill
Instrumentation Laboratory Atomic Absorption Spectrophotometer 157 model with deuterium background correction.
All chemicals were A.R. grade or equivalent from BDH or Fisons.
All glassware was borosilicate glass from Corning, and was acid washed before use.
Element standards of 1000 mg/l were diluted appropriately before use.

Methods

Approximately 1 kg of soil was submitted in a sealed, labelled, polythene bag and each sample was visually assessed on receipt at the laboratory. The difference in physical appearance between topsoil and subsoil was used as an initial sorting test to ensure that samples of topsoil were not contaminated with subsoil and vice versa. The soils of mixed origin were rejected and resampling of that location was carried out. Stones larger than 0.5 cm were removed and the remaining material was air dried at room temperature fro approximately 3 days, when no obvious sign of moisture remained, milled through a 2 mm screen, and stored in labelled polythene bags to await analysis.

As the metallic concentration of the soil was to be expressed on a dry weight basis the residue moisture was determined by drying of air dried soil at 105^0C to constant weight.
The method of analysis has been described in detail previously (3). Briefly the air dried soil was treated with boiling 2M hydrochloric acid to extract the 'total' metals into solution . This solution was then directly used for quantification of the metals using flame AAS.

DISCUSSION

Only 17 of the original 20 samples on the 200ft grid could be taken after respreading. However, because the three missing soils were situated on the extremities of the sampling grid, it was considered that their absence was unlikely to affect the overall conclusions of the experiment.

The ten soil samples taken on the 200 ft grid form the perimeter of the 50 ft grid. The reason for analyzing all 85 samples from the closer grid was to assess, more accurately, the distribution of the metallic contaminants and hence the development potential of the land. Obviously the benefit of sampling from a closer grid must be balanced against the time and cost of obtaining the extra information. However, errors of judgement based on insufficient or inadequate data may invoke cost penalties in terms of additional site restoration costs.

In the case of cadmium, the metal of most concern in the Heatherbrook soil, the increased effort of sampling on the 50ft grid would not have influenced the way in which the site was developed because of the original topsoil from the Heatherbrook area had a reasonably uniform soil metal concentration within the area of the 50 m grid. An area with a more heterogeneous soil metal concentration would have yielded much more benefit .

TABLE 1

Effect of Respreading Topsoil

		200 ft grid		10 samples at perimeter of 50 ft grid		50 ft grid	
No of samples		20 original	17 respread	10 original	10 respread	85 original	85 respread
Cadmium	Range:						
	Minimum	0.6	1.6	1.3	1.6	0.7	0.8
	Maximum	7.2	3.1	4.3	2.9	5.1	4.1
	Mean	3.0	2.5	2.7	2.6	2.7	2.7
	Std.Dev.	1.6	0.4	1.0	0.4	1.0	0.6
	99% upper Tolerance	7.6	3.7	6.0	3.9	5.3	4.3
Lead mg/kgDS	Range:						
	Minimum	35	78	95	78	77	65
	Maximum	424	166	154	150	410	265
	Mean	158	121	125	123	128	137
	Std.Dev.	92	23	17	21	41	38
	99% upper Tolerance	421	188	180	191	236	237
Zinc mg/kgDS	Range:						
	Minimum	89	166	185	166	143	82
	Maximum	569	280	360	279	400	408
	Mean	285	239	265	246	261	275
	Std.Dev.	111	33	56	34	56	53
	99% upper Tolerance	602	335	448	357	409	415

One of the overall objectives of the stripping/respreading exercise, to reduce maximum concentrations of metals in the soil was successful. The 99% upper tolerance limit (the concentration of metal which is only exceeded in 1% of the soil) for cadmium was reduced from 6.0 to 3.9 mg. Cd/kg in 10 samples from the 200 ft grid, 7.6 to 3.7 mg Cd/kg for 20 samples on the same grid, and 5.3 to 4.3 mg Cd/kg in samples from the 50ft grid.

These figures are most significant in that previous guidelines, produced by the D.O.E. (4), suggested a "trigger" concentration of 5. mg Cd/kg of D.S. (later reduced to 3.0 mg Cd/kg. D.S. for small gardens (5)); thus the respreading exercise had reduced the 99% upper tolerance limit of the soil cadmium concentration to such a degree that the use of the soil for small domestic gardens was acceptable.

The reasons for the difference in tolerance limits between the metal concentration of the larger area (the 20 samples on the 200 ft grid) and that the smaller inner area represented by 10 or 85 samples was twofold. First was the uneven geographical distribution of metals; although on Heatherbrook the soil metal concentrations are generally fairly uniform there is a gradient which causes metal levels to increase with increasing proximity to the old sludge spreading area, and hence there was bias in the distribution of the metals in the original soil. When this soil was homogenised during the scraping, stockpiling and respreading this bias was much reduced. Four of the 20 samples of the 200ft grid were closer to the highly contaminated area than any other the other 85 samples, and it was these four samples which contributed most to this bias. The second reason for variation was that these areas with the highest cadmium content (7 mg/kg) were scraped free of the topsoil, which was then isolated and not incorporated back into the Heatherbrook topsoil stockpiles.

Although only cadmium has been detailed, the other metals analyzed show the same general beneficial trend of the stripping/respreading exercise with consequent lowering of the 99% upper tolerance limits over the whole site.

Metal concentrations in the subsoil were very low, and were further reduced after respreading. This reduction was probably due to artifacts, e.g. analytical samples being more closely examined for the presence of topsoil in the resampled subsoils, rather than any real physical mixing effect during respreading of the topsoil.

TABLE 2

Effects of Respreading Subsoil.

		200 ft grid		200 ft grid		50 ft grid
	No of samples	10	10	18	17	85
		original	respread	original	respread	respread
Cadmium	Range: Minimum	0.2	<0.1	<0.1	<0.1	<0.1
mg/kg DS	Maximum	0.9	0.4	0.9	0.5	1.7
	Mean	0.5	0.2	0.5	0.2	0.3
	Std.Dev.	0.2	0.1	0.2	0.1	0.3
	99% upper Tolerance	1.3	0.5	1.1	0.5	1.1
Lead	Range: Minimum	20	6	16	6	6
mg/kg DS	maximum	50	33	50	48	101
	Mean	30	17	32	17	22
	Std.Dev.	10	7	10	10	17
Zinc	Range: Minimum	58	35	40	35	48
mg/kg DS	Maximum	185	86	185	92	206
	Mean	106	65	105	64	76
	Std.Dev.	42	14	37	15	30
	99% upper Tolerance	243	111	212	108	155

CONCLUSIONS

The analysis of Heatherbrook topsoils before the scraping commenced indicated that the metal contamination was widespread over the site and although there was variation in the degree of contamination, this variation rarely exceeded one order of magnitude. however, the process of mixing and respreading did nevertheless result in increased homogeneity and modest reductions in concentrations. It could, therefore, be envisaged that on other sites where the metallic concentration in the soil was more heterogenous, the effect of scraping/stockpiling/respreading could be more dramatic and could significantly assist successful redevelopment of the land. It may also be possible to obtain a significant reduction in development costs in certain sites and with thorough analytical backup, considerable confidence in the success of such an exercise could be achieved.

Quality Assurance of Analytical Data

Soil analyses carried out on Beaumont Leys over twenty years ago, included little analytical quality assurance as it was then essential that the overall pattern of metallic contamination should be determined in a very limited time, occasional results which did not fit into that pattern, were rarely investigated further. Analytical instrumentation was limited, and techniques now considered to be standard, e.g. background correction for AAS were not available on commercial equipment. This project involved fewer analyses and therefore much more credence had to be taken of each individual result.

The Heatherbrook project involved samples being examined before and after respreading, therefore the overriding importance was to achieve analytical comparability between the two sets of data. Samples collected in 1978 from the greenfield site were therefore analyzed alongside the soils submitted after respreading.

Approximately 25 soils were analyzed together, and the quality assurance procedure below was carried out in order to achieve batch to batch consistency.

At the start of the project it was not possible to locate a 'standard' soil with certified values for cadmium and lead, it was therefore decided that ADAS Soil Science laboratory at Derby should collect and prepare six soils containing a range of metal concentrations. The soils originated from various locations but included a number of Beaumont Leys.

Quality Assurance Procedure

With every batch of Heatherbrook samples, a 'standard' ADAS soil (containing a similar concentration of metals) was analyzed for cadmium, lead and zinc. Chromium, copper and nickel were only determined when the Heatherbrook soils themselves required a more comprehensive analysis. (Table 3).

One single extraction of the ADAS' standard soil was re-estimated with each batch of analyses to test the instrumental reproducibility. (Table 3).

TABLE 3

Repeatability of Analytical Determinations

15 separate extractions for Cadmium, Lead and Zinc.(mg/kg DS)

	Range:	Minimum	Maximum	Mean	Std.Dev.
Cadmium		1.4	1.8	1.5	0.1
Lead		85	102	91	4
Zinc		150	190	168	10

Single extraction, 15 estimations for Cadmium, Lead and Zinc (mg/ks DS)

Cadmium	1.5	2.0	1.6	0.2
Lead	89	97	92	2
Zinc	160	200	184	10

Twenty-one of Heatherbrook soil samples were analyzed together in a single batch having been previously analyzed at different time. (See Table 4).

TABLE 4

Batch Repeatability of Analytical Determinations

	Cadmium		Lead		Zinc	
	1	2	1	2	1	2
Range: Minimum	0.5	0.8	74	56	128	82
Maximum	4.6	4.1	221	212	434	408
Mean	3.0	2.9	147	139	303	281
Std.Dev.	0.9	0.8	41	39	76	79

1- soils analyzed together
2- soils analyzed in several batches.

At the time of sampling of the greenfield Heatherbrook site in 1978, ten specimen soils were divided, (with only a cursory attempt made to homogenise them) between the Leicestershire County Analyst's Laboratory and ADAS Shardlow. Each laboratory analyzed the samples using their standard techniques in order to assess interlaboratory comparability. (See Table 5)

TABLE 5

Interlaboratory trial using 10 samples of Heatherbrook topsoil

		ADAS	L.C.C.
Cadmium	Range:		
	Minimum	1.3	1.3
	Maximum	4.0	4.3
	Mean	2.4	2.7
	Std.Dev.	0.8	1.0
Chromium	Range:		
	Minimum	ADAS did not	102
	Maximum	analyze the soils for chromium	390
	Mean		222
	Std.Dev.		86
Copper	Range:		
	Minimum	63	59
	Maximum	110	114
	Mean	86	80
	Std.Dev.	16	17
Lead	Range:		
	Minimum	84	95
	Maximum	164	154
	Mean	115	125
	Std.Dev.	20	17
Nickel	Range:		
	Minimum	41	41
	Maximum	74	80
	Mean	58	55
	Std.Dev.	12	13
Zinc	Range:		
	Minimum	218	185
	Maximum	375	360
	Mean	278	265

ADAS results expressed as mg element/litre of air dried and ground soil.

DISCUSSION

Batch to batch reproducibility was shown to be acceptable. Standard deviations were generally less than 10% of the mean for cadmium, lead and zinc but slightly greater variations were noted for other metals, probably due to the few replicates. the single extraction which was reestimated with each batch provided a similar order of variation to the batch to batch reproducibility.

Analytical variation of the inter-laboratory trial was minimal. Small differences may have been attributable to the samples not being homogenised before being split and also the fact that the metallic content of the soil was calculated by ADAS as a weight/volume relationship with units of milligrams of the element/litre of the air dried and ground soil whereas the L.C.C. analytical results are all calculated as milligrams of the element/kilogram of air dried soil, i.e. weight in weight. However, it is understood that normally the relationship is close to 1:1.

CONCLUSIONS

The analytical quality assurance scheme was devised as an attempt to quantify analytical variations during the project. This was particularly essential during our research project with the D.O.E. which ran from 1978-1984 due to the prolonged timescale together with staff and instrumental changes in the intervening period. About 10% of the analytical work was devoted to AQA.

Generally precision of soil analyses was shown to be excellent using the methods described and although a degree of accuracy was inferred the lack of standard materials prevented the actual confirmation.

BIBLIOGRAPHY

1   Pike, E R, Gram L C and Fogden M W
    JAPA 1975 13, 19-33

2   Pike, E R, Graham L C and Fogden M W
    JAPA. 1975 13, 48-63

3   Musgrove S D. JAPA 1987, 25 113-128

4   Smith M A, Redevelopment of Contaminated Land.
    Tentative Guidelines for Acceptable Levels of Selected Elements in Soils - ICRCL 38-80. (Revised March 1981). D.O.E. London.

5   Guidance on the assessment and redevelopment of contaminated land. ICRCL 59/83. ICRCL. D.O.E., London.

Copyright: Association of Public Analysts - Reproduced with Persmission

LAND CONTAMINATION BY HAZARDOUS GAS - THE NEED FOR A COORDINATED APPROACH

S J EDWARDS, C F C PEARSON
Charles Haswell and Partners Ltd
Ashton House, The Courtyard, Market Place, Belper, Derbyshire DE5 1FZ

ABSTRACT

The 1980's has seen the emergence of land contamination by hazardous gas as an issue in the UK. The classic stages of identification, acceptance, investigation, and reaction have taken place, and the industry is now in a situation where professionals of many disciplines are aware of the potential for gaseous pollutants to present obstacles to development. Methods for identification of contamination and amelioration of hazard are being evolved. The methane gas explosion at Loscoe in 1986, which destroyed an occupied house, served to concentrate the attention of both government and industry on the landfill gas problem, and research and published guidelines have naturally followed. There are, however, many possible sources of gas and these can present a similar or greater risk to that posed by landfill gas. This paper examines guidelines, recommendations and current practices, and illustrates examples of conflict or dual standards. Finally, the need for a coordinated approach to all gas contaminants is discussed.

INTRODUCTION

The presence of hazardous gas within the ground, represents a special case of land contamination. Whereas solid or liquid contaminants generally move slowly and downwards under the force of gravity, gas may move rapidly and, given suitable ground conditions, may move upwards towards the surface. Contamination conveys the meaning of an artificially occurring substance, but naturally occurring gases are also capable of presenting a hazard, and such natural gases are common. Natural methane gas caused the explosion at the Abbeystead pumping station in 1984, in which 16 people were killed. Where naturally occurring gases are the source of pollution, it is often the case that engineering or mining works are responsible for changing the natural equilibrium and increasing the risk of hazardous conditions by allowing gas to move more freely or simply by introducing people to the area of the hazard. This paper considers two of the most common gaseous ground pollutants, methane and carbon dioxide, but it is the case that many other gases may, in certain circumstances, be present in the ground and may represent a hazard. These gases include carbon monoxide, nitrogen, higher hydrocarbon gases, hydrogen and radon. However, it is the landfill gas issue above others that has raised the general levels of awareness of gas in the subsurface. Most of the gases described above may be present in the ground for more than one reason, and depending upon their source, may represent high or low levels of hazard. For these reasons a coordinated approach to measurement, investigation, risk assessment and the setting of guidelines is required.

SOURCES OF METHANE AND CARBON DIOXIDE

Methane

Occurs inevitably in association with coal where it is given the name firedamp. Firedamp typically consists of 90% methane with some higher hydrocarbon gases and other components. Old mine entrances, (shafts, adits etc) may have high methane levels associated with them and where groundwater levels rise on cessation of mine pumping, methane can be forced from the ground surface and possibly into structures. Methane is associated with peat. Where groundwater filters through peat layers and later enters an enclosed space eg a manhole or tunnel, methane may be liberated in sufficient quantity to cause a problem. Natural gas supplied as a fuel can leak from pipes and enter the ground. Where contamination by methane rich gas is suspected, a check should be made for the presence of gas distribution pipes. Marsh gas which is high in methane can be seen bubbling from stagnant water in ponds or boggy ground. This is derived from anaerobic bacterial action on organic material. A very similar gas can be derived from organic rich silts deposited in dock basins or estuaries. Sewer gas is not common but may occur in septic tanks, cess pits or sewer chambers. Landfill gas is high in methane, and may be generated whenever organic wastes are deposited in deep landfills.

Carbon Dioxide

The abundance of methane oxidising bacteria in the ground and in groundwaters means that whenever methane is present in the subsurface, carbon dioxide can be produced by the action of these bacteria.

Blackdamp or stythe is the name given to a carbon dioxide rich gas encountered in some coal measures. Where acidic groundwaters come into contact with carbonate rocks the resulting chemical reaction produces carbon dioxide. Anaerobic or aerobic bacteria acting on landfill or on organic materials within soils produce carbon dioxide. In shallow soils ie within the root zone, carbon dioxide is produced by plant respiration.

Comment

Methane and carbon dioxide, whether from a natural or an artificial source, can represent a hazard to health or a safety risk; methane is flammable and can be explosive when confined, and carbon dioxide is classified as a 'substance hazardous to health' within the meaning of the COSHH Regulations 1988. Either of these gases will act as an asphyxiant when they occur in high concentrations replacing oxygen.

The gases are commonly found together either as a result of methane oxidation or because they are generated together by the same decay process. Due to their multiple sources, it is common that a particular case of contamination can be due to gas from more than one source. There are techniques available which can aid in determining the source of the gas eg bulk gas, trace gas and isotopic analyses, but due to chemical change, dilution and mixing, it will often be difficult to identify positively the source of the contamination and more than one technique will be required. If sufficient gas is present in a sensitive location, a hazardous condition may exist, and this is true regardless of the source. Tables 1 and 2 show some selected occurrences of carbon dioxide and methane and gives concentrations discovered.

TABLE 1
Selected Gas Concentration Statistics - Carbon dioxide

Concentration % v/v	Location	Comment/Attributed Source
6	South Wales	In virgin ground, heavy clay soil above coal measure.[1]
40	Greater Manchester	On landfill site.[1]
9	USA	Marsh gas.[2]
1		Soils overlying Devonian limestone.[3]
8		Mine waste tips.[3]

TABLE 2
Selected Gas Concentration Statistics - Methane

Concentration %$^v/_v$	Location	Comment/Attributed Source
93	Greater Manchester	In borehole approximately 50 metres from gassing landfill site.[1]
55	Greater Manchester	Several boreholes on a gassing landfill site.[1]
7	USA	Marsh gas.[2]
>1	Derbyshire	In general body of unventilated tunnel. Probable mixed source.[4]
9	Lancashire	Developed in 'bubble' above water in distribution tunnel. Probable geological source.[5]
40	South Yorkshire	In shallow soils above coal measures.[6]
2.5	Merseyside	Anaerobic decay of small amount of organic matter/top soil buried by forming of clay bund.[1]
19.5		Peat.[7]
30	Greater London	Organic silt deposits.[8]

GUIDANCE

There is in existence guidance relating to the occurrence of gas in the ground, its monitoring, remedial measures, and planning and building considerations. The guidelines fall into two categories; firstly general documents relating to contamination of all types; secondly, those that offer guidance on gaseous contamination from a specific source, eg landfill sites.

The Building Regulations (1985)[9] These statutory regulations are referred to for building control guidance by many of the documents below, the relevant sections being Part C of Schedule 1. The only guidance included here is: *'C1 The ground to be covered by the building shall be reasonably free from vegetable matter'* and *'C2 Precautions shall be taken to avoid danger to health caused by substances found on or in the ground to be covered by the building'*. The limitations of the Building Regulations as a means for controlling development in areas of potential gassy ground are that they only apply to land directly beneath the structure. An Approved Document (C1/2) carrying practical guidance to meeting the requirements exists and an appendix to this contains some information on gaseous contamination but stresses that the information is *'an introduction to the work of the expert adviser. It is not part of the guidance'*. The appendix includes specific reference to two gaseous contaminants, methane and carbon dioxide, and independently of their source, recommends thresholds for action.

Waste Management Paper 26 'Landfilling Wastes'(1986)[10] A general memorandum which contains an appendix dealing with landfill gas.

Interdepartmental Committee on the Redevelopment of Contaminated Land (ICRCL) Guidance Note 59/83 Second Edition 'Guidance on the Assessment and Redevelopment of Contaminated Land'(1987)[11] Although dealing with contamination in general, this publication draws special attention to the properties of gaseous contaminants and notes that they are capable of migration relatively quickly both laterally and vertically and for considerable distances, and have the ability to accumulate in confined spaces. An important fact included in this note, is that toxic or asphyxiant gases along with harmful liquids are the contaminants most likely to affect site investigation teams and construction workers, especially in excavations, borings and tunnels. Specific examples where gaseous contamination may be likely to be encountered are limited to landfill sites and the only gas named is methane but the note does emphasis that the lists shown are not exhaustive. Reference is again made to the Building Regulations 1985 for building control advice.

Department of the Environment/Welsh Office Joint Circular 21/87 (22/87) 'Development of Contaminated Land'(1987)[12]. This document contains guidance for local authorities in respect of land contamination. Amongst other issues it notes that *'...the emission of landfill gas may be particularly hazardous.'* With regard to development, it states that although few sites are so badly contaminated that they cannot be used, the choice of end use must depend on the nature and extent of the contamination. In relation to building control the circular points to the Building Regulations 1985 for guidance.

'Measurement of Gas Emissions from Contaminated Land'(BRE) (1987)[6] A Building Research Establishment Report, this survey gives much practical advice on gas monitoring procedured but it is also however principally concerned with landfill gas.

Department of the Environment/Welsh Office Joint Circular 17/89 (38/89) 'Landfill Sites: Development Control'(1989)[13] A Circular designed to give local planning authorities information on landfill waste disposal sites and landfill gas as a contaminant. Cross references are made to relevant Waste Management Papers, Government Circulars and Guidance Notes of the Interdepartmental Committee on the Redevelopment of Contaminated Land and the circular refers to the Building Regulations 1985 as containing building control guidance. *'There can be no hard-and-fast rule about the appropriate distance between new landfill sites and existing* *development in relation to the possible migration of landfill gas. A proposal for a site as close as 250 metres to the development will require special attention'.* It is the nature of official guidance that comments such as the above are taken in the absence of alternative information and used as *'de facto'* standards.

Waste Management Paper 27 'The Control of Landfill Gas' (WMP 27) (1989)[14] A technical memorandum containing comprehensive guidance on the monitoring and control of landfill gas. A number of threshold concentrations for methane and carbon dioxide appear within this publication, a new revision of which is expected before the end of this year.

'Monitoring of Landfill Gas'(IWM) (1989)[15] Published by the Institute of Wastes Management. Another very useful document containing much information on the monitoring of landfill gas and including useful notes on the occurrence of gas in the ground from other sources.

Interdepartmental Committee on the Redevelopment of Contaminated Land (ICRCL) Guidance Notes 17/78 Eighth Edition 'Notes on the Development and After-use of Landfill Sites'(1990)[16] This circular contains much useful advice specifically directed at gaseous contamination from landfill sites. A summary of the nature and properties of landfill gas is included and information on the characteristics of gas emission from sites, including non-landfill situations, is given. Gases other than methane and carbon dioxide are briefly mentioned as are protection systems for the control of landfill gas emissions. Most of the information contained within this document is also applicable to gaseous pollution from other sources.

The Building Regulations 1990 Draft[17] The Approved Documents C1/2 are currently at the discussion stage pending revision. It is likely that much more comprehensive advice will be included when the document is approved and that the revised document will become very important guidance for developers and planners. The July 1990 draft shows that more importance is to be attached to the land around any development as well as that immediately below it *'...where there is reason to suspect that there may be gaseous contamination of the site...further investigation should be made to determine what, if any, protective measures are necessary.'* The draft contains broad guidelines relating to methane and carbon dioxide concentrations but also contains the important point that: *'The amount of gas in the ground as well as its pressure relative to the atmosphere also needs to be considered.*

TABLE 3
Summary of Published Methane and Carbon Dioxide Gas Concentration Thresholds

Concentration v/v	Source	Guideline
1% methane 0.5% carbon dioxide	WMP 27	Gas in the ground escaping from landfills should not exceed these limits.
1% methane 0.5% carbon dioxide	WMP 27	When this level is exceeded in an occupied property, evacuation should be instigated.
0.25% methane	WMP 27	Notify under the Reporting of Injuries, Disease and Dangerous Occurrences Regulations 1985 when this is exceeded in occupied property.
0.25% methane	IWM	Install continuous monitors in occupied property.
1% methane	Building Regulations	If this is found in the ground, suggests that remedial measures are needed.
5% carbon dioxide	Building Regulations	If this is found in the ground and buildings where people will normally be present are planned, suggests that remedial measures are needed.
0.05% methane	Building Regulations 1990	Protective measures are not needed where methane levels are below this limit
0.5% carbon dioxide	Building Regulations 1990	Indicates a need to consider possible measures to prevent gas ingress.
0.5% carbon dioxide	COSHH Regulations	Occupational exposure limit 1991 for 8 hour time-weighted average.
1.5% carbon dioxide	COSHH Regulations	Occupational exposure limit 1991 for 10 minute exposure.
1.25% carbon dioxide	Mines and Quarries Act	To be adequate, ventilation must prevent the general body percentage of carbon dioxide from exceeding this level.
2% methane	Mines and Quarries Act	Men shall be withdrawn if the general body of the air exceeds this level.
0.25% carbon dioxide	BRE	May give cause for concern and lead to the designation of the site as 'unsafe on toxic grounds'.
1% methane	British Gas	Evacuation of occupied property at this level.

DISCUSSION

The previous section schedules documents which contain information relating to land contamination by gases. Many of these are either specifically intended to deal with landfill gas, or treat gaseous contamination as a 'minor player' in the range of land contamination. Gases from sources other than landfill receive less attention and where these are from 'natural' sources they are not generally referred to as contaminants. However where such gases occur in sufficient quantities, they must be considered in just the same light when development is proposed. A coordinated approach to the assessment of land contamination by gas is then needed in respect of the following areas: Site investigation; monitoring; setting of thresholds and risk assessment.

Site Investigation

Standard and satisfactory methods for site investigation for gaseous contaminants are in existence (WMP 27, IWM and BRE). These argue for a thorough desk study to be carried out before any physical investigation is executed. The desk study should enable approximate, but minimum budget costs for site investigation to be made. It is important to note that where gaseous contamination is found, further physical investigation will often be needed to quantify its extent and nature, and to assess the potential for the contamination to migrate.

Where a geotechnical site investigation is to be carried out for the purpose of determining engineering properties of ground materials, the additional cost involved in providing the necessary facilities for gas monitoring are relatively small. A coordinated approach to the assessment of land contamination by gas, will include the carrying out of geotechnical and gas contamination site investigations within the same contractor mobilisation, where this is possible. This can be accomplished by providing supervising staff trained in aspects of subsurface gas occurrence, to accompany the site investigation contractors. Boreholes and trial pits can lend themselves to the installation of gas monitoring equipment, but it is important that the installation is carried out satisfactorily. An incorrectly installed gas monitoring standpipe can give misleading results and where this occurs the likely result is a false conclusion. Additional samples of material to those required for engineering tests may be taken at little extra cost during the SI, and specialist laboratory analysis carried out to ascertain the potential for gas contamination. Where cost implications dictate the extent of site investigation, it is probable that gaseous contamination will be missed. Developers and reclamation contractors must accept that the end use for land which suffers from contamination by gas, may often only be determined upon the results of site investigation and that the financial effects of restricted end use may be serious.

Monitoring

An important area in the field of gas contamination where a coordinated approach is required is that of monitoring. WMP 27, IWM and BRE offer much good advice on monitoring and it is vital that this is carried out by suitably trained staff. Areas where mistakes are commonly made are: insufficient time periods for monitoring programmes; inappropriate sampling techniques; use of incorrect or badly suited instruments; failure to measure or otherwise take account of other physical factors which may affect the presence or migration of gas contamination, eg barometric pressure, water level, ground conditions. The published guidance emphasises the need to supplement field measurements with analysis of gas samples in the laboratory. While there are many laboratories which can offer suitable analytical services, these services could be improved by better quality control and improved communication between field workers and laboratories. This communication must be in both directions, so that analytical laboratories can provide tailored services and consultants can appreciate the limitations of analytical methods. A coordinated approach is very important when specialist techniques for sourcing of gas contamination are attempted.

Setting of Thresholds

The setting of trigger concentration for toxic or harmful species, where these are solid or liquid and reasonably static in the ground, has been employed for many years. The basis for calculating such thresholds depends principally on the relative hazard presented by the substance and on the proposed end use of the land.

Thresholds for gaseous contaminants may be based on the same principles but more refinement is required. The discovery of a particular species of gas at a level far below that which could cause any toxic or harmful effect including explosion should not always be ignored. As discussed earlier the mobile nature of gases means that the area of contamination is dynamic. Low levels of gas may be the precursor to dangerous levels.

When considering the current state of official guidelines one must remember that each is developed with a specific aim. Commonly adopted levels for action used in Waste Disposal Licences are 1% by volume for methane and 0.5% by volume for carbon dioxide. Licences often take these levels in boreholes as confirming that landfill gas is migrating to an area. These levels are quoted in WMP 27 as the 'minimum acceptable criterion' for ensuring that gas does not escape from a landfill to present a hazard. By a strict application of these guidelines, land may either be unnecessarily sterilized or may be considered safe for development prematurely. As noted earlier in this paper 'natural' levels of carbon dioxide may be far in excess of 0.5% and methane may have many sources.

Recent draft versions of the latest edition of WMP 27 have recognised the difficulty in setting thresholds where high natural background concentrations of gas occur. Where a background dataset of gas concentrations at a site are in existence, it is probable that gaseous contamination can be detected at lower levels than the ones given in WMP 27. The current approved documents to accompany the Building Regulations includes suggested threshold levels for gases below proposed developments. The documents are for guidance only and the Building Regulations 1985 are restricted to the land directly below the structure and so it is not the case that the guidelines can form a rigorous framework for considering gaseous contamination of structures. However the thresholds for action contained within this approved document ie 1% methane by volume and 5% carbon dioxide by volume would generally be regarded as relaxed. Some workers would specify the most stringent remedial measures possible if gas concentrations as high as this were to be found in the ground under a proposed development. A new edition approved document to Part C of the Building Regulations 1985 is currently under discussion prior to issue and it appears from drafts of this document that the guidelines given for assessment and treatment of gas contamination will be amplified when the document is approved.

Many of the existing advisory thresholds require clarification or amplification; as an example if ten boreholes are sunk over the area of a planned structure, and only one shows methane concentrations in excess of the proposed Building Regulations threshold, does this mean that the whole area of the proposed development site is contaminated? Obviously in all cases, suggested thresholds must be supplemented by judgements made by those with sufficient experience.

Where judgements on contamination or potential contamination and the setting of thresholds are to be made, the points described in the earlier section on Monitoring must be considered. Thresholds should not be set which are non-achieveable or not easily measurable. For example, it has been the practice in the past of at least one Waste Disposal Authority to impose a waste disposal licence condition, setting a threshold concentration for action of 350 ppm of carbon dioxide in a borehole when the atmospheric concentration of carbon dioxide is 300 ppm.

Risk Assessment

Contamination of land by gaseous substances may imply hazards which can range from vegetation distress to explosion. Assessment of the risk in any particular situation depends upon the species of gas present, the volume of gas, the source of gas; the potential for migration; and the sensitivity of areas or structures. A coordinated approach to risk assessment should include full desk and physical studies with reference back to past case histories. Most importantly, risk assessments cannot depend upon gas concentrations alone; gas volumes, pressures and flow rates are also important in evaluating hazard. Every case of land contamination by gas will have factors peculiar to that site and for this reason no guidance or legislation will fully provide a means of risk assessment.

Risks must always be considered with a balance in mind; while much has been written about the risks due to migrating landfill gas and there have been various fires and explosions and indeed some fatalities caused by it, there have been relatively few major incidents. While not seeking to diminish the perception of risk due to landfill gas, it seems that there has been a tendency in some quarters to overplay the hazards. A balanced judgement is always preferable to a hasty reaction.

SUMMARY

This paper discusses aspects of gaseous contamination of land, and presents some arguments in favour of a coordinated approach to that topic. Land contamination is important in the present economic climate, where government, developers and individuals are all keen to see that the best possible use is made of land. In this small and overcrowded country, it is important that further infrastructural development does not take place on green belt land where this is avoidable. This paper has highlighted the special nature of gaseous contamination, in that it is likely to be dynamic, and can leave the ground under some circumstances and enter structures. These properties mean that physical investigation, monitoring, setting of thresholds and risk assessment are more complicated than is the case for liquid or solid contamination.

Where guidelines exist for gaseous contamination they are generally presented either in too general, or in too specific a form. Almost always, they are guidelines to contaminated land in general, or to gas contamination where the source is landfill gas. The interest of recent years in landfill gas has helped to broaden awareness of other forms of gas contamination, and has also allowed research to be carried out into monitoring and remedial work, which has equal application in terms of any gaseous contamination. It is difficult to produce and easy to criticise official guidelines and where these exist, research is continuing to refine them. It should never be the case that advice and guidance become *de facto* standards and regulations, and where land contamination by gas is proved or suspected, specialist advice should be sought.

It is not surprising that given the public and professional interest shown in landfill gas over the past ten years, that the majority of investigations into gas contaminated land are involved with this subject. It is important to remember that more tragedies have occurred due to other gas sources. A coordinated approach to site investigation and monitoring is required and the limitation of techniques and instruments must not be forgotten. The costs of investigations may be considerable and for this reason coordination with any other programme of physical investigation should be encouraged. Saving money by implementing limited physical investigation will almost always prove to be false economy.

The authors are aware that research, sponsored by government and industry is continuing into the area discussed here and believe that the results of this work will prove valuable in future years.

REFERENCES

1. Internal confidential source.

2. Swain, F.M., Composition of Marsh Gases in the Central and Eastern United States. Applied Geochemistry, 1986, 1 301-305.

3. Butterworth, J.S., Methane, Carbon Dioxide and the Development of Contaminated sites. Proceedings of Methane: Facing the Problems, 2nd symposium, 1991.

4. Pearson, C.F.C. and Edwards, J.S., Methane Entry into the Carsington Aqueduct System. Proceedings of Methane: Facing the Problems, 1st symposium, 1989.

5. Health and Safety Executive, A Report of the Investigation by the Health and Safety Executive into the Explosion on 23 May 1984 at the Value House of the Lune/Wyre Water Transfer Scheme at Abbeystead, 1985.

6. Crowhurst, P., Measurements of Gas Emissions from Contaminated Land Building Research Establishment, 1987.

7. Hitchman, S.P., Darling, W.G. and Williams, G.M., Stable Isotope Ratios in Methane Containing Gases in the United Kingdom, BGS Technical Report WE/89/30, 1990.

8. Carpenter, R.J., Building Redevelopment on Disused Landfill Sites - Overcoming the Landfil Gas Problem, ISWA proceedings, 1988.

9. Building Regulations 1985 and Approved Documents, HMSO.

10. Department of the Environment, Landfilling Wastes, Waste Management Paper 26, 1986.

11. Interdepartmental Committee on the Redevelopment of Contaminated Land, Guidance on the Assessment and Redevelopment of Contaminated Land, 1987.

12. Department of the Environment/Welsh Office, <u>Development of Contaminated Land</u> Circular 21/87 (22/87), 1987.

13. Department of the Environment/Welsh Office, <u>Landfill sites: Development Control</u>, Circular 17/87 (38/89), 1989.

14. Her Majesty's Inspectorate of Pollution, <u>The Control of Landfill Gas</u>, Waste Management Paper No 27, 1989, HMSO.

16. Interdepartmental Committee on the Redevelopment of Contaminated Land, <u>Notes on the Development and Afteruse of Landfill Sites</u>, 1990.

15. Institute of Wastes Management, <u>Monitoring of Landfill Gas</u>, 1989.

17. <u>Building Regulations Approved Document C1/2 July 1990 Draft</u>.

A Review of Methods to Control Acid Generation in Pyritic Coal Mine Waste.

I.D. PULFORD.
Agricultural, Food and Environmental Chemistry Section,
Chemistry Department, University of Glasgow,
Glasgow G12 8QQ, U.K.

ABSTRACT

Oxidation of pyrite in coal mine waste leads to two main problems; a very acidic substrate for plant growth and acid mine drainage. The relative importance of these can vary depending on the position, topography, reclamation procedure and final land use of a site. A variety of techniques for controlling acid generation in pyritic spoil have been developed: these methods fall into two broad categories; those which inhibit pyrite oxidation, and those which interact with the oxidation products and render them innocuous. __Barrier methods__ attempt to isolate pyritic waste and prevent oxidation. __Chemical methods__ can be used both to prevent acid generation and to deal with the oxidation products. __Anti-bacterial methods__ are used to inhibit the catalysis of the oxidation by __Thiobacillus ferrooxidans__. __Alternative land use__, particularly the creation of artificial wetlands, is increasingly a means of controlling acid production and of dealing with the oxidation products. The choice of control method depends on the main type of problem faced and the final land use.

INTRODUCTION

Coal and its associated shales are formed under highly reduced conditions which lead to the conversion of sulphur species to sulphide (S^{2-}) and of iron to the ferrous form (Fe^{2+}). Precipitation of iron sulphide occurs; the thermodynamically favoured form being pyrite (FeS_2). This mineral is stable until exposed to air and water, when it oxidizes to produce, among other products, acid. Two oxidizing agents are known to attack pyrite under natural conditions; oxygen and ferric ions. The overall reactions involving these may be written

$$4FeS_2 + 15O_2 + 14H_2O \rightleftharpoons 4Fe(OH)_3 + 8H_2SO_4 \quad (1)$$

$$FeS_2 + 14Fe^{3+} + 8H_2O \rightleftharpoons 15Fe^{2+} + 2SO_4 + 16H^+ \quad (2)$$

These equations hide a series of intermediate reactions which it is important to understand in order to design a treatment to control acid generation.

The first stage of oxidation of pyrite by oxygen is a very slow reaction releasing ferrous ions and elemental sulphur

$$2FeS_2 + O_2 + 4H^+ \rightleftharpoons 2Fe^{2+} + 4S^0 + 2H_2O \qquad (3)$$

Under aerobic conditions and at pH values above 3.5 ferrous ions are oxidized to ferric ions and an iron oxide will precipitate

$$4Fe^{2+} + O_2 + 4H^+ \rightleftharpoons 4Fe^{3+} + 2H_2O \qquad (4)$$

$$Fe^{3+} + 3H_2O \rightleftharpoons Fe(OH)_3 + 3H^+ \qquad (5)$$

Elemental sulphur is oxidized to sulphate ions

$$2S^0 + 3O_2 + 2H_2O \rightleftharpoons 2SO_4^{2-} + 4H^+ \qquad (6)$$

Pyrite oxidation by this pathway is relatively slow, with the rate being determined by the diffusion of oxygen into the surface layers of the spoil [1]. It is the dominant route above pH 3.5, as ferric ions are precipitated and so cannot act as an oxidizing agent on the pyrite.

Below pH 3.5 ferric ions are soluble, but ferrous ions are more stable and the rate of reaction (4) is very slow. Under such acidic conditions, however, the bacterium *Thiobacillus ferrooxidans* is active and can catalyse the oxidation of ferrous ions. This catalytic action has a major effect on the oxidation of pyrite, increasing the rate by more than a million-fold [2]. Sulphur oxidation (reaction (6)) may also be bacterially catalysed, and is also an acid-producing reaction [3].

Above pH of about 4, the oxidation of pyrite is a relatively slow process which is regulated by the supply of oxygen. In more acidic systems, the oxidation rate is very much faster due to bacterial action, and the rate is controlled either by attack on the pyrite by ferric ions or by the rate of oxidation of ferrous to ferric [4].

As reclamation practice has advanced, a variety of techniques has been developed to counter the production of acid, which is the most damaging consequence of pyrite oxidation. These techniques can be divided into two broad categories;

    a) those which inhibit pyrite oxidation

    b) those which interact with the oxidation products and render them innocuous.

To some extent the method used depends on the origin of the spoil and the final land use, which may determine the nature of the acid problem. The two main environmental problems which arise from the oxidation of pyrite are;

    1) a very acidic substrate for plant growth; which results in vegetation not becoming established or, more commonly, the regeneration of acid after reclamation killing off any vegetation which has become established.

2) acid mine drainage; which results in the acidification of drainage waters, groundwater and surface waters, and the precipitation of iron oxides, which are most visible as ochre, and of other metal oxides.

A variety of methods has been used in order to prevent pyrite oxidation;

1) stabilization of pyrite by maintenance of an anaerobic environment

2) limiting contact of air and water with pyrite

3) control of pH to maintain the oxidation pathway by action of oxygen

4) additions of amendments to remove iron by precipitation, to complex the iron or to prevent the oxidation of ferrous to ferric

5) bactericides to inhibit the catalytic role of <u>Thiobacillus</u> <u>ferrooxidans</u>.

Other techniques have been developed to treat the oxidation products;

1) alkali treatment of acidic mine drainage

2) wetland creation.

Some methods may act in more than one of these ways, so four broad groups will be considered; barrier methods, chemical methods, anti-bacterial methods and alternative land use.

BARRIER METHODS

These methods are designed to isolate pyrite from oxygen and water; particularly from flowing water which may contain dissolved oxygen and iron, and which may remove oxidation products.

Specific placement of pyritic spoil is more easily achieved during opencast, or strip, mining operations where the acid-producing potential of the various strata can be assessed. Acid-base accounting techniques have been developed to identify potentially acid-producing material [5], which can then be selectively replaced in order to minimize oxidation. Such a procedure is less easy to use for deep mine wastes, as the whole tip may not be moved, and the properties of unexposed materials are not known.

Although these methods generally try to avoid contact between pyrite and water, one technique which has been used is to place the pyritic spoil below the water table [6]. In theory at least this should maintain pyrite in a reduced form. In practice, however, vertical fluctuations and lateral movement of the water table may mean that the pyrite is periodically exposed to oxidizing conditions and so may produce acid drainage. In general this method has not been widely adopted, as it is feasible only where mining has occurred to or below the water table. It is worth noting, however, that the common practice of flooding and capping old mine shafts and workings relies on this principle. There have been specific incidences (for example in south Ayrshire) where this procedure too has broken down, resulting in contamination of rivers by acid mine drainage.

A much more common approach is to place the pyritic spoil in such a way that it is isolated from groundwater flow, percolating surface water and oxygenated conditions. This method is sometimes referred to as "high and dry placement". Again, this technique is more suited to opencast mining where the replacement of the overburden allows the selective siting of potentially acid-producing material. This pyritic spoil can then be surrounded by overburden strata which are not expected to acidify. Often some specific materials are used to form caps over the zone of pyritic material. These caps may be designed to perform various functions. Skousen et al. [7] referred to the use of plastic films or liners as impermeable barriers to water movement. Their limitations are that they are easily damaged, can deteriorate and are expensive. More commonly, clay barriers have been formed, using a swelling clay such as bentonite. The intention here is that the swollen clay forms an impenetrable barrier to water movement, but experience has shown that the seal is not always perfect. Clay barriers are particularly prone to cracking during dry conditions. Any fault in caps designed to be a seal can allow oxygenated water to percolate through pyritic spoil, resulting in oxidation and transport of the products. Other capping materials could be incorporated in order to alter the hydrological conditions. For example, a coarse gravel layer over the clay would direct most of the percolating water away from the pyrite, and would form a break to the capillary rise of water during dry conditions. Cairney [8] has reviewed the use of soil cover treatments for the reclamation of contaminated land, the principles of which can be applied to coal mine waste.

In addition to isolating pyrite from water, barrier methods aim to limit the supply of oxygen. Erickson [9] measured the oxygen content in unsaturated coal mine waste and found that there were seasonal variations in the oxygen profiles, but that these showed no consistent trends. Overall Erickson concluded that materials used as inert covers over pyritic spoil may not stop the diffusion of oxygen into a tip to depths which would allow buried pyritic material to be oxidized. The extent of oxygen diffusion, and hence the steepness of the concentration gradient, depends on the texture of the waste, and on the development of an oxygen-comsuming rhizosphere in the surface layers, where roots, microorganisms and decaying organic matter all remove oxygen [10]. A similar effect could be obtained by incorporation of an oxygen-consuming material such as sewage sludge, either in the surface layers or as a barrier layer over a clay cap [9].

The use of a topsoil cover, especially over compacted spoil, can also be considered as a barrier method of controlling pyrite oxidation; and is one which can be used for treatment of deep mine waste. Compaction of the underlying spoil will help to reduce the influx of oxygen, but can produce problems of stability of the topsoil cover. The spoil may also be treated, most commonly by liming, to further militate against pyrite oxidation. The establishment of a vigorous vegetation in the soil layer will help to decrease the oxygen content due to root respiration and the water content by evapotranspiration.

In general, barrier methods have been favoured for surface mined coal waste where burial of pyritic material, placement of a capping layer and covering with inert material are more easily achieved. The main problem to be counteracted is the leaching of acid mine drainage, so the overall hydrology of the system should be studied and understood.

CHEMICAL METHODS

The commonest method for treating reclaimed pyritic coal mine spoil is application of lime. This neutralizes any existing acidity, and by maintaining a pH above 4.5 to 5.0 aims to limit the pyrite oxidation pathway to the slow process of attack by oxygen (equation (1)). In many early reclamation schemes the amount of lime applied was calculated by treating the spoil as an agricultural soil and using one of the standard methods for lime requirement [11]. This proved successful in many cases, especially when the final land use was agricultural and aftercare maintenance was applied. As more highly acidic and pyritic spoils were reclaimed, and as the amount of aftercare diminished due to financial constraints and increasing land use for cosmetic or amenity purposes, the problems of acid regeneration became more apparent. Attempts were made to build in to the lime requirement estimate an allowance for potential acidity by measuring pyrite content of the spoil [12]. If pyrite oxidation proceeds through to precipitation of iron oxide (equation (5)) then for each 1% pyrite to a depth of 15 cm, 40 tonnes of limestone are required per hectare [13].

Addition of an estimate for potential acidity to that of existing acidity often resulted in very high lime requirements (> 100 t/ha). There are dangers associated with the use of such high additions of lime, particularly its effect on the growth of clover, an imbalance in the Ca/Mg ratio and the availability of phosphate [14]. It has also commonly been found that even very high rates of lime added at reclamation do not always prevent subsequent re-acidification after a few years [13;15;16]. This arises because of the uneven distribution of pyrite in a spoil [17]. Acid is produced at a point source (a piece of pyrite exposed by weathering) and can easily neutralize any added lime. Thus the pattern observed initially is for a number of small acid patches to develop around individual pieces of pyritic spoil. These patches spread, and merge into larger areas, as the acid water flows over the surface of the spoil. Maclean and Dekker [18] have estimated that spoils containing greater than 4% pyrite will re-acidify even when very high rates of lime additions are used.

Alkali treatment of acid mine drainage presents a different problem. In this case it is the products of the oxidation which are being removed, rather than the prevention of the oxidation. The alkali is added to the drainage water in order to raise the pH and to precipitate metals, notably iron, manganese and aluminium. The chemicals commonly used to treat acid mine drainage are; sodium hydroxide, sodium carbonate, calcium hydroxide and calcium carbonate. Skousen et al. [19] have recently discussed the benefits, drawbacks and costs of their use. The calcium compounds tend to be cheaper, but are less soluble than the sodium ones. Carbonates have less of an effect on pH than hydroxides. Thus the choice of chemical depends on the degree of acidity and metal load of the drainage water. Regardless of which chemical is used, the treated water must be lagooned to allow the precipitated metal salts to settle out. This results in the production of a sludge which must be dredged and disposed of periodically.

Other chemical treatments which have been shown to inhibit acid production by pyrite oxidation all interfere with the chemistry of iron in the spoil. They may act by precipitation, which removes iron from solution, or by complexation, which either keeps the iron in a non-reactive form or inhibits the oxidation of ferrous to ferric.

Phosphate has long been recognized as an inhibitor of pyrite oxidation and was used to prevent acid production in Dutch polder soils [20]. Despite the issuing of a U.S. patent in 1969 [21], it has been only recently that the use of phosphate to inhibit acid production in coal mine spoil has been suggested [22;23;24]. The latter study [24] showed that when a suspension of pyritic spoil was incubated with phosphate the release of acid was inhibited and the iron concentration in solution was reduced almost to zero, suggesting that phosphate was removing iron by precipitation. A number of types of phosphate have been evaluated under field and glasshouse conditions. Renton et al. [23] used rock phosphate ground to various sizes and found that the smaller grains were most effective in controlling acid production; the most effective ameliorant was a clay slurry containing apatite. This result was also obtained by Chiado et al. [25]. Treatment with phosphatic clay also acts as a barrier to water and air, thus isolating pyritic spoil.

Silicate has been tested as an inhibitor of acid production [22;24]. Backes et al. [24] suggested that silicate could act by complexing or precipitating iron, depending on the Si:Fe ratio.

Organic materials are particularly useful at inhibiting pyrite oxidation as they contain functional groups which can chelate ferrous and ferric ions, and can provide a degree of buffering against changes in pH. Wood wastes and manures have been shown to inhibit acid production [24], but only the manure removed iron from solution. This suggests that iron was complexed by the solid manure, while soluble complexing agents were released by the wood waste.

The effectiveness of silicates and organic materials has yet to be proven on a field scale. Both provide the opportunity to use other wastes as amendments for acidic spoil (e.g. pulverized fuel ash, sewage sludge, chicken manure). Care must be taken, however, to ensure that no deleterious factors are introduced with the waste.

ANTI-BACTERIAL METHODS

These methods are aimed specifically at inhibiting the action of Thiobacillus ferrooxidans. Backes et al. [4] showed that a bactericide stopped the oxidation of ferrous ions, thus decreasing the rate of oxidation of pyrite. In the field, however, use of a broad spectrum bactericide would be undesirable as beneficial reactions may also be inhibited. One class of compounds which have been shown to be potentially valuable as anti-bacterial agents are anionic surfactants.

Kleinmann and co-workers in the United States have tested and developed the use of sodium lauryl sulphate (SLS) as an anti-bacterial agent [26;27;28]. SLS attacks the lipid component of the bacterial cell wall, destroying its integrity and allowing acid from the spoil into the cell. The change from a normally neutral pH within the cell to a more acid one can decrease the activity of the enzymes which catalyse the oxidation of ferrous ions. At high surfactant concentrations the bacterial cell can be killed. Kleinmann and Erickson [27] reported a decrease of between 60% to 95% in acid production over a period of 4 to 5 months following application of 55 gallons of 30% SLS solution per acre at a site in West Virginia. This was typical of the results of these early studies; good control of acid production was possible, but only over a limited period. It seemed likely that this short effective period was due to leaching and/or degradation of the surfactant. Attempts were then made to formulate a slow-release form of surfactant [28], and commercial products are now available [e.g. 29].

Another difficulty in this approach is the registering of the use of such chemicals, which may have other environmentally damaging effects. While some surfactant-based products have been registered in the United States, researchers have looked for other, more acceptable, alternatives. Two chemicals which show great promise are potassium sorbate and sodium benzoate [28;30]. As both are used for food preservation they may be more acceptable than surfactants for release into the natural environment.

While the American studies quoted above have been primarily directed towards improvement in the quality of acid mine drainage, there would appear to be no reason why such treatments could not be used to prevent acid production in the root zone. Checks would have to be made to ensure that the chemicals did not adversely affect beneficial bacteria needed for degradative and nutrient cycling processes.

ALTERNATIVE LAND USE

Over the last ten years there has been considerable interest in the United States in the use of artificially created wetlands as a treatment for acid mine drainage [31;and numerous papers in 32]. This approach developed from the observation of sites where acid mine drainage flowed through natural wetlands. Results have been variable and studies are still underway to examine the optimum design of such systems.

Up to now two types of wetland have been used; Sphagnum (a moss) and Typha (cattails). Sphagnum is a moss with a low growth habit, which can tolerate very acid conditions and is most effective at pH values below 4. Typha, on the other hand, is an emergent plant which can tolerate a wider pH range. It has aerenchyematous tissue which allows the transport of oxygen to the roots. A number of studies have reported that concentrations of iron, manganese and sulphate in acid mine drainage have decreased after flowing through a wetland system, and that pH of the water increased. Results have tended to be somewhat variable however. For example, Kleinmann [31] reported that a Sphagnum bed removed between 50% to 70% of total iron from solution, but had no significant effect on acidity. A Typha bog, however, raised the pH of the water from below 6 to near neutrality, and reduced iron concentrations of 20 to 25 mg/l to less than 1 mg/l and manganese concentrations of 30 to 40 mg/l to less than 2 mg/l [31]. In general reported results indicate that iron removal is reasonably efficient, manganese removal much less so and aluminium removal is poor.

A number of processes have been suggested as acting in wetlands to improve the quality of acid mine drainage; oxidation by bacteria, uptake by plants, precipitation as a sulphide and adsorption on to solid phase surfaces. Exactly which of these is important is unclear. Wetlands are characterized by low redox potentials which favour sulphide formation, so removing sulphate and precipitating iron and possibly manganese. In Typha wetlands the oxygenated root zone may create conditions where iron and manganese oxides are precipitated. Recent studies, such as that of Sencindiver and Bhumbla [33] suggest that much more needs to be known about how wetlands work. What is clear, however, is that the design of the wetland is crucial. Such factors as flow rate of water, residence time, length to width ratio and use of limestone outflows are all important [31;34;and numerous references in 32].

Wetlands for treatment of acid mine drainage have created much interest over the last ten years. They provide a relatively low cost method, which also allows a certain amount of variety in the final land use of a reclaimed site.

CONCLUSIONS

Despite the accumulated experience of many years and numerous reclamation schemes, acid substrates and acid mine drainage still present problems. The traditional, and usually relatively expensive, methods of treatment have proven to be unsuitable in many cases where low maintenance systems are required. Use of such amendments as phosphate, silicate or organic materials along with some initial liming may help to maintain non-acidic conditions in reclaimed spoil. Wetland creation may allow some variation in land use and prevent acid mine drainage from entering streams and rivers.

It seems likely that a combination of a number of techniques may have to be employed in order to combat pyrite oxidation, and hence acid production, over the long term. The choice of control method, or methods, depends on the main type of problem faced and the final land use.

REFERENCES

1. van Breemen, N., Soil-forming processes in acid sulphate soils. In Acid Sulphate Soils, ed. H. Dost, International Institute for Land Reclamation and Improvement, Publication 18, Vol.1, 1973, pp. 66-128.

2. Singer, P.C. and Stumm, W., Acidic mine drainage: The rate-determining step. Science, 1970, 167, 1121-1123.

3. Temple, K.L. and Delchamps, E.W., Autotrophic bacteria and the formation of acid in bituminous coal mines. Applied Microbiology, 1953, 1, 255-258.

4. Backes, C.A., Pulford, I.D. and Duncan, H.J., Studies on the oxidation of pyrite in colliery spoil. I. The oxidation pathway and inhibition of the ferrous-ferric oxidation. Reclamation and Revegetation Research, 1986, 4, 279-291.

5. Sobek, A.A., Schuller, W.A., Freeman, J.R. and Smith, R.M., Field and Laboratory Methods Applicable to Overburdens and Minesoils, U.S. Environmental Protection Agency, EPA-600/2-78-054, Cincinnati, Ohio, 1978, 203pp.

6. Kleinmann, R.L.P. and Erickson, P.M., Control of acid mine drainage: an overview of recent developments. In Innovative Approaches to Mined Land Reclamation, eds. C.L. Carlson and J.H. Swisher, Coal Extraction and Utilization Research Center, Southern Illinois University at Carbondale, Southern Illinois University Press, 1987, pp. 283-305.

7. Skousen, J.G., Sencindiver, J.C. and Smith, R.M., A Review of Procedures for Surface Mining and Reclamation in Areas With Acid-Producing Materials. The West Virginia University Energy and Water Research Center, Publication EWRC871, 1987, 39pp.

8. Cairney, T., Soil cover reclamations. In *Reclaiming Contaminated Land*, ed. T. Cairney, Blackie, Glasgow, 1987, pp. 144-169.

9. Erickson, P.M., Oxygen content of unsaturated coal mine waste. In *Control of Acid Mine Drainage*, U.S. Bureau of Mines Information Circular 9027, U.S. Bureau of Mines, Pittsburgh, 1985, pp. 19-24.

10. Erickson, P.M., Kleinmann, R.L.P. and Campion, P.S.A., Reducing oxidation of pyrite through selective reclamation practices. In *Proceedings of the 1982 Symposium on Surface Mining, Hydrology, Sedimentology and Reclamation*, ed. D.H. Graves, University of Kentucky, Lexington, 1982, pp. 97-102.

11. Doubleday, G.P., Soil forming materials: their nature and assessment. In *Landscape Reclamation* Vol. 1, University of Newcastle-upon-Tyne, IPC Business Press, London, 1971, pp. 70-83.

12. Dacey, P.W. and Colbourn, P., An assessment of methods for the determination of iron pyrites in coal mine spoil. *Reclamation and Revegetation Research*, 1979, 2, 113-121.

13. Costigan, P.A., Bradshaw, A.D. and Gemmell, R.P., The reclamation of acidic colliery spoil. I. Acid production potential. *Journal of Applied Ecology*, 1981, 18, 865-878.

14. Costigan, P.A., Bradshaw, A.D. and Gemmell, R.P., The reclamation of acidic colliery spoil. III. Problems associated with the use of high rates of limestone. *Journal of Applied Ecology*, 1982, 19, 193-201.

15. Doubleday, G.P., The reclamation of land after coal mining. *Outlook on Agriculture*, 1974, 8, 156-162.

16. Pulford, I.D., Walker, T.A.B., Devlin, J.G. and Duncan, H.J., Amendment treatments for acidic coal waste reclamation. In *Environmental Contamination*, CEP Consultants Ltd., Edinburgh, 1984, pp. 483-487.

17. Pulford, I.D., Backes, C.A. and Duncan, H.J., Treatments to prevent acid production by pyrite oxidation in coal mine waste. In *Innovative Approaches to Mined Land Reclamation*, eds. C.L. Carlson and J.H. Swisher, Coal Extraction and Utilization Research Center, Southern Illinois University at Carbondale, Southern Illinois University Press, 1987, pp. 637-654.

18. Maclean, A.J. and Dekker, A.J., Lime requirement and availability of nutrients and toxic metals to plants grown in acid mine tailings. *Canadian Journal of Soil Science*, 1976, 56, 27-36.

19. Skousen, J., Politan, K., Hilton, T. and Meek, A., Acid mine drainage treatment systems: Chemicals and costs. *Green Lands*, 1990, 20, 31-37.

20. Quispel, A., Harmsen, G.W. and Otzen, D., Contribution to the chemical and bacteriological oxidation of pyrite in soil. *Plant and Soil*, 1952, 4, 43-55.

21. Flyn, J.P., Treatment of earth surface and subsurface for prevention of acidic drainage from soil, 1969, U.S. patent 3,443,882.

22. Watkin, E.M. and Watkin, J., Inhibiting pyrite oxidation can lower reclamation costs. *Canadian Mining Journal*, 1983, **104**, 29-31.

23. Renton, J.J., Stiller, A.H. and Rymer, T.E., The use of phosphate materials as ameliorants for acid mine drainage. pp. 67-75 in ref 32.

24. Backes, C.A., Pulford, I.D. and Duncan, H.J., Studies on the oxidation of pyrite in colliery spoil. II. Inhibition of the oxidation by amendment treatments. *Reclamation and Revegetation Research*, 1987, **6**, 1-11.

25. Chiado, E.D., Bowders, J.J. and Sencindiver, J.C., Phosphatic clay slurries for reducing acid mine drainage from reclaimed mine sites. pp. 44-51 in ref 32.

26. Kleinmann, R.L.P., Crerar, D.A. and Pacelli, R.R., Biogeochemistry of acid mine drainage and a method to control acid formation. *Mining Engineering*, 1980, **33**, 300-306.

27. Kleinmann, R.L.P. and Erickson, P.M., *Control of Acid Drainage From Coal Refuse Using Anionic Surfactants*. U.S. Bureau of Mines Report of Investigations 8847, U.S. Bureau of Mines, Pittsburgh, 1983, 16pp.

28. Erickson, P.M., Kleinmann, R.L.P. and Onysko, S.J., Control of acid mine drainage by application of bactericidal materials. In *Control of Acid Mine Drainage*, U.S. Bureau of Mines Information Circular 9027, U.S. Bureau of Mines, Pittsburgh, 1985, pp. 25-34.

29. Shellhorn, M.A., Sobek, A.A. and Rastogi, V., Results from field applications of controlled release bactericides on toxic mine waste. In *Proceedings of the 1985 Symposium on Surface Mining, Hydrology, Sedimentology and Reclamation*, ed. D.H. Graves, University of Kentucky, Lexington, 1985, pp. 357-360.

30. Onysko, S.J., Erickson, P.M., Kleinmann, R.L.P. and Hood, M., Control of acid drainage from fresh coal refuse: Food preservatives as economical alternatives to detergents. In *Proceedings of the 1984 Symposium on Surface Mining, Hydrology, Sedimentology and Reclamation*, ed. D.H. Graves, University of Kentucky, Lexington, 1984, pp. 35-42.

31. Kleinmann, R.L.P., Treatment of acid mine water by wetlands. In *Control of Acid Mine Drainage*, U.S. Bureau of Mines Information Circular 9027, U.S. Bureau of Mines, Pittsburgh, 1985, pp. 48-52.

32. U.S. Bureau of Mines, *Mine Drainage and Surface Mine Reclamation, Vol.1 Mine Water and Mine Waste*, U.S. Bureau of Mines Information Circular 9183, U.S. Bureau of Mines, Pittsburgh, 1988, 413pp.

33. Sencindiver, J.C. and Bhumbla, D.K., Effects of cattails (*Typha*) on metal removal from acid mine drainage. pp. 359-366 in ref 32.

34. Stillings, L.L., Gryta, J.J. and Ronning, T.A., Iron and manganese removal in a *Typha*-dominated wetland during ten months following its construction. pp. 317-324 in ref 32.

RANKING OF "PROBLEM" SITES AND CRITERIA FOR CLEAN-UP OF CONTAMINATED SITES

Michael A Smith
Clayton Environmental Consultants
288 Windsor Street, Birmingham B7 4DW

ABSTRACT

The Environmental Protection Act requires Waste Regulation Authorities to identify sites that threaten public health or the environment, and also provides for local authorities to set up registers of potentially contaminated sites. In addition, the National Rivers Authority has announced an intention to set up a register of problem sites. In other countries, such surveys have identified hundreds of "high risk" sites. The United Kingdom can be expected to be similar. Once sites are identified, it will be necessary to assign local and national priorities so that limited resources are spent cost-effectively on enforcement, site investigation and remediation. Criteria will be needed to aid assessment of hazards and risks and decisions about the standards to be achieved in any remediation. These must refer not only the potential hazards to users of the site, but also to the potential to pollute ground or surface water, or to damage the environment.

INTRODUCTION

Various bodies in the United Kingdom are about to embark on surveys and other activities that are will to lead to the identification of sites that pose an immediate or imminent threat to public health or to the environment. The resources required to investigate, and in due course, remedy such sites, are certain to exceed the resources that are immediately available. It will thus be necessary to decide priorities at several stages through the process from identification/discovery to remediation. It will also be necessary to decide on the criteria to be applied when deciding whether remedial action is required and the standards to be achieved in the remediation. These two needs can be addressed by the establishment of "Hazard Ranking Systems" and "Clean-up criteria." The nature of these criteria will be explored in more detail below but they can be briefly defined as follows:

A **Hazard Ranking System** is a means of indicating the relative importance of a site in terms of actual or potential hazards to public health or the environment in order to establish priorities for further investigation or remedial action.

Clean up criteria are intended to assist in deciding whether remedial works are required and in defining the standards to be achieved as a result of the remedial action.

The Environmental Protection Act 1990 requires the Waste Regulation Authorities (WRAs), once established, to identify sites which may pose a threat to public health, the environment or to natural resources due to migration of contaminants off-site. The wording of the Act appears to embrace any "site" not just those used for waste disposal. However, even if the task were to be restricted to "landfills" there are many thousands of these that must be located (identified), assessed and investigated and then, possibly remedied either by the responsible party or by the NRA.

If the task is taken seriously each WRA must go through a lengthy and iterative process along the following lines:

identification/location of sites
- for example all sites of a particular type (e.g. former landfills, former gasworks sites) by use of current records, maps and other historic sources

appraisal of sites
- collection of site specific information and site visits

staged investigation(s)
- what is the contamination, where is it, how much is there, and is there evidence that it is migrating, together with information on hydrogeology, geology etc.

site-specific risk assessment

health studies of potentially affected populations and other possible targets

immediate remedial works

long-term remediation

long-term monitoring of site and populations etc

It is obvious that throughout this process it will be necessary to make judgements about the urgency of action and the most cost-effective allocation of resources.

The principle in the Act is that the land owner or other responsible party should be required to meet necessary costs but this can only happen if the WRA first allocates resources to identify the "problem sites" and to take necessary technical and legal actions to achieve this end. There will always remain a number of sites where there is either no identifiable owner or other responsible party, or where these do not have the financial resources to carry out necessary investigations or remediation. Such sites are generally known as "orphan" sites in other countries. The Act does not really address the question of how works on such sites are to be funded.

The Environmental Protection Act also makes provision for the establishment of registers of potentially contaminated sites by local authorities. Authorities will need to decide whether to attempt to carry out a fully comprehensive survey of their areas immediately or to first seek to identify sites that can be regarded as "high risk" because of their past uses, location or current status in development terms.

It is highly probable that most local authorities will identify "suspect" sites which have already been developed for sensitive uses such as allotments, housing and schools. They will also identify sites, such as former gasworks, where there is a high probability of groundwater pollution and therefore a risk that contaminants will have migrated off-site to affect adjacent houses. Local authorities will have to decide which sites to pursue in more detail following a similar staged and iterative process to that already described. The possible magnitude of the task that some authorities may have to faced is shown by the results of a survey of a small area in the West Midlands presented in Table 1.

TABLE 1
Results of survey for potentially contaminated sites of
an area in the West Midlands

Classification	% of area surveyed
Contaminated	24.5
Probably contaminated	17.2
Potentially contaminated	9.2
Pre-1930s housing	10.0
Mixed land use	7.4
Use unknown	0.5
Uncontaminated	15.1
Contaminated but in sensitive use	16.1

The National Rivers Authority (NRA) will face a similar task if it does indeed set up its own register of "problem" sites.

In the absence of a comprehensive national survey to date it is difficult to estimate how many contaminated sites there are in the UK but comparisons with other countries (see Table 2) suggest that there could be about 100000 of which several thousand could be regarded as "high risk" in terms of actual or potential risks to public health or the environment.

In practice the United Kingdom has still not set up a national programme for the identification of "high risk" sites. Something which all other countries that have taken the issue seriously have felt a need to to do so that the "worst" sites can be identified and remedied.

Unfortunately, the responsibility for identifying these "high risk" sites is divided between a number of different parties. In England and Wales alone these include District Councils, WRAs, and the NRA, with no national coordination or allocation of priorities or money. It could be a recipe for confusion and duplication of effort. There is a serious danger of different standards being applied in different areas.

TABLE 2
Results of surveys for contaminated sites in various countries

Country	Total number of sites	"High risk" sites	Estimated clean-up costs
Canada	10 000	1000	
Czechoslovakia	15 000**		
Denmark	20 000		£700M
Finland	20 000	100-1000	
France		100***	
Germany	100 000		
Netherlands	110 000	6000	£15 000M
Norway	2 441	61	
USA	33 000	1200*	
United Kingdom	100 000 ?	?	

* 33000 sites screened under "Superfund" programme. >1200 sites on National Priorities list. For total of US sites must add State sites and private sites

** known landfills - 5000+ with groundwater pollution

*** orphaned sites for national government action over next 5 years

Note.   Figures are not on a common basis but provide a general indication of the magnitude of the problem in each country

Canada has recently set up a formal national programme to identify contaminated sites and then to remedy them. Each Province has agreed to introduce legislation requiring the identification of sites and their remediation on the basis of the "polluter pays" principle. But Federal and Provincial funds (a total of $200M) are to be provided for dealing with "orphan sites." A national hazard ranking system and "decommissioning criteria" are to be introduced to assist in deciding the priorities for tackling sites [1].

The US Environmental Protection Agency (EPA) uses a formalized hazard ranking system to decide which of the tens of thousands of sites brought to its attention should go on the "National Priorities List" (NPL) so that they become eligible for funding under the "Superfund" programme for detailed investigation and remediation in due course. Of the approximately 32755 sites screened by the EPA by September 1990, 1207 currently have achieved the dubious status of being put on the NPL. A substantial number of sites are removed from the NPL when investigation leads to a revision of their hazard ranking. Many of the sites that do not make it on to the NPL will be dealt with in parallel State programmes [2].

The US EPA Hazard Ranking System (HRS) has been recently substantially revised to take account of deficiencies revealed since it was first introduced in about 1982. Among other changes has been to give greater weight to actual as opposed to potential risks, to observed effects/damage, and to risks to sensitive ecosystems [3].

HAZARD RANKING SYSTEMS

A hazard ranking system is simply a systematic way of doing what must be done in deciding on priorities in the allocation of resources of money, manpower, and possibly equipment. It may be very simple or quite complex in terms of the amount of information that is used to rank sites and how various factors are weighted in arriving at the final ranking. The system should be no more complex than is necessary to meet the immediate need. The "ranking" should be kept under constant review as new information becomes available.

It is to be assumed that some local authorities and Waste Disposal Authorities have been applying some form of ranking system in addressing the problems posed by the potential migration of gas from operating and closed landfill sites. Having identified the sites, one assumes that they have considered such factors as the age of sites and the proximity of houses and other buildings to the sites, together with geological settings, and any record of gas problems, in deciding the order in which they are to be investigated.

Initial ranking is done from readily available information (maps, site records etc). Subsequently it moves in to a more detailed assessment of documentary evidence, moving progressively towards detailed site-specific risk assessments. Essentially the system has to provide a means of comparing sites on the basis of potential or actual risks, either separately or collectively to: public health, the environment, and natural resources.

A range of factors are scored on arbitrary scales, the scores weighted as seems appropriate and then combined to give an overall score which provides a ranking when comparison is made with other sites. In the first instance, some of the scales might have only two points e.g. there is or is not housing on the site, there is or is not surface water within or close to the site (say within 100 metres).

In practice, it is difficult to devise a scheme, other than for a first broad appraisal, that can provide a single ranking covering all three aspects i.e. parallel listings may be required.

Factors that may be considered include:

the nature of the contaminants (e.g. toxicity, persistence, mobility in the environment)

probability of migration off-site in the groundwater, as gas and vapours, or by wind or water action (surface condition of the site, presence of drums, surface deposits, exposure, fill density, surrounding geology are some of the relevant parameters as far as the latter are concerned) targets at risk (people, flora, fauna, water courses, drinking water aquifer, aesthetic quality etc)

proximity of targets and numbers potentially at risk

potential routes of exposure, both direct and indirect.

These are all ingredients to be fed subsequently into a site-specific risk assessment.

Models/systems developed to assist in the evaluation of the suitability of new waste disposal sites may be adaptable to this application provided the different aims are taken into account.

The ranking given to the site will be based on an as objective assessment as possible of various technical factors. However, it is only a guide to action. It can never be a precise process and when the ranking of an individual site, or group of sites is considered, is quite proper for other considerations of a socio-political or economic nature to be taken into account. For example, once high-risk sites have been rendered safe, perhaps by placing a high and secure fence around them or controlling gas migration, it might be considered more cost-effective to clean up three sites at modest cost than to put all available funds immediately into a more highly ranked site (in any case the emergency works should have lowered its ranking). Or a site might be the key to redevelopment of an area which would then allow some other improving works to go ahead that might permit the relocation of a population affected by other, more hazardous sites and thus justify added priority. Similarly, where litigation is involved, it might be judged more cost-effective to pursue three cases that are winnable in the short-term, than the most difficult case which will tie up people and money for many years.

If one considers the case of a local authority that has completed its register of contaminated sites it might like first to place sites into a few simple categories. For example based on the development status of the land. Suitable categories might be:

- sites currently being developed

- sites already developed for "sensitive" uses

- sites already redeveloped for less sensitive uses

- sites in areas scheduled for redevelopment

- sites unlikely to undergo a change of use in the next few years

- "undeveloped" sites

It is important to recognise that the latter are not the same as "uncontaminated" sites because undeveloped farmland may be affected by migrating contaminants or may have become contaminated through the use of agro-chemicals or soil amendment materials such as sewage sludge or still-bottoms from the distilling industry.

The authority will need to have regard not only to sites which are themselves polluted, but those which may have been affected by migrating contamination.

Having categorised the sites as above, other factors can then be considered such as:

the history of the site (the reason why the site is on the register but more detailed information may be required),

the current use of the land - is it residential and if so is it houses or flats,

the number of people that might be at risk,

the presence of a specially at risk population - travellers through their way of life may be at greater risk than a more stable population,

the use of neighbouring land,

whether the land liable to flooding,

whether the water table stable or is rising (a problem in a number of our major cities),

presence of obvious signs of damage to vegetation, water courses, property,

the age of the houses or other buildings and infrastructure - (what condition are the services such as water supply pipes and sewers likely to be in ?),

who owns the site,

whether the pollution likely to be associated only with the site in question or is it one of a series in the same area,

what "environmental" targets are at risk, and

whether a sensitive ecosystem (eg a Site of Special Scientific Interest) at risk ?

In reviewing the available information and deciding how to proceed the authority will need to recognise the need to bring in extra expertise and to consult with other bodies such as the National Rivers Authority.

Such schemes are not new to the UK. Many years ago a survey was carried out of metalliferous mining sites in Wales for the Welsh Development Agency with the intention of identifying those sites which were, or might, pose a threat to water courses etc. Many of these are in fairly remote locations but flooding could cause loose tailings to be washed downstream resulting in damage to river life, or to fields lower downstream that were subsequently flooded [4].

CLEAN-UP CRITERIA

Clean-up criteria are required for two purposes:

>   to help a decision on whether some form of (further) investigation or remediation is required (action criteria)

>   and to provide guidance on the standards to be achieved when such work is carried out (clean up targets)

They are required for soils and for groundwater as sources of contaminants, and may be required for targets such as surface waters adjacent to a site, or drinking water aquifers. Criteria are needed that address risks from soil and groundwater to immediate targets (e.g site occupants) and also to targets likely to be affected by migration i.e. that proportion of the contamination that is, or may become, mobile in the environment.

It might be concluded that ideally these should always be set on a site-specific basis taking into account the targets at risk, the routes of exposure etc. i.e. after a site specific risk assessment. But this will be rarely practicable or cost effective, except perhaps, for the most problematic sites and more general criteria will generally be sufficient for most contaminated sites.

The distinction between the two types of criteria (action and target) can be seen by reference to the Dutch "ABC" values for soil and groundwater. The "B" value indicates a need for caution and further investigation. If the "C" value is exceeded action must be taken to clean-up to achieve the target condition defined by the "A" value.

The Dutch system [5] sets the same target whatever the intended use of the land. In contrast the British Columbian system [6] sets different targets for soils for different land uses, and for waters depending on whether the concern is drinking water quality or "environmental."

The British Columbian system allows that for a particular site, either site-specific criteria may be used, or the Provincial guidelines applied. The latter are deemed to reduce risks to an "acceptable" level in the majority of situations. Although the Canadian "National Decomissioning Criteria" have not yet been published they are likely to follow a similar pattern.

Environment Canada, with support from the US EPA, have been developing an expert system (AERIS) as an aid to developing site-specific criteria [1,7].

Criteria on their own are not enough. Guidance is also required on sampling strategy and sampling methods both at the assessment stage and for validation of the clean-up process, bearing in mind that the latter may be undertaken either in-situ, or after excavation/extraction with the intention that the soil/fill or groundwater should be returned to the site, otherwise returned to the environment, or put to beneficial use. Both the target criteria and the (statistical) basis for judging that they have been achieved must be in place before remediation is started.

In the UK the only guidance that we have available is the Department of the Environment/Interdepartmental Committee on the Redevelopment of Contaminated Land (ICRCL) guidance values which provide us with "threshold trigger" values for some contaminants, and both "threshold" and "action" trigger values for a further limited number of contaminants. They tell us when to worry but not necessarily when to take action. They do not tell us directly what we are to achieve if we do take action, although the threshold values could be taken to define a "clean enough" target.

The ICRCL guidelines do not, of course, cover groundwater, either in terms of in-situ suitability, or potential for remote harm. In addition they suffer from a number of other deficiencies [8] not least of which the limited number of contaminants that are covered. The Department of the Environment has announced [9,10] an intention to extend the guidelines but has not been able to indicate when they might appear.

Additional to the clean-up criteria to be applied to a site, are the criteria, limits and statistical evaluation, that should be applied when the intention is to try to evaluate a technology for wider application [11,12].

CONCLUSIONS

If the need to identify contaminated sites, in particular those that might be regarded as "high risk" or problem sites is taken seriously, substantial manpower and money resources will have to be deployed. The agencies responsible for identifying sites and then pursuing a remedy to the identified problems (WRAs and local authorities), either by "persuading" someone to pay or paying themselves, vary greatly in size and in their technical capabilities. They will require guidance on how to assess the hazards posed by sites and how to decide their priorities for allocation of what are very limited funds in relation to the costs of comprehensive site investigations, lengthy litigation, or major clean-ups.

It is to be assumed, that those responsible for administering Derelict Land Grant and similar grant systems, will also need to be able to judge where best to spend money on an equable basis, which is based on the hazards posed by a site, irrespective of where it is, rather than on local unemployment figures or historic factors.

Guidance is also required on other aspects of these site discovery programmes. Such as how and when to inform and involve the public, how to involve local general practitioners in health surveillance programmes [13,14], and how to evaluate the options for remedial action.

REFERENCES

1. Schmidt, J. W., Contaminated land: a Canadian viewpoint. In <u>Proc 2nd Conf. Contaminated Land - Policy, Regulation and Technology</u>, London 1991, IBC Technical Services, London 1991

2. Hansen, P. and Previ, C., The Superfund site assessment process: a status report. In *Proc. Superfund '90*, Washington DC 1990, Hazardous Materials Research Institute, Silver Spring MD, 1990. pp. 77-79

3. Wells, S. and Caldwell, S., Overview of the revised Hazard Ranking System (HRS), In *Proc. Superfund '90*, Washington DC 1990, Hazardous Materials Research Institute, Silver Spring MD, 1990. pp. 71-76

4. Survey of metalliferous mining sites in Wales, report for Welsh Development Agency, Liverpool University Environmental Advisory Unit.

5. Keuzzenkamp, K., Dutch policy on clean-up of contaminated soil. *Chemistry and Industry*, 1990, (3), 63-64

6. Criteria for Managing Contaminated Sites in British Columbia. Ministry of the Environment, Victoria BC, 1989

7. Ibbotson, B. G. and Powers, B. P., AERIS - An expert system to aid the establishment of clean-up guidelines. In *Proc 4th Conf. Petroleum Contaminated Soils*, Amherst, Mass. 1989

8. Smith, M. A., Identification, investigation and assessment of contaminated land. In *Proc. Symposium on the Redevelopment of Contaminated Land*, Birmingham 1990. Institution of Water and Environmental Management, London 1990.

9. Department of the Environment. The Government Response to the First Report of the House of Commons Select Committee on the Environment: Contaminated land. HMSO, London 1990

10. Denner, J., Contaminated land: Policy development in the UK. In *Proc 2nd Conf. Contaminated Land - Policy, Regulation and Technology*, London 1991, IBC Technical Services, London 1991

11. Smith, M. A., International study of technologies for cleaning up contaminated land and groundwater. In *Proc. LAND REC 88*, Durham 1988. Durham County Council, Durham 1988, pp. 259-266.

12. Smith, M. A., Options and criteria for the remediation of contaminated sites. In *Proc 1st Conf. Contaminated Land - Policy, Regulation and Technology*, London 1990, IBC Technical Services, London 1990

13. Smith, M. A., The social etc impacts of polluted sites: public involvement. In *Proc. LAND REC 88*, Durham 1988. Durham County Council, Durham 1988, pp. 25-34

14. Smith, M. A., An international study on the social impacts etc of contaminated land. In *Contaminated Soil*, ed. K. Wolf, W. J. van den Brink and F. J. Colon, Kluwer, Dordrecht, 1988, pp. 1653-1572.

RECLAMATION OF THE FORMER LAPORTE CHEMICAL WORKS
ILFORD, ESSEX

MARTIN SURY CENG MICE AND ALAN SLINGSBY
Johnson Poole and Bloomer
53 Moor Street, Brierley Hill, West Midlands, DY5 3SP

ABSTRACT

This paper describes the design and execution of the decontamination works required to prepare this 12 hectare site for a new residential development. Considerable physical and chemical restraints associated with the sites former use as a chemical works were overcome. Local authority planning conditions prevented movement of materials via local roads and so a temporary rail sidings was constructed. A cost effective reclamation strategy was employed by retaining low level contamination on site and covering with a layer of clean imported material. The lack of capacity of existing foul sewers on the boundary of the site required a new outfall sewer to be designed to negotiate a tidal river and a main road. A small bore tunnel was chosen to overcome this problem.

INTRODUCTION

The former chemical works was purchased in 1988 by Trust Estates, a subsidiary of London & Edinburgh Trust. Soon after, Johnson Poole and Bloomer were asked to advise on the reclamation of the former chemical works site for residential redevelopment. The brief was to prepare a scheme which would provide a site suitable for new homes economically and at the earliest possible time. Advice was sought from Peter J Rice Associates, environmental chemists. In addition, due to the site's association with radiochemical processes, Nuclear Services Group were appointed to undertake identification of areas affected by radioactive materials. The client was represented by Robert Fish John Savage, Project Managers.

SITE DESCRIPTION AND HISTORY

The 12 hectare Laporte chemical works was built close to Ilford town centre and bounded by the tidal River Roding, a main line railway, and a pre-war residential area. Chemicals had been manufactured on the site since the end of the last century. A range of processes were used in the production of materials from citric acid to radioactive thorium. These operations had left a legacy of contamination in the ground that created a considerable hazard to be overcome.

Ground conditions in the main consisted of varying levels of made ground overlying terrace gravels which in turn overlay London Clay. In addition alluvial deposits were found in the river bank area.

A length of river wall had failed and chemical contamination levels were hazardous in places. A large tip of organic waste existed at the northern end of the site up to 6m deep. This was composed largely of cinchona bark residue, the waste generated from the quinine extraction process. Other tipped material included calcium sulphate, a fine white powder, particularly aggressive to concrete. Also, organic solvents producing noxious odours and a whole range of other chemical substances.

The plant had been closed in 1980 after considerable public and political pressure had been brought to bear. There had been much public concern for many years concerning both the smell emissions and the safety of the plant. Two explosions had already taken place on the site and it was common knowledge that cyclohexane, the chemical that caused the Flixborough disaster, had been stored there. After the closure of the plant, the site was the subject of various development initiatives, all of which had foundered. The presence of radioactive deposits on the site made an emotive issue and the local residents were naturally keen to see them removed in a safe and controlled manner.

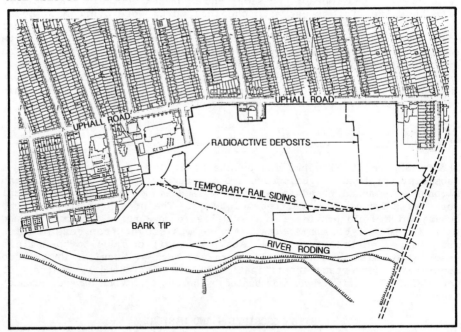

Figure 1. Site layout.

FEASIBILITY STUDY AND DESIGN

A comprehensive site investigation was carried out which systematically examined both the physical and chemical conditions over the whole site. This was carried out in accordance with the ICRCL 59/83 guidelines (1). Major restraints consisted of:-

(a) Made ground varying from reworked terrace gravels to soft silts and organic materials. Underground tanks, foundations, pipework and other obstructions were also present.

(b) Chemical contamination generally consisted of toxic organic chemicals, both heavy and phytotoxic metals, and sulphates.

In addition an obnoxious odour was given off from the bark tip when the material was disturbed. Some of the organic vapours were found to be toxic.

(c) Deposits of the radioactive compounds radium and thorium.

A reclamation philosophy was formulated which involved:

(a) the removal of radioactive materials
(b) minimum disposal of contaminated materials off site and
(c) the import of a layer of clean granular material to cover retained contaminated materials.

The bark tip presented a particular challenge. By using piled foundations it would have been possible to support new buildings and infrastructure over this material. However, the risks associated with gases and odours from the material were considered unacceptable and it was decided to remove the material from the site.

On the main body of the site not affected by the bark tip, most of the chemical contamination was limited to 0.7m below ground level. It was therefore decided to remove this material and cover underlying low level contamination with a layer of clean imported fill. After separation of the seriously contaminated soils, the more granular excavated material was considered suitable for controlled filling and compaction in the bark tip area, once the organic waste had been removed.

The comprehensive chemical analysis had identified areas of the site where conditions were more hazardous than others. Less contaminated areas warranted only 200mm thickness of clean cover. However up to 1 metre of cover with a capillary break were required in other areas, particularly where gardens were planned. This site specific philosophy allowed cost effective reclamation of the site.

Design of the outfall sewer presented a particular challenge. The route required negotiation of both the tidal River Roding and the North Circular Road, an overall distance of some 180m. The feasibility of various methods was assessed. A pumped system would have the lowest construction costs, but when maintenance costs and a commuted sum payable to the Water Authority were taken into account, it was found to be no cheaper than other methods.

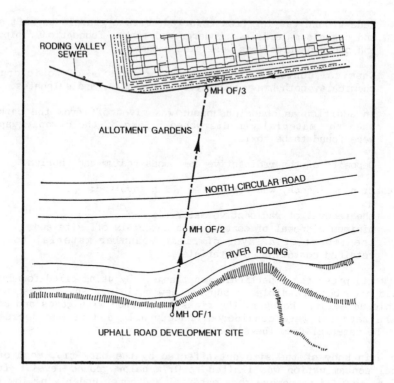

Figure 2. Route of outfall sewer.

Difficult ground conditions required careful consideration of available tunnelling methods. Up to 4m head of groundwater and variable soils meant that compressed air and grouting would have been necessary in a man-entry size tunnel. The cost of this option would have been considerable and the working conditions would have been hazardous. In addition, much of the tunnel capacity would have been wasted since only a small diameter pipe was required. Considerable advances recently in small bore tunnelling technology provided a viable alternative. Of the two main types of system available, a slurry machine was preferred to an auger mole because of its ability to support a hydrostatic head at the tunnel face. Following discussions with specialist contractors, an optimum diameter of 500mm was chosen, based on the mole's ability to deal with hard material up to 30% of its size. Hence any cobbles up to 150mm could be overcome.

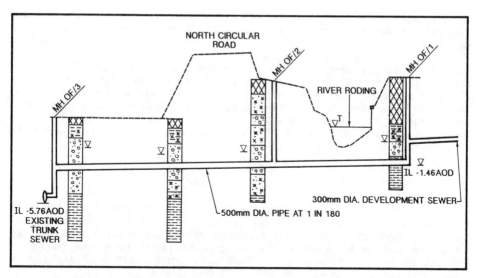

Figure 3. Section of outfall sewer.

The planning and design of the reclamation works had to overcome:-

(a)   the need to avoid known "hot-spots" of radioactive deposits until the material had been removed from within containment structures by Nuclear Services Group.

(b)   the requirement to monitor all excavations for further unknown deposits of radioactive materials.

(c)   the restriction applying to transportation of materials on and off site. Due to local planning restrictions the local highways could not be used for this purpose.

The feasibility of transporting materials by barge on the River Roding was investigated. However, the duration of high tide was insufficient to enable the works to be carried out in a reasonable time scale, and the option was discounted.

An adjoining mainline railway provided the opportunity for transport of materials on and off the site using trains. British Rail were consulted, along with disposal sites and quarry operators possessing railheads. British Rail were able to provide two trains per day into and out of the site to take contaminated waste to Shanks and McEwans site in Bedfordshire. In addition, quarried stone fill could be imported to the site on one train every day. In order to satisfy British Rail's timetable requirements, an assessment of on site materials handling was carried out. For maximum economy, the longest length of sidings that could be laid into the site was considered. A 250m track length could accommodate 12 rail waggons each carrying 3 containers of 20t gross capacity. The maximum tonnage that could be removed from the site on this basis in one train load was 720t.

By considering the production rates of excavators and dump trucks, it was concluded that it was feasible to plan for 2 export and 1 import trains per day.

Contract document were prepared on the above basis on a 5th Edition ICE Conditions form, and with remeasurement. A significant allowance was made for contingencies. Tenders were invited, and A F Budge were appointed as reclamation contractor with Henry Boot Railway Engineering as nominated subcontractor for the railsiding.
Miller Markham were appointed for the mini-tunnel with Barhale Construction as service contractor.

RECLAMATION WORKS

In order to facilitate an early start to the works, the client arranged for the supply of specialist plant for the project before the contractor had been appointed. The containerisation of the waste meant that haulage vehicles had to be adapted to carry containers from the point of excavation to the railsiding. A fork lift truck capable of lifting 20 tonnes was provided as well as 180 ½ ISO containers.

A chemists laboratory was established on site for the monitoring of samples of soil, water and atmosphere. Air monitoring stations were established around the site perimeter to monitor for the emission of radioactive dust particles and toxic vapours and gases. A Permit to Work system (PTW) was established to ensure the safety of site personnel. This was a proforma detailing the work and the permitted work location for each machine. This was circulated in advance to NSG's health physics supervisor ;and to the site chemist (PJRA). The PTW's were issued by the Resident Engineer from Johnson Poole and Bloomer for each machine on each and every day of the contract.

The works were divided into five specific aspects of the scheme. These were:-

(a) Identification and clearance of radioactive waste
(b) Identification and clearance of all hazardous chemical contamination in the near surface fills.
(c) Removal of old foundations and general earthworks
(d) Construction of rail siding
(e) Waste disposal and transportation

Each of these were let as separate contracts as follows:-

(a) Clearance of radioactive waste - Nuclear Services Group Ltd (NSG)
(b) Earthworks and clearance of chemical contamination - A F Budge (Contractors) Ltd (AFB)
(c) Waste disposal and transportation - Shanks and McEwan (SMc)

The rail siding installation was to be carried out as a nominated subcontract under the general earthworks contract and the transportation element of the waste disposal contract was sub-let to British Railfreight.

Railsidings

Works started in October 1988. The first operation on the critical path was the establishment of the railsidings since reclamation relied on the disposal of contaminated waste and import of clean material.

The railsiding was designed as a twin track to allow a locomotive to pull in one train, to shunt onto the other track and push out the loaded train on the other track. In order to allow sufficient room for the fork lift truck to manoeuvre, a 15 metre wide level strip had to be established on either side. Preparation for track laying included removal of all oversite concrete and buried foundations, on site crushing and screening of the broken concrete for re-use as hard core and railway ballast. Also the excavation of all remaining contaminated fills over the area of the track together with a 15m wide strip on each side of the siding.

The operation was hampered by the discovery of deposits of unknown radioactivity material on the western side of the siding. This discovery prevented trains being loaded by forklift truck from this side. It was concluded that the only way to maintain an output of two trains per day was to load both trains using a crane. This would need to have a large enough capacity to lift 20 tonnes at a radius of some 8-10m enabling it to operate from the eastern side of the track only. All excavated materials were checked for radioactivity, using hand held monitors, both at the point of excavation and the point of deposition.

Figure 4. Loading of rail wagons.

Bark tip

Excavation of the bark tip commenced with the start of rail movements in early February 1989 and was substantially completed around the end of September that year.

During this period approximately 83,000 cubic metres of waste was removed from the tip, loaded into ½ ISO containers and taken by rail to Shanks and McEwan's waste disposal facility at Stewartby, Bedfordshire.

The nature of the waste was such that monitoring the material at the face or in the machine bucket for deposits of radioactivity was out of the question. In many places the material was not competent to support the weight of a man and the fumes within the excavation area were often unbearable. Excavator drivers were provided with emergency oxygen cylinders and all personnel entering or working in the vicinity were required to wear half face masks equipped with organic filters. A gantry was built onto which sensors were fixed and haulage vehicles carrying full ½ ISO containers were driven underneath. The vehicle was required to stand under this gantry for 2-3 minutes whereupon the driver would be given a green or red light to proceed to the rail siding or not depending on the level of radioactivity detected.

Figure 5. Radiological monitoring gantry.

Permission was granted by Thames Water Authority to discharge leachate into the public sewers under very stringent conditions. Discharge was to be permitted only between the hours of midnight and 5 am. The rate of discharge was strictly limited and the permitted concentration of certain chemicals was to be carefully regulated. Holding tanks were installed to store the leachate and the liquid was pumped from the bark tip using two 150mm Univac pumps operating in tandem. Samples were taken for analysis which revealed that a dilution rate of 20:1 would be required before discharge could take place. The leachate was discharged through an oil interceptor into the existing foul water sewer off the site, the flow being carefully regulated by a system of gate valves. Clean water was discharged into the existing foul water sewer simultaneously, at a rate measured to provide the required level of dilution.

Following the placing of a layer of 500mm of broken concrete at the base of the tip, fill materials excavated from the main works area where placed on top of the crushed concrete. This material was placed in layers and compacted.

A radiological clearance survey to a depth of 2m was carried out after refilling and when cleared, clean cover material was laid over the area. The filling operation was monitored with compaction and sand replacement tests. Plate load tests were carried out on the final surface.

OUTFALL SEWER WORKS

The tunnel was driven in two parts from a drive pit located on the west bank of the river. The pit measuring 5m x 3m was sunk to a depth of 9m by first driving interlocking steel sheet piles to form a cofferdam then excavating inside using a 22RB crane with a grab attachment.

The tunnelling machine, designed by the Okumura Company of Japan, comprised a remote controlled crushing head and shield and was lowered into position on a special cradle at the base of the pit. A hole was cut through the sheet piles at the tunnel entrance after cement/bentonite grout had been injected into the ground behind to stabilise the soils and prevent ingress of groundwater. The alignment of the machine, the speed of the cutting head and the thrust of the rams were all computer controlled from a control station at the surface. Groundwater pressure was neutralised at the cutting head by pumping a bentonite slurry through the tunnel machine at a pressure slightly above groundwater pressure and the excavated material, after being crushed in the crushing head, was mixed with the bentonite slurry and returned through the machine to two large settling tanks at ground level.

Special temporary steel pipes with flush bolted joints were jacked into the face behind the cutting unit. This system had the benefit of greater accuracy and control of line and level because the steel pipes were made to better tolerances. In addition, the system allowed the cutting unit to be withdrawn if an obstacle could not be negotiated. After retrieval of the machine at the reception pit the permanent concrete pipes were jacked into the tunnel from the drive pit forcing the special temporary steel pipes out at the reception end.

Both the reception pits were constructed using 3m dia precast concrete caisson rings. The upstream pit was sunk using bolted segmental rings, while the downstream pit used complete 3m diameter one piece units. The caissons were sunk under gravity by excavating the material from inside using a Poclain 90 with a grab attachment. The final downstream link to the trunk sewer was achieved using a combination of ground treatment and pipejack techniques. At the request of Thames Water a 300mm pipe was laid inside the pipejack, since this was the maximum size of connection allowed.

The 16 week contract involved 2 tunnel drives of 110m and 66m. The 110m length was completed after only 8½ days setting a new U.K distance record. The 66m length taking only 3½ days to drive. The tunnelling machine's accuracy was comfortably within the maximum specified tolerance of + or - 20mm. The machine performed well through the Terrace gravels as expected but was hindered by the alluvial clay.

CONCLUSIONS

With advance knowledge of the development layout a cost effective strategy was possible. Non-building areas were not required to be reclaimed to such a high specification and cost savings resulted.

In the main it was possible to compact both excavated fills and imported cover material to achieve a bearing capacity of $80kN/m^2$. On this basis shallow raft construction was possible. However, where 3 or more storey buildings were planned, less suitable fill material was placed since the foundations were to be piled.

A Completion Report provided details of the types of materials used and also recommendations for redevelopment. Depending on the type of retained contamination and the end use, various thicknesses of clean cover were to be maintained. Although costly to establish, the railsiding had provided an alternative to road haulage of materials and avoided the associated inconvenience and nuisance to local residents. The outfall sewer was installed without any serious inconvenience to local residents and road users.

The housing development, which started before reclamation works had finished, is now well underway, and the project has successfully brought a tract of polluted and derelict land back into beneficial use.

REFERENCES

Inter-departmental Committee on the Redevelopment of Contaminated Land. ICRCL59/83 : Second Edition July 1987.

TOPSOIL FROM DREDGINGS:
A SOLUTION FOR LAND RECLAMATION IN THE COASTAL ZONE

B R THOMAS
Stanger Consultants Limited, Cardiff

M S DE SILVA
Applied Environmental Research Centre Limited
Colchester, Essex

ABSTRACT

Ever increasing land use pressures and demands for development sites has led to additional reclamation of marginal lands in the coastal zone. These sites are typified by poor ground conditions, they may be contaminated and have often remained derelict, requiring engineered land reclamation solutions to return the sites to beneficial use. Traditionally, this reclamation has required the stripping of agricultural land for topsoil, often involving considerable transportation costs and depletion of a valuable land resource.

This paper describes research completed at the University of Strathclyde which has established the technological and economic feasibility of rapidly producing (in 2-3 months) commercial quantities of a soil product 'Clydesoil' from River Clyde dredgings. The project has determined the physical, chemical and plant growth characteristics of the dredgings, and has established methods for handling, dewatering, desalination and amelioration prior to use in land reclamation schemes.

INTRODUCTION

Large quantities of maintenance dredgings are removed annually in the UK from harbours, estuaries, rivers and deep water channels, amounting to 38,870,000 tonnes of in-situ material in 1987 (1). This material, which comprises clay-silt-sand sized material is derived from soil erosion processes. Historically, this material has generally been considered a waste, and is mainly sea-dumped. It has been estimated that if all this material were suitable for reclamation it would represent a land gain of 500ha/pa in the UK.

Land reclamation from dredgings is minimal in the U.K. with only three ports, Manchester, London and Medway disposing of approximately 5% of dredged material to land. Reasons for the limited number of initiatives include inherently poor geotechnical properties of fine grained material, concern over possible in-situ contamination of the sediments, problems of economical management and jurisdictive issues.

Anticipated sea-level rises due to global warming may place at risk much of the low lying coastal areas of the UK (areas below +7m OD which may lie below existing sea defences). It has been suggested that these areas be reserved for low density use - agriculture, conservation, recreation. Additionally the demise of the UK shipbuilding industry has been highlighted by the conversion of former dockyards to public amenity use in the form of the Liverpool, Glasgow and Newcastle Garden Festivals. This demise has also resulted in the dereliction of support and complementary industries close to coastal and riverside sites. Much of this industry had been established in proximity to large urban conurbations, and many of these sites are now proving an eyesore. An increase in leisure time places ever increasing pressure on the provision of recreative space and amenity land for public use including waterside developments, barrages, marinas and dockside redevelopments. Such areas are prime targets for reclamation and restoration utilizing dredged materials although their proximity to sources of suitable restoration materials is often poor.

TOPSOIL FROM DREDGINGS - PROCESSES

Considerable advances in the understanding of the behaviour of dredged materials have been made in the last twenty years, particularly in the Netherlands and the USA. The Dutch experience relates to reclamation of low lying lands and the disposal and recycling of dredged materials (2,3). The US Army Corps of Engineers assessed the beneficial uses of dredged material and the environmental impacts of dredgings disposal (4) with industrial, commercial and amenity area land uses most prevalant. Experience of onshore dredged material in the UK is very limited and has generally concerned the consolidation of hydraulically placed material in disposal lagoons. In most cases and when undertaken, the recycling of dredgings to form a soil product has taken several years.

A three year study at the University of Strathclyde has demonstrated that material removed from an industrial harbour can be successfully recycled to rapidly form a landscaping soil (5). Four areas were identified as requiring substantial research effort including a resource assessment, suitability for use, determination of manufacturing processes and the economic feasibility of the recycling of dredgings. This paper concentrates on the dewatering, desalination, contamination and nutrient status aspects. The other processes and potential problems involved in rapidly converting dredged saline sediments into a soil product are described elsewhere (5, 6, 7).

Geotechnical Properties
The following types of River Clyde dredgings are referred to in this paper:

Fines - typically medium to coarse organic clay silts.
Coarse - fine to medium coarse sand with occasional gravel and silt
Mix   - 50% fines: 50% coarse by weight.

It should be noted that specifications relating to the end product - topsoil - now need to be considered rather than the more familiar engineering characteristics. The mix material complies with accepted standards for topsoil (8) which can be assessed using soil classification charts.

DEWATERING

The inherently poor geotechnical properties of fine dredgings, (various sources):

Placement moisture content	60 to > 220%
Void ratio	2-6
Bulk density	< 1.3 Mg/m^3
Permeability	10^{-5} to 10^{-9} m/s
Liquid limit	60-150
Plastic limit	20-70
Plasticity Index	15-100
Liquidity Index	0.2-6.0

coupled with lagoon disposal heights of up to 10 metres (e.g. Manchester deposit grounds) results in large volumes of material which has consolidated to approximately the liquid limit under a thin dessication crust (<1m). This behaviour was confirmed by a full scale trial (7) involving approximately 6,000m^3 of suction-dredged silt placed in a lagoon at about 1.2m height (see Figure 1). Consolidation settlement was essentially complete after 37 days. After 400 days moisture contents remained above the liquid limit below 0.20m depth, and further dewatering could not be achieved without rotivation and mixing with coarse material. In comparison, 0.40m of grab dredged fines placed upon 0.40m of sand on a Clyde quayside dewatered from moisture contents of 136% to 50% in 50 days.

Previous studies (2, 4, 9) examined physical, chemical and biological techniques for dewatering contained fine grained dredgings. Given the requirement to rapidly dewater dredgings to form soil, the most favourable dewatering methods seemed to be the use of under-drained thin layers with evaporative drying, surface trenching and rotivation, although large areas will be required for dewatering. The further research

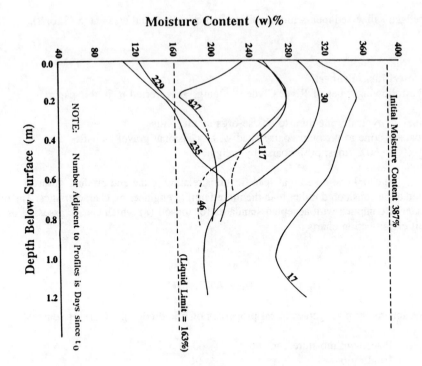

Figure 1. Moisture content depth profiles vs. time for full scale trial.

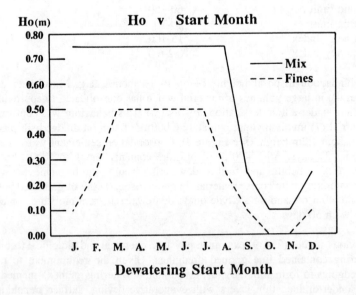

Figure 2. Predicted best monthly dewatering options for fines and mix dredgings.

therefore concentrated on field dewatering trials using thin layers and laboratory tests. This represents a reversal of previous research trends.

The dewatering behaviour of fine, coarse and mix dredgings was examined under constant and variable evaporative potential to determine dewatering characteristics and identify the optimum dewatering conditions for drying throughout a calender year. This involved examination of material behaviour in scaled field plots and full scale trials. The plots indicated greatly improved dewatering behaviour for the mix compared to the fines. Evaporative dewatering was not effective for plot heights of over 0.50m for the fines, but up to 0.75m height could be used for the mix. The provision for sand under-drainage aided dewatering for both material types.

A constant rate evaporation model developed by Benson and Sill (10) was successfully calibrated against 86% of drying tests indicating that the model gives a good description of the evaporative drying process in the laboratory.

The need to predict field dewatering behaviour led to the development of a variable evaporation dewatering model for use under field conditions (7). The model was verified mathematically and calibrated against 80% of field plots using derived potential evaporation for the Glasgow area. The next step was to use the dewatering model as a predictive tool to determine optimum dewatering conditions i.e. production of the maximum material mass in the minimum time to a moisture content and textural condition which enables handling by standard earthmoving plant throughout the calendar year. Prediction runs were undertaken using the following test permutations:

- Initial dredgings height = 0.10, 0.25, 0.50, 0.75m
- Initial moisture content, Fines = 136%-252%; Mix = 44%-193%
- rotivation/no rotivation (where applicable)
- dewatering starting at monthly intervals throughout the calender year

The best monthly dewatering options thus obtained are summarised in Figure 2.

Consideration of all test results indicated that the mix material is a more efficient dewatering medium that the fines under the same evaporation potential.

DESALINATION

The desalination of soils derived from estuarine materials, which although different in physical characteristics from 'natural' soils poses problems which are similar to those experienced in the leaching of salts from saline and alkaline soils. These include problems of soil-structure and plant-salinity relationships.

The desalination behaviour of River Clyde dredgings was assessed using laboratory column tests and large scale field trials (5, 6, 7). Initial observations indicated a major conflict of interests in that dewatering was essential for successful handling, whereas wetting with fresh water was required for desalination. The tests showed that salt removal in the wet state is not possible due to the very low to practically impermeable nature of the dredgings coupled with net upward evaporative flux and salt movement during drying (secondary salination). Rapid desalination could however be achieved using initially dry dredgings rotivated to produce a loose and open soil structure i.e. dewatering-desalination is a two stage process.

Classic wetting/salt fronts observed in the desalination columns allowed mapping of salt movement through the dredged material. Desalination efficiency decreased as initial soil moisture increased or if movement of water through the dredged material was too rapid. The addition of sand to fines improved hydrodynamic dispersion. Most of the test column salinity was removed by adding the equivalent of no more than 330mm of rainfall. In the summer months (May - June), irrigation is necessary to achieve desalination, while in winter adequate drying and drainage must be provided. Time of rotivation is critical to produce an open structure suitable for leaching and care has to be taken to ensure that redistribution of salt from lower layers does not occur.

Removal of residual salinity takes several months but good mixing of material ensures that no salinity 'high spots' are present, which may prove detrimental to plant growth. Natural rainfall can, with time, reduce this 'residual' salinity.

CONTAMINATION

UK guidance on metal levels in the environment is largely taken from DoE ICRCL document 59/83 (11). These guidelines must be interpreted with care, and informed professional judgement is a pre-requisite of decision making processes resulting from their use. The application of guidelines to dredged materials is particularly important in view of the poorly developed soil structure, and potentially high available metal content.

Careful evaluation of dredged materials to determine contaminant concentration is necessary to assess their suitablility for use on land. Numerous studies, particularly in Europe and the USA, have examined the propensity for pollutant uptake from dredged materials disposed of to land (12, 13, 14, 15); however, relatively little work has been conducted in the UK. Furthermore, many of the studies have been conducted on passive disposal areas containing sediments known to have high levels of contaminants.

The economic use of dredged materials must be based on evaluation of the sediment system prior to dredging. This requires an extensive survey of the river, estuarine or coastal system prior to dredging operations. Selective disposal to land for restoration purposes can then be made. This in-situ assessment was made for River Clyde sediments prior to landing of material ashore (6, 16).

Although the studies listed earlier have indicated intolerable contaminant uptake into crops grown on dredged material, there are many examples in the literature of restoration and reclamation schemes with soft after-uses including parks, sports fields and other areas for recreational pursuits. River Clyde dredged materials were evaluated primarily for this purpose with assessment for total and extractable metals and contaminant uptake in vegetation. Field and laboratory experiments utilizing a homogenised mix of Clyde dredgings with organic waste materials including sewage sludge, mushroom compost and tree bark were used.

All soil mixes were sown with a seeding mixture recommended for restoration of derelict land, adapted to give a hard wearing, closely knit sward, with fast establishing low maintenance *L.perenne*. In comparison with a locally derived topsoil control, variation was seen in total metal contents. Dredged material zinc, copper, nickel, lead and manganese concentrations were all found to be significantly lower. Although chemical extractant techniques developed largely for agricultural soil use also indicated lower plant available concentrations of these metals, uptake into vegetation was found to be greater for zinc, nickel and manganese. These metals have important implications on phytotoxicity, however, no visual signs were apparent and analysis showed no statistical difference in biomass over three vegetation harvests in the field. Comparisons between unamended and amended dredged materials exhibit a decrease in soil total concentrations of zinc, nickel and copper, which are all primarily phytotoxic. Significant reductions are also noted for soil total concentrations of cadmium and mercury (Table 1).

Careful selection of material in relation to proposed afteruse can minimise metal uptake into vegetation to levels which are not detrimental to growth. However, a soil must provide other functions such as enabling biological and chemical processes and it is important to evaluate dredged materials in their ability to perform these functions.

NUTRIENTS

Preliminary studies by Roberts and Roberts (17) on alternative cover materials for landfill sites evaluated the nutrient levels of River Thames dredged materials. This work indicated that there is a largely soluble nitrogen pool in dredged material which is either taken up rapidly by plant roots or lost from the system by leaching. This resulted in rapid establishment of vegetation in the first season followed by inability to sustain this acquired biomass.

Although nitrogen was the first nutrient to be specifically recognised as necessary for plant growth and largely accepted as the most important and generally most deficient soil factor limiting growth, the importance of the other macronutrients must not be underestimated. Recent research on pasture grasses and legumes deprived of single mineral nutrients demonstrated that nitrogen is the most limiting nutrient followed by potassium, magnesium and phosphorus (18). The long term success of any restoration project is largely determined by the ability of the restored site to adequately cycle nutrients. For this reason extensive nutrient application experiments were conducted on dredged materials from the River Clyde. Experiments were conducted on

TABLE 1
Summary of comparative behaviour of dredged material for various treatments.

		R.Clyde Dredged Material	D M + S S	D M + S D	D M + PFA	D M + BARK	D M + M C
TOPSOIL higher in	soil Pb	***	***	***	***	***	***
	soil Zn	***	***	***	***	***	***
	soil Fe			***	***	***	***
	soil Cd		***	***	***	***	
	soil Cu		***	***	***	***	***
	soil Ni	***	***	***	***	***	***
	soil Hg		***			***	
	soil Cr					***	
	veg Cu						**
TOPSOIL	soil Cu	***					
	soil Cd	***					***
	soil Cr		***			***	
	soil Hg	***					
	veg Zn	***	***	***	***	***	
	veg Mn	***	***	***	**	***	***
	veg Pb				**		
	veg Cd		**			***	

Left hand column indicates status of the topsoil control in relation to each of the other treatments. All ameliorants added at 20% V/V

DM -	dredged material	MC -	mushroom compost
SS -	sewage sludge	SD -	sawdust
**	Significant at p < 0.01	***	Significant at p < 0.001

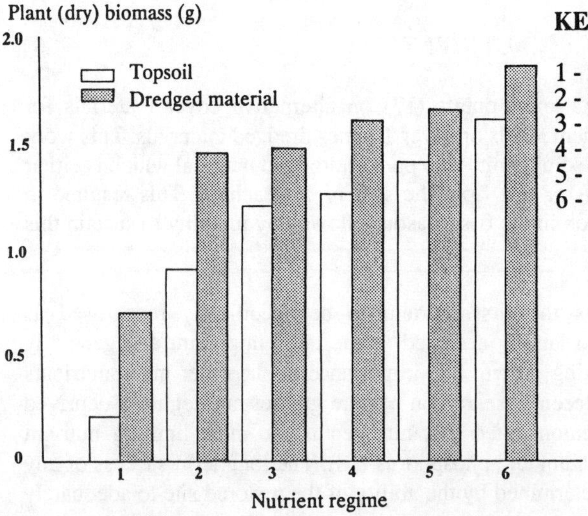

KEY TO NUTRIENT REGIMES

1 - no nutrient
2 - 50:12.5:25   NPK
3 - 100:25:50   NPK
4 - 100:25:50   NPK + 1000 lime
5 - 200:50:100   NPT (kg/ha)
6 - 400:100:200   NPT

Figure 3. Response to nutrient addition in dredged material.

L.perenne and *T.repens* to assess the nutrient requirements of River Clyde dredgings in order to support plant growth. These two species were used because of their ecological importance and their ability to co-exist due to different uptake strategies (19). Dredged material from the River Clyde was found to compare favourably wth the topsoil control in macronutrient content. Dredged material generally had higher total and extractable nitrogen potassium and phosphorous. Additions of 20% organic waste generally improved the nutrient content of the material particularly total and extractable nitrogen. Increases in biomass were noticeable, particularly in the sewage sludge amended dredged material.

Nutrient addition experiments conducted on River Clyde sediments included factorial increases in NPK fertilizer and lime. Further experiments assessed the effect of individual nutrients on growth. Additions of 100:25:50 Kg/ha NPK showed increases in plant biomass. Further nutrient increases of up to four times this applicaton did not result in increased biomass in newly dredged sediments (Figure 3). However, in older sediments nutrients are found to be limiting. There is also a notable response to lime addition, with treatments receiving lower NPK and lime producing greater biomass than in higher NPK treatments. This is of particular importance in the older material, indicating the development of pH limiting conditions due to oxidation processes converting sulphides to sulphuric acid. The addition of all individual nutrients was also found to be statistically significant, although nitrogen and lime are found to have the greatest effect on biomass for *L.perenne*. Phosphorus was found to be the most limiting factor for *T.repens*.

These experiments have indicated the importance of nutrient and lime additions to ageing dredged material as part of their successful management. Long term success requires the development of soil processes in order to recycle nutrients within the system. Preliminary investigation of desalinated dredged material has indicated the ability of the soil to maintain uninhibited nitrogen mineralization. This is clearly an important step in developing a self-sustaining nitrogen economy.

CONCLUSIONS

The potential of dredged soil for land reclamation in the coastal zone is considerable given the current net increases in land dereliction in the UK. Natural topsoil is expensive, takes many years to develop and is generally in short supply in areas where demand is greatest - urban renewal, docklands development areas and landfills etc. The use of dredged material for reclamation brings benefits to all involved. Dredging disposal costs are reduced and sale of the product generates a revenue. There are several technical and logistical difficulties to overcome, many of which are site specific requiring the correct technical input and management procedures before, during and after reclamation.

The use of 30,000 tonnes of Clydesoil in the Glasgow area and the establishment of a 100,000 p.a. soil factory on the River Clyde proves the technical and economic feasibility of the process. The success of the research has also been recognised by the project winning a commendation in the 1987 Better Environment Awards for Industry.

Further research into the economic conversion of dredged material to a soil medium is required if the technology is to be used on a large scale. Increasing pressures on land use, the presence of derelict land in the coastal zone, the threat to our coastline from tidal action and predicted sea level change, combined with legislation to ban the dumping of dredged materials at sea suggests that the opportunity and incentive for further research into dredged material has never been better.

ACKNOWLEDGEMENTS

The Authors gratefully acknowledge the support of the Scottish Development Agency and the Clyde Port Authority which enabled the research to be undertaken. The advice and guidance of Professor G. Fleming, Dr J. Riddell and Dr P. Smith is acknowledged.

REFERENCES

1. Ministry of Agriculture,Fisheries and Food.Report on the disposal of waste at sea, 1986 and 1987.HMSO, London,1989.

2. de Glopper, R.J. and Smits, H., Reclamation of land from the lakes in the Netherlands. Outlook on Agriculture ,1974, **8** No.3, 148-155.

3. Van Driel, W., Use and disposal of contaminated dredged material. Land and Water International ,1984, **53**, 13-18.

4. U.S.Army Engineer Manual (1986). Beneficial uses of dredged material. Report EM 1110-2-5026. Department of the Army, Corps of Engineers, Office of the Chief Engineers, U.S. Army, 1986.

5. Fleming, G., Smith, P., Riddell, J.F., de Silva, M. and Thomas, B., Feasibility study on the use of dredged material from the Clyde estuary for land renewal - Final Report. Dept. of Civil Engineering, University of Strathclyde, Glasgow, 1988.

6. de Silva, M.S., Chemical and biological aspects of using dredged River Clyde sediments for land restoration. PhD Thesis, Dept. of Civil Engineering, University of Strathclyde, Glasgow, 1990.

7. Thomas, B.R., Clyde sediments: physical conditioning in relation to use as a topsoil product for land reclamation. PhD Thesis, Dept. of Civil Engineering, University of Strathclyde, Glasgow, 1990.

8. British Standard 3882: Topsoil. British Standards Institution, London, 1965.

9. Haliburton, T.A., Best ways to dewater dredged material. Civil Engineering, ASCE Proceedings, 1979, 57-61.

10. Benson, R.E.and Sill, B.L., Modelling the drying of dredged material, at "Dewatering and dredged material disposal", Proc. ASCE Conference, Florida, 1984.

11. Guidance on the assesment and redevelopment of contaminated land. Interdepartmental Committee on the development of contaminated land. ICRCL 59/83, (2nd edition), 1987.

12. Groenwegen, H.J. and Nijssen, J.P.J., Redevelopment of former dredged material disposal sites. Heavy metals in the Environment, International Conference, Athens, 1985.

13. Smilde, K.W., van Driel, W. and van Luit, B., Constraints on cropping heavy metal contaminated fluvial sediments. Sci. Total. Environ., 1982, **25**, 225-44.

14. van Luik, A., Mined land reclamation using polluted urban navigate waterway sediments. i. Trace metals, ii.   organics. J.Environ.Qual., **13**, 3, 410-21.

15. Fanning, M.C.B., Growth of and uptake of Zn,Cu,Mn,Cd,Ni and Pb by radishes (*Raphanus sativus L.*) on two Baltimore harbour dredged materials in the greenhouse as affected by pH,composted sewage sludge and fertiliser. M.S.thesis,University of Maryland,USA,1983.

16. de Silva, M.S., Smith, P.and Fleming, G., A study of the distribution of pollutants in sediments from the River Clyde. 2nd International Conference on Environmental Contamination, Amsterdam,September 1986.

17. Roberts, D.R.and Roberts, J.M., Selection of cover materials for reclamation of landfill sites. Proceedings, Reclamation '83.

18. Fenner, M. and Lee, W.G., Growth of seedlings of pasture grasses and legumes deprived of single mineral nutrients. J. Appl. Ecol.,**26**, (1),pp223-232.

19. de Wit, C.T., Tow, G.P. and Ennik, G.C., Competition between legumes and grasses. Agric. Res. Rep.,1966, Waginegen, 697,pp1-30.

SLURRY TRENCH CUT-OFFS FOR GAS AND LEACHATE CONTROL

Dr. Stephan Jefferis
European Centre for Pollution Research
Queen Mary & Westfield College, University of London
Mile End Road, London E1 4NS, UK

ABSTRACT

Slurry trench cut-off walls are now widely used for the control of gas and leachate from landfill sites. The slurry trench process is a very flexible method for forming cut-off walls. The paper discusses the suitability of the various types of cut-off wall for gas and leachate control and gives an overview of some of the parameters which ought to be considered when designing containment systems.

1. THE SLURRY TRENCH PROCESS

If a trench is excavated in the ground to a depth of more than about 2 metres it is likely to collapse. However, if filled with an appropriate fluid the trench can be kept open and excavated to almost any depth without collapse. For this purpose the fluid must have two fundamental characteristics: it must exert sufficient hydrostatic pressure to maintain trench stability and it must not drain away into the ground to an unacceptable extent. These requirements can be met with a bentonite clay slurry, a bentonite-cement slurry or a polymer slurry.

2. TYPES OF SLURRY TRENCH CUT-OFF

There are now a significant number of different types of slurry trench cut-off and as demands for pollution control increase further developments must be expected. Currently the principal types of cut-off wall are:

2.1 Soil-active clay cut-offs

These are formed by excavation under an active clay slurry (almost invariably bentonite) and backfilling the trench with a blend of soil and the active clay. The soil needs to contain of order 20 to 40% of material finer than 75 microns (0.075 mm) and yet be reasonably friable so that it can be mixed with the active clay to produce a uniform high slump material. The best economy can be achieved if the excavated material is used as the backfill. However, it is often necessary to add some extra fines. The bentonite may be the used slurry from the excavation but this will lead to a rather low bentonite content in the blended material. It is usual to add some bentonite in powder form as well.

Design procedures for soil-bentonite backfills are given in D'Appolonia, 1980.

The backfill of soil plus active clay must be at a moisture content such that it will self-compact when placed in the trench. The combination of a relatively high moisture content and a swelling clay makes the system potentially sensitive to chemical attack. Any change in the chemical environment may lead to change in the interactions between clay particles and shrinkage of the clay structure with increase in permeability. For good chemical resistance dense non-swelling clays systems are required but these cannot be installed with the conventional slurry trench process (see Section 2.5). Soil-bentonite cut-off walls have been used for groundwater control at a number of gravel workings. They have not found general application for pollution control in the UK.

2.2 Clay-cement cut-offs
Bentonite cement cut-offs are effectively the only type of slurry trench regularly employed for groundwater and pollution control in the U.K. For bentonite-cement slurries the typical range of mix proportions is:

Bentonite	20 - 60 kg
Cement	100 - 350 kg
Water	1000 kg

To prepare the slurry the bentonite powder must be mixed with the water and allowed to hydrate (swell) for at least 4 hours and preferably 24 hours before the cement is added. If less than 20 kg of bentonite is used the bentonite-cement slurry is likely to show excessive bleed (separation of free water on standing). Above 60 kg the hydrated bentonite slurry will be very thick. Furthermore when cement is added to hydrated bentonite slurry there is always a rapid and substantial thickening. The effect lasts for only a few minutes and after this there is some thinning. However, at high bentonite concentrations the addition of cement may cause excessive thickening and an unworkable material.

Clay-cement slurries usually are used as self hardening slurries and simply left in the trench to set at the end of excavation. Details of materials and specifications for self hardening cut-off walls are given in Jefferis, 1990 and a discussion of durability is given in Jefferis, 1988.

If a semi-rigid wall is required the slurry may be displaced by a clay-cement-aggregate plastic concrete prepared by blending the displaced slurry with a suitably graded aggregate.

2.3 Clay-cement-aggregate cut-offs
The addition of aggregate will reduce the quantities of clay and cement required to form a wall. However, the economy achieved by the use of a reduced quantity of slurry may be more than offset by the costs of additional materials handling (especially if the excavated spoil cannot be used in the backfill). For clay-cement-aggregate mixes the bentonite and cement concentrations of the slurry phase generally will be towards the lower end of the ranges listed in the above Table. Despite this the mixes tend to be much stronger and stiffer than simple bentonite-cement systems and the strain at failure may be low

especially if the aggregate particles are in grain to grain contact (this may be difficult to avoid unless the slurry is thickened so that the aggregate particles can be held in suspension prior to set). For pollution control the reduced content of clay and cement and the potential for cracking at the slurry-aggregate interfaces tends to militate against their use except for systems using specially formulated aggregate gradings.

2.4 Cut-offs with membranes

The use of membranes may be appropriate where high levels of pollution exist or very aggressive chemicals are present as membranes can be obtained to resist a wide spectrum of chemicals. For gas migration control membranes may be essential. Most slurry based cut-off materials have a relatively high water content. If this water is lost they may become gas permeable and thus the potential for such materials for gas control must be regarded as still not proven.

The major problems with membranes relate to the sealing of the membrane to the base of the excavation and the joints between membrane panels. The base seal is often achieved by mounting the membrane on a frame and hammering the whole assembly a pre-determined distance into the base layer. This operation has to be undertaken with some care to avoid damaging the membrane and the inter-panel joints.

A number of different systems are available to form the inter-panel joints but all require considerable care in use. Typically a membrane panel may be over 10 metres high by 5 metres wide and will be mounted on a heavy metal frame during installation. The panel joints therefore need to be robust and simple to put together. Details of jointing systems are given in Krause, 1989.

Typically a bentonite-cement trench will be of order 600 m wide. The insertion of a membrane will substantially disturb the behaviour of the slurry in the trench and the membrane may act as a sliding plane and encourage cracking. Therefore if a membrane is used it should be treated as the sole impermeable element and the cement-bentonite regarded as providing only support and mechanical protection.

2.5 High density walls

These are relatively new cut-off materials developed specifically for polluted ground. The material is based on a graded soil, calcium bentonite, sodium silicate and a number of proprietary admixtures. The hardened material is hydrophobic and the water permeability may be very substantially lower than with conventional soil-bentonite or cement-bentonite walls. Indeed the permeability may be so low that diffusion becomes the most important transport process. Chemical resistance to aqueous pollutants may be better than for conventional slurry walls. Material costs are higher and special excavation plant may be necessary because of the high density of the slurry (this may be over 1800 kg/m^3) which follows from the high solids content. Details of the system are given in Hass & Hitze, 1986.

2.6 Drainage walls

Pollution migration (liquid or gaseous) may be controlled by gravel filled drainage trenches. These trenches must extend to a sufficient depth to intercept all the pollutants. For liquid pollutants the trench must be permanently drained so that it is not bridged by polluted liquid. For gas, vent pipes must be installed at regular intervals as any exposed gravel surface is likely to be rapidly clogged by vegetation etc.

Deep drainage walls may be formed by excavating a trench under a degradable polymer slurry. On completion of excavation the trench is backfilled with the gravel drainage material and the slurry broken down with an oxidising agent or left to degrade naturally. In principle the drainage wall system could be useful for the formation of gas venting trenches. However, such trenches need not extend below the water table and unless this is exceptionally deep direct excavation will be cheaper than a slurry trench procedure.

2.7 Future Developments

It is clear that the current level of activity with barrier systems will ensure that there is a continuing flow of new developments. Many of these will be basically improvements to existing systems. However, it is to be hoped that some will represent lateral approaches to the problem. Two examples of such approaches are outlined below:

The Bio-barrier: In oil wells bio-barriers have been developed to clog undesirable permeable regions (Howsam, 1990). The bare bones of the procedure are as follows: a sample of the natural bacterial population is obtained and cultured and then starved. Certain bacterial species on starving may remain viable but reduce very substantially in size and develop an electrically neutral cell wall (which reduces attachability). These bacteria are termed ultra-micro bacteria. The formation to be blocked is permeated with a culture of the ultra-micro bacteria and a slow acting feed. The bacteria then develop and expand and so block the formation. At the present time the technique is limited to materials within a relatively narrow band of permeabilities. It would be most elegant if a system could be developed which selectively clogged a leachate or gas migration path.

Active barriers: The fundamental aim of barriers systems is to prevent all flow from the landfill, repository etc. This is an ideal requirement and strictly unachievable in a passive system. However, it is possible that active systems may be developed where the flow is dynamically controlled. In many areas of science if a no flow condition is required then a guard ring procedure is used. In electrical measuring circuits guarding is a well established. For heat experiments adiabatic conditions may be achieved using a guard dynamically driven to match the sample temperature. Jefferis, 1989 has developed the guard ring concept for the elimination of edge leakage errors when measuring the permeability of concrete. Thus the prevention of flow by use of an active guard driven to the sample potential (electrical, thermal, hydrostatic etc.) is well established. Unfortunately it is difficult to develop a guard ring for a landfill site as are many different potentials which can influence flow through a barrier (see Section 4) and there are also many different chemical species involved. Mitchell, 1991 has shown that the migration of cations may be retarded by applying a reverse potential for short periods during a permeation test. However, the procedure accelerates the migration of the anions and it may have little benefit for organics. There may be some scope for the development of electrical sandwich layers designed to retard (or recover) anions and cations separately.

3. THE REQUIREMENTS FOR A CUT-OFF

A cut-off should represent a sufficient barrier that the rate of escape or transit time through the barrier (the break-through time) of the pollutant is such that it will not adversely affect the surroundings. Thus in principle the barrier performance ought to be related to the local geology, hydrogeology, the nature of the contained materials and the land use adjacent to the site.

For the control of liquids there seems to be some international consensus that the barrier should have a permeability equivalent to 1 metre of 10^{-9} m/s material. Despite this for slurry trench cutoffs thinner than 1 metre (shallow cut-offs often will be only 0.6 m thick) the specified permeability is not usually reduced to compensate for the reduced thickness. Thus there is some flexibility in the figure. In fact it would seem that 10^{-9} m/s has been selected not because it represents the level at which any escape is insignificant but rather it is the level at which permeation ceases to be the most important vector for escape. As the permeability drops below 10^{-9} m/s diffusion becomes progressively more important though as will be shown below there are many other driving forces which may influence the escape of the contained species.

For the control of gases the Waste Management Paper 27 says "reworked natural clay or calcium bentonite linings are probably the most suitable commonly available materials for gas barriers. They should be laid and compacted to achieve a maximum water permeability of 10^{-9} m/s". Thus a gas barrier is to be designed to achieve a specified water permeability though it is noted that "the effectiveness of liners in preventing gas migration is not yet fully understood and their permeabilities to gas have yet to be determined. It would seem likely that they would be more permeable to gas by several orders of magnitude compared to leachate". With the current state of knowledge to design gas control measures on water permeability is unsatisfactory. For bentonite-cement materials the gas permeability will be very sensitive to the moisture content and even for a clay there may be no useful correlation between the air and water permeabilities.

If a high density polyethylene (HDPE) membrane is used the gas permeability might be of order 50 ml/m^2/day per atmosphere of differential pressure for methane or 200 ml/m^2/day/atm for carbon dioxide (the permeability is sensitive to the gas type). These are very low figures. It does not follow that simply because they can be achieved with an intact membrane that such low values are actually required. In practice because there will be joints both at the toe of the membrane and between sections the overall permeability will be higher than the above figures. Clearly when designing membrane systems careful attention must be given to the joints and mechanical protection of the membrane.

4. POTENTIALS TENDING TO CAUSE FLOW

The following potentials ought to be considered when assessing the magnitude and direction of flow through a barrier (Mitchell, 1991).

Hydraulic potential due to difference in water levels, the potential for flow may be inward or outward depending on the relative water levels on either side of the barrier.

Chemical potentials tending to cause diffusion. In general the diffusion potential will tend to be outwards as concentrations of chemical species will be higher inside the landfill than outside.

Osmosis, in general this will promote inward flow as concentrations within the landfill will be higher than outside.

Electrical potentials, these may be generated by the difference in chemical conditions between the landfill and the surrounding soil. For example the environment within a landfill may be significantly more reducing than outside. The effect of electrical potential will depend on the charge of any migrating species (eg cations may be accelerated and anions retarded or vice-versa). There also will be electro-osmotic effects tending to induce flow of the water itself

Temperature differentials may drive flow. Such differentials may result from the generation of heat by microbiological activity within the landfill,

Gas flow also may be influenced by changes in atmospheric pressure.

Thus there are many different potentials and as shown above they may act in different directions. Furthermore in a multi-barrier system each component of the barrier will have different response to the potentials. Clearly a great deal more work needs to be done on multi-barrier systems. However, it is apparent that fluid could accumulate between layers of a multi-barrier system. Equally, there could be a net removal. Thus for a waste underlain by a membrane - compacted clay - membrane sandwich, the clay layer might be desiccated by osmotic flow.

At the present time permeability seems to regarded as the controlling parameter for barrier design. clearly the other driving forces need to be considered and the chemical as well as the physical boundary conditions need to be established.

5. GAS PERMEABILITY

As noted above a water permeability of less than 10^{-9} m/s is suggested as possibly suitable for gas control. The gas permeability of cement-bentonite cut-off materials is complex. If the material is fully saturated there can be no gas flow and methane etc can move only by diffusion or advection with the water. However, if the material dries out it may shrink substantially and thus become highly permeable. In the partially saturated state if the water forms a continuous system throughout the pore space then the flow may be effectively blocked unless the pressure is sufficient to force the water from the pore space and establish a flow path. The necessary pressure may be large and thus the partially saturated material may still behave as an effective barrier. However, a small further loss of water (from the 'blocked' partially saturated state) may open up flow paths of quite significant permeability. For example a 1 metre thickness of good quality structural concrete when air dried might show a gas flow rate of 50,000 ml/m^2/day/atm (which may be compared with perhaps 50 ml for a membrane, see Section 3). Clearly the permitted gas flow rate through a membrane must be related to the venting arrangements etc. In most situations there is probably sufficient moisture in the ground to ensure that most fine grained materials such as clay or a cement-bentonite are of low permeability to gas. However, droughts do occur and these have on occasion lead to the shrinkage of clays to significant depths. The author therefore prefers to regard the in-situ performance of cement-bentonites for gas control as unproven. Unless or until it has been shown to be satisfactory a membrane always should be included when designing a gas control cut-off wall unless the full depth which may be subject to drying is adequately protected by vent trenches (but these will exacerbate drying and thus the problems become circular).

6. WATER PERMEABILITY

As noted above 10^{-9} m/s has come to be regarded as the necessary in-situ permeability for barrier materials. However, the permeability of cement-bentonites develops with time. Figure 1 shows a typical permeability time plot. It can be seen that the permeability drops by almost two orders of magnitude from the start of the test (sample age 50 days) to 400 days. There is a tendency to require the permeability of 10^{-9} m/s to be achieved at 28 days. Clearly this is unnecessary as within a relatively short time (in terms of the life of the barrier) it would seem that the permeability will have dropped by at least one order of magnitude. The drop in permeability seems to be a function of not only the age of the sample but also the time under permeation. Work is in progress to try and identify the mechanisms associated with the permeability reduction.

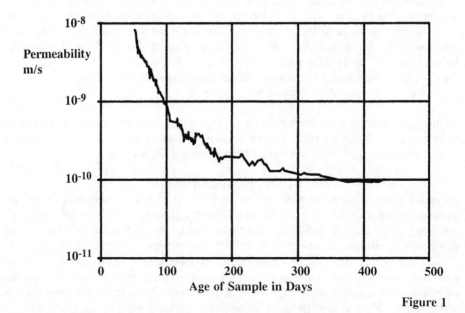

Figure 1

The reduction in permeability with time could be regarded as merely a bonus providing a further factor of safety against liquid escape. However, a specification of 10^{-9} m/s at 28 days may lead to an unnecessarily brittle material.

7. STRESS-STRAIN BEHAVIOUR

Most specifications require that the cement-bentonite should have a strain at failure of at least 5%. This may be compared with concrete which may fail at of order 0.2%. In general cemented materials are brittle or quasi brittle and thus to require 5% strain at failure of a cement based materials is to seek a rather unusual system. For cement-bentonites this strain can be achieved only under confined drained conditions with a confining stress comparable to the unconfined compressive strength of the material. Similarly plastic behaviour will be achieved in the trench only if the in-situ stresses are comparable with the unconfined compressive strength of the cut-off material. As stresses in a trench may be low (especially near the top) it follows that the material must be designed to be of low strength. The strength of a material with a 28 day permeability 10^{-8} m/s may be substantially less than that of a 10^{-9} m/s material. Thus an unnecessarily tight specification for early permeability may lead to a wall which although of lower intact permeability is more sensitive to cracking if there are any ground movements.

8. CONCLUSIONS

The required permeabilities for barrier systems are sufficiently low that chemical, electrical and possibly thermal potentials need to be considered when assessing the magnitude and direction of pollutant flow. In a multi-barrier systems it does not follow that the flows through all elements of the barrier will be in the same direction. This could lead to damage though probably only in the very long term as flows should be small. Design of gas barriers on the basis of liquid permeability is not sound. Droughts and drying to considerable depths can occur. For cement-bentonite materials gas permeabilities may change by many orders of magnitude if severe drying occurs. With the present state of knowledge it would seem that if gas is to be controlled then a membrane must be included in the cut-off. However, the gas permeability of the membrane itself is possibly lower than strictly necessary (a 2.5 mm membrane may equivalent to a 1000 metre thickness of air dried structural concrete). It is very difficult to assess what is required by way of permeability control especially for gas. More quantitative data are necessary. If a membrane is included then great care must be taken with the design of the joints and the assembly of these joints in the trench.

9. REFERENCES

D'Appolonia, D.J., Soil-bentonite slurry trench cut-offs. J. Geotech. Eng. Div. ASCE, 106(4), pp399-417, 1980.

Hass, H.J. & Hitze, R., All-round encapsulation of hazardous wastes by means of injection gels and cut-off materials resistant to aggressive agents, ISME3 Seminar on Hazardous Waste, Bergamo, Italy, 1986.

Haxo, H.E., et al, Liner materials for hazardous and toxic wastes and municipal solid waste leachate, Noyes Publications, 1985.

Her Majesty's Inspectorate of Pollution, Waste Management Paper 27, The Control of Landfill Gas, 1989.

Howsam, P, Editor., Microbiology in Civil Engineering, Cranfield Institute of Technology, E & F.N. Spon, 1990.

Jefferis, S.A., The design of cut-off walls for waste containment, In land disposal of hazardous waste, Eds. Gronow,J.R., Schofield, A.N., Jain, R.K., Ellis Horwood Publisher, 1988.

Jefferis, S.A., and Mangabhai, R.J., The divided flow permeameter, Materials Research Society, Symposium on Pore structure and permeability of cementitious materials, Vol 137, 1989.

Jefferis, S.A., Bentonite-cement cut-off walls for waste containment: from specification to in-situ performance, Symposium on Management and control of Waste fill sites, Leamington Spa, 1990.

Krause, R., New developments and trends in ground water protection with flexible membrane liners, 1989.

Mitchell, J.K. Conduction Phenomena, 31st Rankine Lecture, to be published Geotechnique, 1991.

BURNING CHEMICAL WASTE SITE: INVESTIGATION, ASSESSMENT AND RECLAMATION

D L Barry BE, CEng, MICE, MIHT

Head of Wastes and Contaminated Land Department
WS Atkins Environment
Epsom, Surrey

ABSTRACT

A series of underground fires on a foreshore site in Flint (Fig 1), which was previously used for disposal of chemical wastes from the alkali and nylon industries, was causing an environmental nuisance as well as restricting the commercial development of the site (4 ha) and adjacent areas. Atkins were commissioned by Delyn BC to define the key combustion and environmental problems, recommend a practical solution and manage the necessary rehabilitation works.

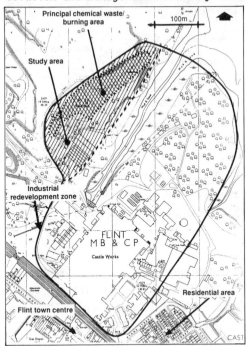

Figure 1 Location

A phased investigation strategy, commencing with a review of site history, allowed a progressively more accurate judgement to be made on the actual conditions, in terms of the extent of combustion, the materials involved, and the resultant products. Samples of solids, gases and vapours were collected, including pyrolitic and combustion products, using a range of techniques such as bulk sampling from trial pits and boreholes, and condensate traps. A selection of the samples were analysed using conventional and more sophisticated analytical techniques.

The principal relevant wastes were viscose wastes, gelatinous cellulose polymers, Leblanc wastes and process sludges. One area had approx 20,000m³ of viscose wastes, up to 10m deep in a valley formed by Leblanc wastes and presented profound physical and environmental problems. Other areas had more inert wastes mixed with some organics.

In two of the three combustion zones identified the seats of combustion were essentially in thin layers of carbonaceous material (small coke fractions) and were 'cooking' the overlying organics. The most critical combustion zone had developed a refluxing system (Fig 2) in which high concentrations of less volatile condensible organic compounds collected within the groundmass; the more volatile fractions escaped to atmosphere causing an odour problem.

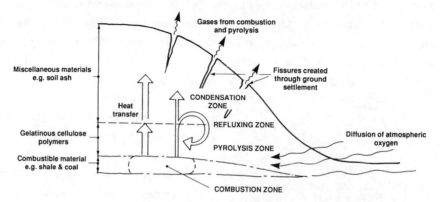

Figure 2 Simplified interpretation of Combustion Zone 1

The critical compounds identified were higher aliphatic hydrocarbons and polycyclic aromatic hydrocarbons. Airborne problems from these compounds could be avoided if temperatures could be lowered to about 20°C ie vapour pressure would be low and excavation hazards would be greatly reduced.

The adopted solution to the combustion problem was to dynamically compact (DC) the relevant ground, thereby reducing air ingress to the combustion zone with a consequent fall in temperature. Rigorous environmental measures were adopted during the subsequent excavation and removal of the relevant materials, which was undertaken when thermocouples installed in ground probes showed the DC process to have been effective.

The project was funded by WDA and ERDF, the latter for the assessment phase only.

REDEVELOPMENT OF A GAS WORKS CONTAMINATED SITE

JULIAN BROMHEAD and DAVID ROCHE
Frank Graham Geotechnical Limited, Consulting Engineers
22 Waterbeer Street, Guildhall Centre, Exeter EX4 3EH

Abstract
This technical note considers the redevelopment of a site contaminated by materials from a former gas works. It reviews how the degree of contamination was investigated and assessed, and describes the engineering measures adopted to develop the site for new offices. It also addresses certain difficulties encountered during construction.

The Site
The site comprised a semi-derelict area of about one hectare located adjacent to a former town gasworks. Prior to redevelopment it was partly occupied by a factory unit with hardstanding and grass. A river and a millstream are located adjacent to the site.

Initial Investigation
The initial site investigation included some five boreholes and six trial pits. It identified up to about 1.5m of made ground with some tarry contamination overlying alluvial soils, underlain by mudstone bedrock. Some visual contamination of the alluvial soils was noted. Eight samples of the made ground were analysed for typical gas works contaminants.

A geotechnical report prepared by the site investigation contractor identified the made ground to be contaminated with significant levels of coal tars and phenols; locally high levels of cyanides and sulphates were also reported. Monitoring in two standpipes indicated only trace levels of methane, hydrogen sulphide was not detected. The contamination was believed to be derived from the adjacent former gas works.

Preliminary Engineering Design
Frank Graham became involved with the project after the initial site investigation had been completed when the client became aware of the very difficult nature of the site. A preliminary geotechnical appraisal and engineering design to facilitate site redevelopment was prepared. This concluded that encapsulation of the majority of contaminated material would be the most appropriate remedial solution involving minimum disturbance. Encapsulation would isolate construction personnel from contaminated material and coupled with hard landscaping would minimise any further leaching of contamination. Large scale excavation was considered inappropriate due to the risks of disturbing areas of contaminated material near to the groundwater table and adjacent watercourses, and

because there was no local landfill facility for special wastes.

The preliminary encapsulation design consisted of a 300mm layer of coarse limestone rockfill to provide a capillary break, overlain by a geotextile separator and sand blinding which in turn was overlain by suitable inert fill. To avoid the problems of contaminated arisings associated with bored piles, it was proposed to emplace small section driven piles through the encapsulation layer and to construct ground beams within the inert fill above the capillary break. A suspended ground floor with an integral odour/methane barrier was also proposed.

Detailed Investigation

The detailed investigation was subsequently carried out as a proving survey during the early phase of site clearance and groundworks. It included trial pits with sampling on a 25m grid. Some 68 samples were analysed and the key results are compared against the trigger values for hard cover and buildings (from ICRCL 18/79 and 59/83) as follows:

Coal Tar (PAH)	-	73% >threshold	16% >action
Phenol	-	14% >threshold	2% >action
Sulphide	-	32% >threshold	4% >action
Cyanide	-	9% >threshold	0% >action
pH	-	38% >threshold	22% >action

Detailed Engineering Design

The detailed investigation proved overall levels of contamination very similar to those identified by the initial investigation and so the design philosophy remained virtually unchanged. Certain identified 'hotspots' of contamination were excavated and removed from site prior to encapsulation. The design was subsequently approved by the Building Regulations Authority, Environmental Health Authority, National Rivers Authority, Waste Regulation Authority and Health and Safety Executive and development at the site proceeded satisfactorily.

Difficulties During Construction

Despite the successful outcome of this project, some major difficulties during construction can be highlighted. These can be attributed primarily to the pressures of the 'fast track' development programme, which can be especially problematic for contaminated sites unless suitably experienced engineers are involved at an early stage to provide integrated assessment and engineering design. The initial investigation was carried out and reported by a contractor with no integral link to the engineering design, and no desk study was undertaken. Subsequently, the client turned to Frank Graham when the need was identified to engage consulting engineers with specialist experience of ground contamination and with structural design resources to meet the fast construction programme. However, despite recommendations that a more detailed contamination assessment was necessary, involving desk study, more exploratory holes and contamination testing, and that the development proposals should be discussed and agreed with the relevant statutory authorities, the client opted to proceed directly with construction. Considerable delays were subsequently incurred during construction to accommodate the detailed investigations and design to satisfy the requirements of statutory consultees, in particular the environmental health authority who initially objected to the construction activities.

LANDFILL GAS PROBLEMS

DAVID HOBSON B.Tech C.Eng MICE MIHT
W A Fairhurst & Partners, Environment Division
1 Arngrove Court, Barrack Road, Newcastle upon Tyne NE4 6DB

INTRODUCTION

The problem of gas from landfill is now widely recognised, however, in the past the potential difficulties were generally underestimated. This has led to a large number of sites which have the potential to generate gas, but for which no designed mitigation measures are in place. Guidance on the design of new sites are contained within Waste Management Papers nos 26 and 27, but measures for controlling gas from older completed sites are less clear.

CHARACTERISTICS OF OLD LANDFILLS

All sites which have been filled since 1960 must be suspected of producing gas. These will fall into two main categories.

1. Land used specifically for municipal wastes.

2. General landfill sites.

The former would be more likely to be high gas producers as they will contain a large proportion of degradable materials. Sites filled since 1974 under regulations imposed as a result of the Control of Pollution Act will tend to be better compacted and contain layers of inert material. Anaerobic conditions generating flammable gas will be prevalent and layering will cause a tendency for gas pressure build up and a lateral migration into surrounding land.

Other tips containing a wider range of materials with a lower degradable content will often have been placed under less demanding controls. The result is lower gas production and often large parts of the tip remain in an aerobic stage producing carbon dioxide rather than flammable gases. Gas pressures tend to be low or non-existent and movement of gases outside the site is slow and mainly due to diffusion effects.

MONITORING OF GASES

Waste Management Paper no.27 provides recommendations for monitoring, however, each site requires individual consideration. Monitoring will have a number of different objectives which should be understood.

Firstly conditions within the tip can be monitored. This will be a useful an exercise in the early stages after completion of a new tip or during any works which are being carried out that will disturb an older landfill. Monitoring should be on a regular basis with the frequency appropriate to the anticipated rate of change within the tip. Often weekly, falling to monthly frequency is adopted until the conditions within the tip have been determined. To establish the gas characteristics within a tip a minimum of twelve months monitoring is necessary as considerable variations will occur during the seasons. Normally gas production is higher in the warmer part of the year than in the cooler. The results of monitoring are best presented in a graphical format for ease of interpretation. It will be useful to note the depth to groundwater, temperature and barometric pressure at the time of each reading so the effects of these parameters can be assessed. Once a "steady state" has been reached within the tip there is little point in frequent monitoring and annual readings (normally taken in the summer) should be adequate. Once the tip has ceased gassing more frequent monitoring will again be necessary for about a 2 year period to confirm that activity has ceased.

The strategy for monitoring conditions outside the tip will depend on ground conditions and the significance of any lateral migration. If there are no sensitive targets then anything more than occasional monitoring is unnecessary. Where sensitive targets are present within a reasonable distance (Waste Management Paper 27 seems to suggest 250 metres) then regular monitoring will be advisable and provide reassurance. However, once high gas conditions are established, unless remedial action is taken and it often is not, then frequent monitoring again becomes somewhat academic. Monitoring of gas outside the site should be taken at varying distances. Readings taken close to the perimeter of an unprotected old landfill will provide little information about the extent of migration through the surrounding ground. Finally it should be remembered that many areas contain several landfill sites and in such conditions the soil atmosphere within the entire locality may have elevated levels of gas, particularly carbon dioxide. There will be no pressure associated with the gas and little hazard. Rigid application of the interest levels contained within Waste Management Paper 27 should therefore be avoided.

CONCLUSION

The problem of gas from infilled land is extremely variable. Serious hazards are possible under certain conditions, but the threat from a large number of sites is probably very small. In assessing any site the characteristics of the tip and the surrounding ground should be carefully researched and the sensitivity of targets assessed. The level of monitoring should be consistent with the information required and a separate strategy should be adopted for on site and off site installations. Guidelines for acceptable gas levels should not be rigidly applied, but should be considered alongside other relevant factors.

RECLAMATION OF GAS WORKS SITE
FOR HOUSING DEVELOPMENT AT SALISBURY: A CASE STUDY

EDWARD J WILSON
E J Wilson & Associates
Consulting Engineering Geologists
8 Brunswick Square, Gloucester GL1 1UG

St Paul's Gas Works in Salisbury is a classic case. Built in the early 19th century outside the city, it became engulfed by the tide of late 19th century housing, in the middle of which it became an environmental problem and after its closure, a potential hazard. Thus there was a particular desire by Salisbury District Council to see this site redeveloped with housing, rather than a less sensitive end use which would have been the easier option. In 1986 the 0.65 ha site was purchased by Vine Homes Ltd for a residential development of 51 units.

THE SITE

The site is underlain by the Upper Chalk, with a patchy cover of sandy gravel and colluvial clay. This was overlain by a mainly ashy fill averaging 2m but varying between 0 and 4m in thickness.

The site contained a substantial brick retort house with a large battery of horizontal ovens and 3m deep basement, together with many smaller buildings, the older carbonisation plant, used as miscellaneous garages, a large gas holder and two batteries of purification tanks. There was visual evidence of contamination in the spent oxide covering the ground surface of the upper part of the site around the purification tanks, and in the many pipes within and around the buildings which were lagged with asbestos, then loose and flapping in the wind.

INVESTIGATION

Investigation was in three parts, archival research, the excavation in several phases of 51 trial pits and trenches with associated laboratory testing, and close and frequent control inspection throughout groundworks.

Although only one gas holder was still visible, old maps showed a total of five and these were investigated by careful trial pitting. One had disappeared without trace beneath the retort house, the remaining three were found to be still open below ground level, the older two capped by brick barrel vaulting and the third by a reinforced concrete slab supported by pillars and ring beam. All were filled with a separated mixture of tar and barely contaminated water, with a small air void filled largely with hydrogen sulphide, which was released when the system was pierced.

REMEDIAL GROUND WORKS

All buildings on site were demolished and all clean and suitable concrete and brickwork were crushed to hardcore on site using a portable crushing plant, and stockpiled. This amounted to 8050m^3.

All visible substructures and any surrounding contaminated ground were removed. The gas holder wells were pumped out, and all contamination removed. The water was found fit for disposal via the public sewer, but the tar, 463m^3, was tankered together with 1210m^3 of heavily contaminated solids, to licensed tip in Bedfordshire.

Simultaneously the level of fill was reduced, and the arising fill sorted into reusable and unsuitable. In the upper levels, contamination was general over no more than about one third of the site area.

The remaining fill was stripped or turned over down to natural ground level, removing all substructures and significantly contaminated material, or material unsuitable for recompaction. 7523m^3 of less contaminated solids, soil and masonry, were taken to licensed tip in Swindon.

Mechanically suitable granular fill was returned as a controlled fill, the marginally contaminated materials being placed first. The crushed hardcore was filled first into the 4-5m deep excavations left by the gas holder wells, the sides of which had been battered back, and then as a general granular fill over the site as a whole, once all ground-derived material had been placed. By this stage the site had been shaped to a restoration of the original natural ground profile, with level benches for buildings. All granular fill was laid in horizontal layers not exceeding 200mm in thickness, compacted with a tandem vibrating roller.

Foundations were designed as ring beams, and were cast in shuttering on heavy duty polythene sheeting over the compacted hardcore. The remaining clean hardcore was then used to achieve final levels. The site was finished with concrete pavior hard surfacing, all ornamental soft landscaping being in raised beds incorporating clean imported soil.

COMMENTS AND ACKNOWLEDGEMENTS

The selective and carefully targetted investigation including a close control on remedial groundworks, produced a safe and very satisfactory development, with site investigation costs amounting to a mere 0.32% of total sales value of the development.

The cost of all excess ground works, investigation and associated fees, were almost completely covered by a derelict land grant.

Acknowledgements are expressed to Vine Homes Ltd, to Dr B Baker of the City of Birmingham Industrial Research Laboratory for close collaboration throughout the project, and to the Officers of Salisbury District Council for a positive approach throughout the development.

SECTION 4 : BIOTECHNOLOGY

EVALUATION OF COMPOSTED SEWAGE SLUDGE/STRAW FOR THE REVEGETATION OF DERELICT LAND.

S.L. ATKINSON, G.P. BUCKLEY AND J.M. LOPEZ-REAL[*]
Department of Agriculture, Horticulture and Environment,
[*]Department of Biochemistry and Biological Sciences,
Wye College, Wye, Ashford, Kent. TN25 5AH

ABSTRACT.

Composted sewage sludge was evaluated as a material suitable for the revegetation of derelict land. Field trials carried out at three contrasting sites demonstrated that the addition of composted sewage sludge/straw to derelict land increased the dry matter yield of a grass/clover mixture compared to the unamended controls. At one site two different methods of applying the compost were compared (blanket layer versus incorporated layer) and were found to have no significant effect on the yield.

INTRODUCTION.

The major factors limiting the revegetation of derelict land are the adverse physico-chemical properties [1]. Derelict land is typically deficient in nutrients, especially nitrogen and phosphorus and has little capacity to retain inorganic nutrients. Physical problems, e.g. lack of water holding capacity, arise from the lack of organic matter [1]. The natural development of vegetation on derelict land is slow [2] and one of the simplest solutions to hasten revegetation is to bring topsoil to the site, but this is costly [3]. In recent years, considerable research has been carried out to find alternative approaches to revegetating derelict land [4].

Sewage sludges contain large amounts of organic matter, nitrogen and phosphorus. They give up their nutrients slowly and enable vegetation to establish without regression [5] and thus complement the needs of derelict land [1], however, they are unpleasant to handle [6]. Composting radically alters the physical nature of sewage sludge and eliminates many of the objectionable factors [6,7]. Between 1985 and 1990, the Sewage Treatment Works at Canterbury in Kent (an activated sludge plant) composted the sewage sludge with straw by means of the forced aeration static pile process. The sludge and the straw were mixed and windrowed over perforated pipes. Aerobic conditions were maintained by blowing air through the piles for 21 days and the temperature of the piles was controlled by feedback. After composting, the product was remixed and windrowed for several months, during which time it matured. The matured compost is easy to handle, relatively odour free and rich in useful plant nutrients. Composted sewage sludges are ideal for creating large pools of nitrogen for the revegetation of derelict land [8].

MATERIALS AND METHODS.

Field trials were established at three contrasting sites to evaluate the use of composted sewage sludge/straw to revegetate derelict land. At each site the compost was weighed out and applied to plots using a front end loader and a grass/clover mixture was sown by hand (table 1). The experimental layout at each site was a randomised blocks design with four replicates.

TABLE 1.
Composition and sowing rate of grass mixture 1.

Lolium perenne v. melle	25.0 %
Lolium perenne v. bravo	17.5 %
Festuca rubra v. rubra	15.0 %
Festuca ovina	15.0 %
Poa compressa v. reubens	20.0 %
Agrostis capillaris	5.0 %
Trifolium repens	2.5 %
Sowing rate	80.0 kg ha^{-1}

The Trial at Betteshanger Colliery Spoil Tip.

The trial carried out at Betteshanger colliery spoil tip was on one of the more sheltered parts of the tip. The shale in this area was well weathered and had a relatively low pyrite content (0.3-0.7%). The composted sewage sludge/straw was applied at rates of 0, 32, 64 and 128 tonnes of dry solids per hectare (tds ha^{-1}) to plots measuring 2 x 4 m and incorporated to a depth of approximately 5 cm using a tine plough. The grass was sown on 11th July 1988 and one representative quadrat (0.5 x 0.5 m) of herbage was harvested from each plot on 14th July 1989 and 20th July 1990.

The Trial at Westbere Gravel Quarry.

The trial was established on a gravel quarry which had been infilled with washings (a mixture ranging from fine silt to gravel) and partially capped with topsoil. The composted sewage sludge/straw was applied at rates of 0, 40, 80 and 160 tds ha^{-1} to plots measuring 2 x 4 m. For each rate of compost application, the compost was applied as a blanket layer (mulch) and an incorporated layer (to a depth of approximately 20 cm using a tine plough). The grass was sown on 11th July 1988 and one quadrat (0.5 x 0.5 m) of herbage was harvested from the centre of each plot on 10th November 1988 and 16th July 1990.

The Trial at Shelford Landfill Site.

An area being used to stockpile London clay was made available for the field trial. The composted sewage sludge/straw was applied at rates of 0, 70, 140 and 280 tds ha^{-1} to plots measuring 2 x 12 m. The compost was left as a blanket layer. The grass was sown on 21st September 1989 and one representative quadrat (0.5 x 0.5 m) of herbage was harvested from each plot on 11th July 1990.

RESULTS.

The Trial at Betteshanger Colliery Spoil tip.

The composted sewage sludge/straw was difficult to spread and formed a cloddy seed bed. The grass seed germinated within two weeks of sowing, although the growth of the grass was slow due to dry weather. The results of the two harvests are shown in figure 1. The composted sewage sludge/straw had a very highly significant ($p<0.001$) effect on the dry matter yield of the herbage for both harvests and there were very highly significant linear trends ($r = 0.91$ and 0.83 respectively), with the dry matter yield of herbage increasing in proportion to the rate of compost applied. The application of compost also increased the growth of weeds.

The Trial at Westbere Gravel Quarry.

The results of the trial are shown in figure 2. Difficulty in spreading the composted sewage sludge/straw was experienced, as for the trial at

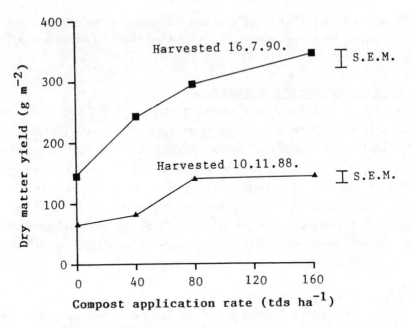

Figure 1. The yield of herbage grown on colliery shale amended with different rates of composted sewage sludge/straw.

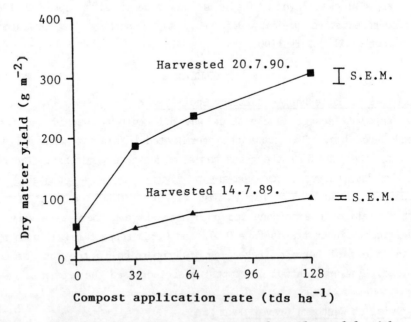

Figure 2. The yield of herbage grown on gravel spoil amended with different rates of composted sewage sludge/straw.

Betteshanger colliery tip. The grass seed germinated within two weeks of sowing. The composted sewage sludge/straw significantly ($p<0.05$) increased the dry matter yield of herbage at both harvests. There was a highly significant linear relationship ($r = 0.44$) between yield of herbage and the rate of compost application for the harvest in November 1988 and a very highly significant linear trend ($r = 0.72$) for the harvest in July 1990. There was no difference in yield due to the different methods of applying the compost for either of the harvests, therefore the results shown in figure 2 are the means of the two methods of application. Weed growth was also increased by the application of composted sewage sludge/straw.

The Trial at Shelford Landfill Site.

The composted sewage sludge/straw was easy to spread and formed a fine tilth. The germination of the grass seed was slow, but appeared to be improved on the plots amended with composted sewage sludge/straw. The application of composted sewage sludge/straw had a highly significant ($p<0.01$) effect on the yield of the herbage and there was a very highly significant linear trend ($r = 0.73$). The results are shown in figure 3. Weed growth was increased by the application of composted sewage sludge/straw.

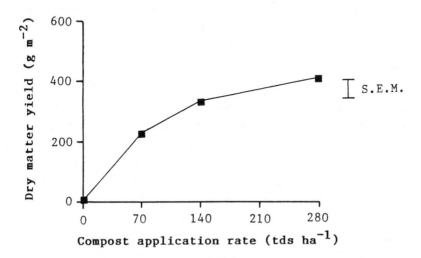

Figure 3. The yield of herbage grown on London clay amended with different rates of composted sewage sludge/straw.

DISCUSSION.

The results from the three field trials (figures 1 to 3) demonstrated that composted sewage sludge/straw increased the yield of herbage grown on spoil materials compared to the unamended control. In all the trials there was strong evidence of a linear relationship between the yield of herbage and the application rate of the compost, with an increase in the yield of herbage as the rate of compost application was increased. However, figures 1 to 3 show that the yield would probably not continue to increase as the rate of compost application was increased further. The increased yield was probably due to the composted sewage sludge/straw improving the nutrient status and the physical properties of the spoil. The trials on the colliery shale and the gravel spoil show that the composted sewage sludge/straw has a residual effect on the yield of herbage, as it was able to sustain the growth of herbage over more than one growing season. For both of these trials, the dry matter yield of herbage was greater at the second harvest and this is probably due to an increased availability of nutrients as the compost mineralised.

The composted sewage sludge/straw used in the field trials on the colliery shale and the gravel spoil produced a cloddy seed bed. This was due to the compost being stored out in the open prior to the trials which resulted in it becoming wet, anaerobic and sticky. The composted sewage sludge/straw used in the trial on the London clay, however, had been stored under cover and produced a dry, friable seed bed.

There was no evidence of inhibition of germination at either of the three sites and the composted sewage sludge/straw appeared to improve the germination of the seed on the London clay. This was thought to be due to the compost protecting the seeds from the weather and improving the water status as it was noted that the germination on the unamended clay was restricted to the troughs of ruts (caused by the front end loader) where the seeds were more protected and moisture collected.

The application of composted sewage sludge/straw also increased the weed growth at all three sites, particularly in the trials on the colliery shale and the gravel spoil where the compost was thought to be the source of the weeds.

CONCLUSIONS.

The revegetation of derelict land can be improved by the application of composted sewage sludge/straw. The composted sewage sludge/straw was found to have no adverse affects on the germination of seeds. In general, the yield of herbage increased in proportion to the amount of composted sewage sludge/straw applied, but the method of applying the compost (blanket layer of incorporated) had no effect on the yield. The storage of the compost was shown to be important as the compost used in two of the trials had been stored in the open resulting in it becoming wet and difficult to handle. It was also contaminated by weeds. The compost used in the third trial had been stored under cover and was easy to handle and did not contain weed seeds.

REFERENCES.

1. Hall, J.E. and Vigerust, E., The use of sewage sludge in restoring disturbed and derelict land to agriculture. In Utilisation of sewage sludge on land: rates of application and long-term effects of metals. Proceedings of a seminar held at Uppsala June 7-9, 1983, ed. S. Berglund, R.D. Davis and P. L'Hermite, D. Reidel Publishing Co., Dordrecht, 1984, pp. 91-102.

2. Bradshaw, A.D., The landscape reborn. New Scientist, 1982, 30 September, 901-904.

3. Bradshaw, A.D., The quality of topsoil. Soil Use and Management, 1989, **5(3)**, 101-108.

4. Roberts, D.R. and Simpson, T.G., The development of soil physical properties in restoration materials. Journal of the Science of Food and Agriculture, 1987, **40**, 221-222.

5. Marrs, R.H., Nitrogen accumulation, cycling and the restoration of ecosystems on derelict land. Soil Use and Management, 1989, **5(3)**, 127-134.

6. Griebel, G.E., Armiger, W.H., Parr, J.F., Steck, D.W. and Adam J.A., Use of composted sewage sludge in revegetation of surface-mined areas. In Utilisation of municipal sewage effluent and sludge on forest and disturbed land, ed. W.E. Sopper and S.N. Kerr, The Pennsylvanian State University Press, 1979, pp. 293-305.

7. Hall, J.E., Soil sludge management. Treatment and use of sewage sludge - concerns and options. WRc Technology Day, 14 June 1990, Water Research Council, Medmenham.

8. Crites, R.W., Land use of waste water and sludge. Environmental Science and Technology, 1984, **18(5)**, 140A-147A.

IN SITU BIOLOGICAL TREATMENT OF CONTAMINATED LAND - FEASIBILITY STUDIES AND TREATMENT OF A CREOSOTE CONTAMINATED SITE

PETER BARRATT and PAUL HAROLD
Biotreatment Limited,
5, Chiltern Close,
Cardiff CF4 5DL, UK

ABSTRACT

This paper describes the development of a microbially based system for the in situ treatment of creosote contaminated ground. Results indicate that the use of microorganisms capable of degrading creosote components, particularly polynuclear aromatic hydrocarbons, and surfactants to increase the solubility of those components, provide a good basis for soil clean up. A system installed at a creosote contaminated site is described. This has been in operation over a period of approximately 5 months, and significant reductions in the concentrations of creosote within both the groundwater (52.5%) and the soil (64%) have already been demonstrated. The system will be operated over a further period of 2 years.

INTRODUCTION

Sites contaminated with creosote derived from the fractional distillation of coal tar pose a particular risk to the quality of both ground and surface waters, as well as to that of the soil itself. Problems are due to the complex mixture of toxic organic components that constitute coal-derived creosotes. These components range from highly soluble and mobile compounds such as phenols, to comparatively insoluble compounds like polynuclear aromatic hydrocarbons (PAHs). The adsorption of organic contaminant molecules to soil particles, coupled with the low solubility of such compounds (1), is of particular relevance to the biological remediation of creosote contaminated soils. When considering in situ techniques for such remediation the latter is particularly true, as mobilisation of the target contaminants into the soil water phase is a prerequisite for a successful treatment programme.

Creosote present at the Blekholmstorget site in Stockholm is typical of many coal tar creosotes in chemical terms, containing high concentrations of PAHs, as well as phenols and oil hydrocarbons. In addition to the elevated concentrations of these anthropogenic residues in the soil, the extent of the contamination at Blekholmstorget demonstrated infiltration of the creosote to depths of up to 3m in the soil profile. Thus, most of the creosote contamination was present in highly anaerobic soil zones, where natural degradation of recalcitrant organics is expected to be much slower, and the mineralisation of contaminants may be incomplete (2).

This paper describes a research and development programme carried out to investigate the feasibility of treating creosote contaminated soil at the Stockholm site, in order to achieve concentrations of creosote below 200 mg kg^{-1}.

Based upon the findings of this study, an in situ system has been designed and constructed for full-scale reclamation of part of the Blekholmstorget site. This treatment system is currently in operation, and has now completed the first seasons treatment. Initial results of the in situ treatment programme are summarised.

MATERIALS AND METHODS

Site and soil description

The section of the site to be decontaminated by in situ microbial treatment was defined as two distinct areas. The volumes of contaminated material beneath these areas was partitioned from the adjacent ground by sheet piles driven deep into the underlying clay.

The two areas were of 3600m^2 (area A) and 1020m^2 (area B) respectively. The heaviest creosote contamination within the confined material extended to a depth of approximately 3m from the surface, below which lies relatively impermeable clay. The volumes of creosote contaminated ground to be treated were therefore 10800m^3 (area A) and 3060m^3 (area B) respectively. In addition to the sheet pile partitions, because of its dimensions, and the requirement for the maintenance of adequate flows through the soil, area A was further subdivided into four quadrants by impermeable geotextile membranes positioned within narrow trenches.

The subsurface ground to be treated consisted largely of made ground, comprising sand and clay, with some wood chip wastes. The creosote contamination lay within this layer of mixed fill.

Analysis of soil samples taken from the contaminated areas during a site investigation showed creosote contamination to be in the order of 20000 mg kg^{-1} in many places. The soil pH was generally slightly alkaline (x_1 = 7.8 ± 0.3). Phenol concentrations ranged from 6 - 154 mg kg^{-1}. Total nitrogen and phosphorous were in the ranges 1900 - 24000 and 29 - 50 mg kg^{-1} respectively.

Permeability and leaching studies

Laboratory studies were undertaken to define the leaching characteristics and permeability of the contaminated fill materials, in order to ascertain the suitability of the ground for in situ treatment.

Homogeneous samples of the contaminated fill were generated by passing representative materials through a 10mm sieve, and mixing the sieved sample thoroughly. Wood chip wastes were also collected for testing.

Soil permeability was determined by using a constant hydrostatic head technique. The volume of liquid passing through a soil column over a fixed time period was determined by maintaining a fixed head of liquid above the soil. The hydraulic conductivity of the material was established by using the equation $K = QL/(A+(h_2 + L)$, where K = hydraulic conductivity, Q = volume of collected liquid, A = column cross sectional area, h_2 = height of hydrostatic head, L = column height and t = time.

Leaching tests used both water and surfactant solutions to determine the removal of creosote from the sandy loam soil matrix. Two surfactants, synperonic PE/P94 (ICI, UK) and ethylan CD916 (Lankro Chemicals, UK) were individually screened against a distilled water control by shaking 50g of homogenised field soil with 200ml of liquid for 3 hr (180 rpm). The soil suspension was then centrifuged (10000 rpm; 10min) and the soil creosote concentration determined and compared with that of an unleached soil.

In addition, a soil column leaching test was performed using both water and the two surfactants. This was undertaken by a static head method similar to that used in the permeability assessments, by passing 6000 ml of liquid through the 300g soil column. The soil was then drained and analysed for creosote.

Analytical methods

Creosote determinations were undertaken by either GC or HPLC analysis. These methods were as follows.

For analysis by GC with Flame Ionisation Detection (FID) 25g wet soil was extracted for 3h with 25ml acetone and centrifuged (10000rpm; 15min). A 2ml aliquot of the supernatant was then evaporated under nitrogen, and redissolved in 2ml of dichloromethane. This was dried with anhydrous sodium sulphate and analysed by GC (FID) using a non-polar capillary column for the separation of creosote components.

Quantification of eleven priority peaks against the corresponding peaks from a Dutch standard creosote (Cindu Chemicals, Holland) was established as a measure of creosote concentrations in mg kg^{-1}.

For analysis by HPLC against the standard creosote, 25g of fresh soil was extracted as for the GC analysis and the supernatant analysed directly by reversed phase HPLC. The mobile phase consisted of acetonitrile:water gradient elution and the stationary phase that of octadecyl silane. Detection was by ultra violet absorbance (254mm). Ten PAH peaks and the peak corresponding to dibenzofuran were quantified individually.

Liquid samples containing creosote were extracted using 10 parts of sample and one part dichloromethane (v/v).

Microbiological studies

In order to achieve enhanced degradation of the creosote within the full-scale in situ treatments at Blekholmstorget, it was planned to amend the groundwater with microorganisms capable of degrading creosote components.

Five bacteria isolated from laboratory enrichment cultures of variously contaminated soils were selected and tested for their efficacy in degrading both Blekholmstorget creosote and a Dutch standard creosote in liquid culture. Two of these bacteria were isolated from contaminated Blekholmstorget soil (S001 and S003), and the other three (E001, E002 and E003) from coal tar contaminated soils in the UK. All 5 isolates were cultured at 20°C on agar media (3) containing 1000 mg kg^{-1} creosote as the sole source of carbon.

Batch liquid cultures of each isolate were established in the laboratory at 20°C. Each flask contained 100ml nutrient medium (3) supplemented with 500 mg l^{-1} of either Dutch standard creosote or creosote extracted from the Blekholmstorget site. The latter was found to contain other non-creosote components as witnessed by the lower control concentration in the Blekholmstorget flasks (Table 3).

Control flasks remained uninoculated, whereas those established as microbial treatments were inoculated with the individual isolates. Triplicate flasks were prepared for each treatment.

All flasks were incubated aerobically for 10-15 days after which each was analysed for optical density (620 mm), total viable count on creosote agar, and creosote concentration.

Continuous culture system

An experiment to ascertain the degradation of creosote within a continuous culture system (simulating a field treatment) was established. A continuous flow of liquid media including inorganic nutrients (100 mg l^{-1}) creosote (1000 mg l^{-1}) and surfactant (100 mg l^{-1}) was passed through a culture of isolate S001, then through a mixed culture of creosote degrading microorganisms and finally through a column of charcoal (Grade GAC; 12-22 USS).

Over a period of 16 days, flow rates through the system, as controlled by calibrating a peristaltic pump (Watson and Marlow, UK), were varied from 0 to 3 ml min^{-1}, and measurements of creosote in the influent, the two microbial treatment tanks and the final effluent were made at intervals.

Full-scale treatment system

The in-situ system for decontamination of the Blekholmstorget site was constructed during 1989-90, and consists of a series of perforated subterranean pipes; one set (infiltration) buried approximately 10cm below the surface, and the other (extraction) at approximately 350cm depth. Both sets of pipework, which are at right angles in gravel surrounded runs, are independent of one another, and the system relies on the passage of groundwater downwards through the contaminated fill, from the infiltration to the extraction pipes. From here, it is drawn out of the ground through vertical connecting pipework by vacuum pumping, and into two sparge aerated treatment tanks (each $5m^3$ capacity) situated within a portable cabin on the site. The flow rate of leachate through the whole system is maintained between 15 and 20$m^3 h^{-1}$. Flow rates through each of the five infiltration pipes were approximately 5$m^3 h^{-1}$, and those from each of the two extraction networks in the order of 10$m^3 h^{-1}$.

From the galvanised steel treatment tanks, leached groundwater, contaminated with creosote, is periodically treated prior to its return to the infiltration network for recirculation. The groundwater within each treatment area is maintained at a level just below the soil surface, and so the in situ treatment volume is essentially flooded.

Chemical and microbial treatment of the circulating water includes nutrient supplementation to maintain a C:N:P ratio in the order of 50:5:1, hydrogen peroxide dosing to maintain high influent oxygen levels (approximately $20 mgO_2.l^{-1}$) for aerobic microbial activity, and inoculation with known creosote degrading microorganisms. To date, microbial amendments have included the application of freeze dried preparations of bacterial isolates S003 and S001 (Pseudomanas fluorescens and P. cepacia). Inoculations were estimated to provide approximately 10^{10} cells per m^3 leachate.

Table 1 summarises the in situ groundwater conditions during 1990.

The treatment season is limited by ambient temperature, and is thus from April to November. Treatment was initiated in August 1990, and this paper describes the results obtained from then until November 1990. The treatment is to continue until November 1992.

TABLE 1

Summary of leachate conditions within the in situ treatment system during 1990

Analytical Parameter	MONTH			
	August	September	October	November
Temperature ($^\circ$C)	16.5	14.0	10.5	7.0
Oxygen - infiltrating (mg l^{-1})	5.6	7	15.0	20.0
Oxygen - extraction (mg l^{-1})	3.0	<0.1	3.6	5.0
Total Nitrogen (mg l^{-1})	<100	50	37	ND
Total Phosphorous (mg l^{-1})	1	ND	1.9	ND
Total Phosphate (mg l^{-1})	0.2	1.6	1.6	ND
pH	7.0	6.7	6.9	7.0

Monitoring the in-situ system

Flow rates, dissolved oxygen (influent and effluent), pH and temperature are all routinely monitored on site, and an autosampler (Epic Products Limited, UK) periodically removes water samples for creosote and nutrient analysis.

Although construction of buildings over the treatment areas severely impedes soil sampling, soil and water samples are removed at intervals from area B. Leachate samples only are withdrawn from Area A.

Contaminants in the emergent leachate from the two treatment areas were analysed following the abstraction of monthly samples.

RESULTS AND DISCUSSION

The soil within the in situ areas ranged from impermeable clay ($<10^{-9}$ m s^{-1}) to relatively permeable sandy loam and wood chip matrices with permeabilities of 7.6×10^{-4} and 2.0×10^{-3} m s^{-1} respectively. Since the clay mostly underlies the other contaminated material permeabilities were viewed as suitable for in situ treatment. The addition of 1% surfactant (ethylan CD916) to the permeating water was found to increase permeabilities in the sandy loam and wood chip soils by a maximum of 93 and 80% respectively.

Shake flask and column leaching studies produced results as summarised in table 2.

TABLE 2

Creosote concentrations (mg kg^{-1}) in soil before and after leaching in shake flasks and soil columns. Surfactant added at 1% (w/w) wet soil.

SHAKE FLASK	Before leaching	After leaching	% Creosote reduction
Water	2846	1704	40
Ethylan CD916	2846	1418	50
Synperonic PE/P94	2846	1485	48
COLUMN			
Water	6114	5261	14
Ethylan CD916	6114	4566	25
Synperonic PE/P94	6114	5253	14

These results illustrate the benefit of performing column-based leaching tests to discern the most effective surfactant treatment, as although there was little difference between creosote removal by the two surfactants in shaken flasks, a clear difference was seen in the column study. In this respect ethylan CD916 was viewed as the most successful surfactant for the mobilisation of creosote from the soil into the water phase.

Table 3 shows the results of batch microbial treatment of creosote, using the five selected microbial isolates.

TABLE 3

Batch liquid culture studies of microbial creosote degradation. Creosote added initially at a concentration of 500 mg l^{-1}

Isolate	Optical Density (600nm)	Viable count (colony-forming units ml^{-1})	Final Creosote Concentration	% reduction in creosote over control
DUTCH STANDARD CREOSOTE				
Control	0	0	336	-
S 001	0.25	1×10^6	310	15
S 003	ND	ND	ND	ND
E 001	0.35	1×10^4	261	29
E 002	0.01	0	270	27
E 003	0.08	1×10^4	283	23
BLEKHOLMSTORGET CREOSOTE				
Control	0	0	200	-
S 001	0.23	1×10^7	83	60
S 003	0.13	1×10^6	137	31
E 001	0.05	0	154	23
E 002	0.01	0	140	30
E 003	0.02	0	142	29

ND = Not Determined.

There are no clear differences in the extent of creosote reduction by each isolate and the type of creosote to be degraded. Although batch cultures of P. cepacia degraded 60% of the creosote extracted from contaminated soil, it was the least effective degrader for the standard creosote. However, as this bacteria removed approximately 100% more of the creosote components than any of the other isolates, it was selected for use in the full scale treatment system.

Data obtained from the continuous laboratory treatment of a creosote contaminated waste stream, using two microbial treatment tanks and a charcoal filter, revealed an optimum flow rate for the system (Figure 2).

The system was operated for 16 days, and at approximately 4 day intervals the liquid passing from one treatment stage to the next sampled and analysed. Flow rates were increased from 0.5 during the first 4 days, to 1.5, 2.0 and 3.0 ml min^{-1} at similar intervals. The discrepancy between the starting levels of creosote at each flow rate setting is explained by the gradual settling of some of the heavier creosote components in the primary holding tank. The data summarised in Figure 1 indicate that the greatest microbial degradation, after primary treatment by S001, and after secondary degradation by the mixed culture, occurred at 1.5 ml min^{-1}, the proportions degraded at each stage being 84 and 91% respectively.

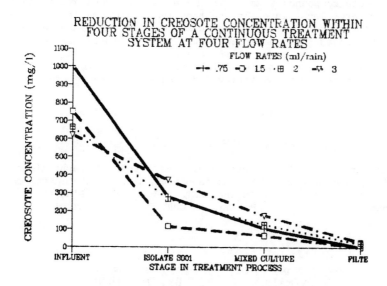

Figure 1.

Gradual saturation of the activated carbon polishing unit (the tertiary treatment) is indicated by gradually increasing breakthrough of creosote between day 8 (2.5 mg L^{-1}) and day 16 (32 mg L^{-1}).

Figures 2 and 3 show a marked decline in the concentration of creosote within the in situ groundwater in areas A and B respectively.

Decreases in the levels of individual PAHs and dibenzofuran (a three-ringed, heterocyclic aromatic hydrocarbon) are clear in the in situ groundwaters. In area B the only exception to this is acenaphthene, which increased from 61 to 386 g l^{-1}. This could be due to a gradual release of acenaphthene, a water-insoluble, clay binding PAH (4), into the groundwater from the soil creosote. Microbial transformation of such a three-ringed PAH is slow relative to the smaller PAHs (5).

The results for soil creosote component concentrations in area B (Figure 4), showed a similar pattern of decline to those in the in situ groundwater (Figure 3), indicating the successful operation of the in situ clean up system in area B to date.

The data for groundwater within area A are rather different from those in area B, the starting concentrations being an order of magnitude lower, whilst the higher molecular weight three- and four-ringed PAHs were present in proportionally greater quantities. The area A leachate results indicate disappearance of these larger PAHs in proportions similar to those seen in area B. However, in area A there were appreciable increases in the lower weight PAHs e.g. 1-methyl naphthalene, between August and November, which were the reverse of the changes witnessed in area B. Although no soil data exists to clarify these results, they are assumed to be due to the gradual release of the more soluble creosote compounds from the soil matrix into the circulating in situ groundwater. Thus, later in the treatment programme it is to be expected that the the levels of these components in the leachate will gradually decline as that remaining in the permeable soil reaches a minimum, and that in the water phase is microbially degraded. The enhancement of release of the remaining creosote hydrocarbons into the groundwater is expected to be achieved during the following season by the addition of surfactant.

KEY TO FIGURES 2, 3 AND 4

CRE = Total Creosote, NAP = Naphthalene, 2-MN = 2-methyl naphthalene, 1-MN = 1-methyl naphthalene, ACEN = Acenaphthene, DIBE = Dibenzanthracene, FLUO = Fluorene, PHEN = Phenanthrene, ANTH = Anthracene, F'AN = Fluoranthene, PYR = Pyrene, CHRY = Chrysene.

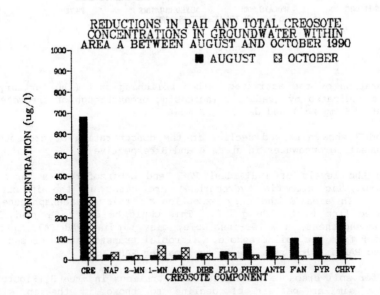

Figure 2.

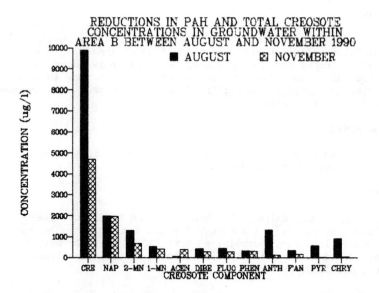

Figure 3.

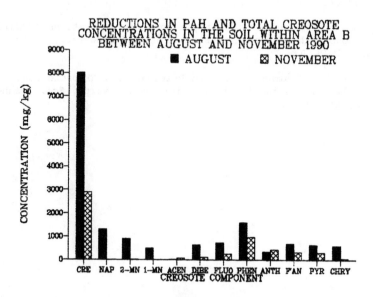

Figure 4.

CONCLUSIONS

Results obtained during this two stage study have provisionally indicated the successful development and operation of an optimised in situ biodegradation system for the treatment of creosote residues in a contaminated soil. Reduction in the concentrations of one to four-ringed PAHs has been demonstrated during the first of a three season treatment at the Blekholmstorget site.

ACKNOWLEDGEMENTS

Thanks to Skanska AB and the SNV (Swedish Environmental Agency) without whom this work would not have been possible.

REFERENCES

1. Weissenfels, W. D., Walter, U., Beyer, M. and Klein, J., PAH-degradation by bacteria - bioassay for the evaluation of microbial soil remediation. In Contaminated Soil '90, eds. F. Arendt, M. Hinsenveld and W. J. van den Brink, Kluwer Academic Publishers, Netherlands, 1990, pp. 483-484.

2. Lee, M. D., Thomas, J. M., Borden R. C., Bedient, P. B., Ward, C. H. and Wilson, J. T., Biorestoration of aquifers contaminated with organic compounds. In CRC Critical Reviews in Environmental Control, 18, 1988, 29-89.

3. Stanier, R. Y., Palleroni, N. J. and Doudoroff, M., J. Gen. Micro., 1966, 159-271.

4. Meyers, P. A. and Oas, T. G., Comparisons of associations of different hydrocarbons with clay particles in simulated seawater., Environ. Sci. Technol., 1978, 12.

5. Mueller, J. G., Chapman, P. J. and Pritchard P., Creosote contaminated sites - their potential for bioremediation., Environ. Sci. Technol., 1989, 23, 1197-1201.

SITE FACTORS AFFECTING TREE RESPONSE ON RESTORED OPENCAST GROUND IN THE SOUTH WALES COALFIELD

N.A.D. BENDING, A.J. MOFFAT and C.J. ROBERTS
Forest Research Station, Alice Holt Lodge, Farnham, Surrey GU10 4LH, UK

ABSTRACT

The physical and chemical properties of spoils resulting from opencast coal mining were studied at a number of sites in the South Wales coalfield, in conjunction with a survey of the health and performance of Japanese larch (*Larix kaempferi* (Lamb) Carr) planted on them.

Tree response on restored ground lacking soil was found to be extremely variable. Analysis of foliage indicated that nitrogen deficiency was widespread, but the poorest trees also suffered from a range of micronutrient deficiencies. Excessive spoil magnesium concentrations caused toxicity locally. Physical spoil conditions also adversely affected tree response. Problems of ground compaction, winter waterlogging and summer drought were identified.

The study has identified important areas where further research is required, for example the assessment of the suitability of overburden types for soil forming materials, methods of redressing nutrient deficiency, and the development of effective ground preparation techniques.

INTRODUCTION

The afforestation of opencast coal workings in South Wales began in the early 1950s and to date some one thousand hectares of ground have been restored to this afteruse. The majority of sites have been reclaimed without top or sub-soil materials.

Growth of the variety of coniferous species planted on opencast coal spoils has been extremely variable. Better grown trees are of acceptable size and habit, but the poorest are small and stunted. It seems likely that many of the plantations will not form a commercial timber crop.

The purpose of the present study was to identify site factors affecting tree response on existing restored opencast ground, so that future restoration could be based upon those site conditions which produced the best tree performance. An investigation was conducted into the relationships between spoil properties and the growth of Japanese larch. This species has exhibited the greatest commercial potential of all tree species used in opencast planting programmes.

METHODS

Studies were conducted at four former opencast workings in the South Wales coalfield restored between 1968 and 1981. Spoils used in the reclamation of each of the sites were derived from the Upper Carboniferous Lower and Middle Coal Measure Series (Westphalian A and B Stages). Lithologically the overburden consisted of shale, sandstone, siltstone, mudstone, coal and additional minor constituents. The growth of Japanese larch was extremely heterogeneous. The spatial variability of tree vigour was systematic rather than random.

Ten study sites were identified from within the four restored workings. Each study site was composed of either three or four arbitrarily defined sub-sites representing good, medium, poor (and very poor) tree growth. Within each sub-site five assessment trees were selected, to represent the average health and growth of trees surrounding them.

Detailed investigations were carried out in each sub-site including assessments of tree growth, foliar nutrition, spoil chemical, hydrological and physical conditions, and rooting habit.

The top height and diameter at breast height of assessment trees were measured. Extreme diversity in response over small areas and the comparatively young age of stands rendered assessments of yield class using conventional techniques problematical [1]. A technique modified to accommodate a smaller population than is normally employed in such assessments was carried out.

Samples of foliage were collected from the first lateral below the leader on each assessment tree (September, 1988). Samples were processed and analysed by standard laboratory procedures [2]. Concentrations of nitrogen, phosphorus, potassium, calcium, magnesium, iron, manganese, copper, zinc and boron were measured.

Spoil samples were collected systematically from beneath assessment trees at 0-15cm depth. Samples were air dried and sieved and the <2mm fraction retained for analysis. Spoil pH was determined in a 1:1 soil solution suspension and soluble salts by measuring the electrical conductivity of a 1:5 soil water extract. Organic matter content was determined by loss-on-ignition at 450°C, and cation exchange capacity was determined by the Chapman method [3]. Total nitrogen was determined by the Kjeldahl method [4]. Phosphorus was extracted by the Bray 2 method [5], and potassium and magnesium by standard ADAS methods [4].

Boreholes, gypsum blocks and tensiometers were installed at seven sites to study soil hydrology. Water table levels at depths up to 80cm and soil water potentials at a depth of 30cm were recorded at seven day intervals from October 1988 to October 1989.

Soil pits were excavated at one study site from each of the four former opencast workings. The size of excavations was determined by the depth and width of root penetration and extension. Bulked samples were removed at 20cm increments through the profile for assessment of spoil bulk density and percentage volume stone content [6]. Rooting in the frontwall face of each soil pit was assessed by detailed counts of root number, classifications of root diameters and the mapping of root distributions [7].

RESULTS

Growth Performance

Stands examined were between 9 and 22 years of age and their top heights ranged from 35-1325cm. Estimates of yield class for all sites ranged from 1.0 to 13.9, with a mean of 5.9 (SD 3.1). The maximum yield class of the stands examined exceeds that which would normally be expected for Japanese larch at such elevation, (300 metres OD) within the district (yield class 10-12). Mean growth rates were however, at least four yield classes lower than those of undisturbed ground adjacent to workings.

Foliar Nutrition

The macronutrient content of foliar samples are summarised in Table 1. Details of normal and deficient concentrations of macronutrients in Japanese larch are shown for comparison.

TABLE 1
Concentrations of macronutrients in foliage of Japanese larch on restored opencast ground (percentage dry weight)

	FOLIAR ANALYTICAL DATA				
ELEMENT	RANGE	MEAN	SD	DEFICIENT*	NORMAL RANGE*
Nitrogen	0.59-2.04	1.22	(0.41)	<1.80	1.80-2.50
Phosphorus	0.09-0.78	0.45	(0.16)	<0.18	0.30-0.75
Potassium	0.47-2.53	1.59	(0.46)	<0.50	0.75-1.41
Calcium	0.26-0.81	0.47	(0.16)	<0.24	0.28-0.67
Magnesium	0.15-0.46	0.27	(0.07)	<0.10	0.20-0.24

SD: Standard Deviation. Population Size = 31
* SOURCE: [8].

Nitrogen deficiency was widespread, with less than 10% of stands sampled containing adequate concentrations. Concentrations of phosphorus and calcium were within normal ranges for over 90% of stands. Magnesium and potassium concentrations extended from marginally deficient to excessively high, with only 26% and 36% of samples respectively, containing concentrations within the normal range. Foliar concentrations of iron, zinc and boron were low and approached deficiency levels. Concentrations of copper were within the normal range. High concentrations of manganese in foliage were identified at older workings.

Multiple regression analysis was used to determine a model of response between yield class and macro and micro nutrient content. Four elements contributed to the model, which produced a highly significant regression ($P<0.001$). The relationships between nitrogen, phosphorus and copper and yield class were positive, while that between magnesium and yield class was negative.

Spoil Chemistry

Results of analyses of selected spoil chemical properties are summarised in Table 2.

TABLE 2
Chemical properties of spoils on restored opencast ground

	RANGE	MEAN	SD
Total nitrogen (% dry weight)	0.06-1.06	0.20	0.19
Available phosphorus ($mgkg^{-1}$)	0.5-11.8	5.2	2.6
Available potassium ($mgkg^{-1}$)	84-212	136	40
Available magnesium ($mgkg^{-1}$)	384-890	586	128
pH	4.5-8.2	6.7	1.2
Electrical conductivity ($\mu s\ cm^{-1}$)	48-194	146	35
Cation exchange capacity ($me100g^{-1}$)	5.7-11.3	8.1	1.3

SD: Standard Deviation. Population size = 21.

Spoil pH ranged from strongly acid to moderately alkaline. Total nitrogen and available phosphorus contents were extremely variable. Exceptionally high spoil magnesium concentrations were determined in a large proportion of the samples.

Yield class was negatively correlated with available magnesium concentrations ($r = -0.53$ $P<0.01$) and pH ($r = -0.51$ $P<0.05$). No significant correlations between foliar macronutrient content and concentrations of total nitrogen, available phosphorus, potassium and magnesium in spoil were identified.

Site age was negatively correlated with pH ($r = -0.64$ $P<0.01$) and electrical conductivity ($r = -0.63$ $P<0.01$), and a positive correlation existed between cation exchange capacity and site age ($r = 0.47$ $P<0.05$). These results suggested some amelioration in spoil chemical conditions with time.

Spoil Hydrology

Borehole data was expressed in terms of the number of days water table levels occurred above 40cm and 70cm depth (W40 and W70 days: [9]). Waterlogging appeared to be broadly correlated with tree performance: in general medium and good sub-sites experienced waterlogging to within 70cm of the surface for less than 30 days per annum compared to poor sub-sites where waterlogging was often recorded for 90-180 days per annum.

Figure 1 shows water table levels in dipwells at the Abercrave 3 site. The poor sub-site was subject to permanent waterlogging, which was most severe between late autumn and late spring. The good sub-site was subject to intermittent winter waterlogging, while the medium sub-site (not shown) was exceptionally well drained.

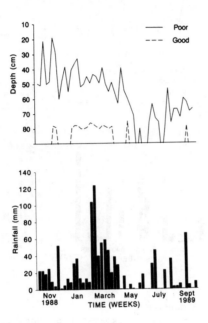

Figure 1. Weekly water table levels in dipwells, and rainfall at the Abercrave 3 site.

Soil water potentials indicated that drought conditions (ie soil water potentials less than -2 bar for in excess of 60 days) were widespread during the summer of 1989. Consistent patterns between soil water potential and tree performance were exhibited only at sites at the oldest of the four workings.

Figure 2 shows soil water potentials at the Abercrave 3 site. Soil water potentials below -2 bar were maintained for extended periods throughout the summer and autumn months at the good and medium sub-sites. Large weekly fluctuations in soil water potential were common. At the poor sub-site soil water potentials below -2 bar were infrequent, and weekly fluctuations in potential were relatively small.

Figure 2. Weekly soil water potentials and rainfall measured by
tensiometers and gypsum blocks at the Abercrave 3 site.

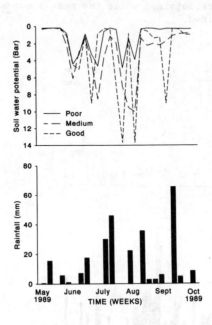

Spoil Bulk Density and Stone Content
Mean values of bulk density for the 19 excavations increased consistently with depth, from 1.54 gcm^{-3} (SD 0.14) at a depth of 0-20cm to a value of 1.77 gcm^{-3} (SD 0.25) at a depth of 20-40cm, and a mean of 1.83 gcm^{-3} (SD 0.21) at a depth of 40-60cm. The mean stone content of samples removed from 0-20cm were 16.2% (SD 5.48). Stone contents from 20-40cm and 40-60cm classes were 25.1% (SD 16.0) and 24.0% (SD 11.5) respectively.

Rooting Habit
Most consistent patterns in root system development were exhibited by trees at the Abercrave working. Figure 3 shows rooting sections for good and poor trees at the Abercrave 3 site. Trees at the good sub-site had greater lateral and vertical root development, a larger total number of roots, an improved size distribution of roots within diameter classes and a more uniform distribution of roots with depth, than those at the poor sub-site.

The incidence of shallow rooting was widespread. The mean maximum rooting depth of trees at all sites was 41.0cm (SD 11.8), while the total number of roots within 10cm and 30cm of the spoil surface were 35.1% (SD 14.0) and 88.8% (SD 10.4) respectively.

Figure 3. Rooting sections of good (top) and poor (bottom) trees at the Abercrave 3 site.

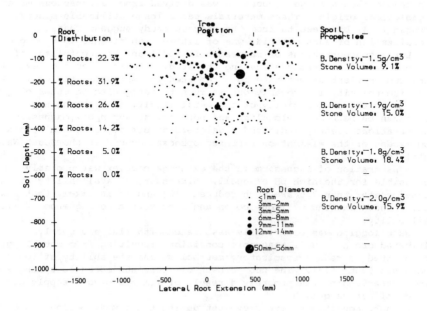

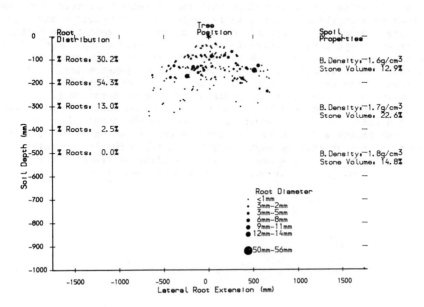

DISCUSSION

Severe nitrogen deficiency was found to be a principal cause of the poor response of Japanese larch on restored opencast ground. Although total spoil nitrogen levels were similar to a range of temperate soils [10], it is probable that a large proportion was derived from carbonaceous material of geological origin. These materials can release utilizable mineral nitrogen [11], but results from the present study suggest that mineralization of such material was of little significance to tree growth.

Large concentrations of magnesium in spoil and in the foliage of Japanese larch were recorded during the study. Toxic effects of excess magnesium on plant growth have been reported for serpentinitic soils [12]. Magnesium toxicity is probably due as much to antagonism to other elements as it is to the phytotoxicity of the metal itself. The strong negative relationships between yield class and both foliar and spoil magnesium concentrations clearly indicated a deleterious effect to growth arising from excess of the element on restored opencast ground in the South Wales coalfield.

A saturation of magnesium in the exchange complex was probably responsible for the high pH of spoil. Micronutrient availability at pH values greater than 7.5 are much reduced [12] and it is likely that low foliar concentrations of iron, boron and zinc were a consequence of limited availability.

Waterlogging was principally associated with flat and poorly cultivated ground. Soil anaerobic conditions resulting from waterlogging are reported to reduce respiratory metabolism and the ability of roots to absorb nutrients [13]. The poor growth of trees subjected to spring and summer waterlogging was probably a result of limited oxygen supply during periods of active growth.

Drought conditions were prevalent during the summer period. Oldest and most vigorous stands were most severely affected, with large soil moisture deficits. It is possible that the increased moisture requirements of trees in such stands was partially responsible for spoil desiccation.

Waterlogging was identified as a major cause of restricted vertical root growth, with rooting commonly observed to cease at the depth of the permanent winter watertable. High bulk density and large stone contents were also identified, and probably inhibited root penetration. Bulk densities of the order 1.8 gcm^{-3} have been reported to prevent root penetration due to mechanical impedance and reduced aeration [14]. The effects of compaction on root system morphology and root form have been described [15]. Physical conditions influenced the development of root system morphology and root form in the present study and major roots were observed to run at shallow depth parallel to the direction of cultivation. These results have implications for the stability of trees on restored opencast ground, especially on elevated and exposed sites.

CONCLUSIONS

The results of the study indicated that the poor performance of Japanese larch on opencast land restored between 1968 and 1981 is attributable to the type of spoil material and the ground cultivation techniques employed during site restoration.

The study suggested that in the future the suitability of overburden strata intended for use as soil substitutes should be assessed. Classifications of overburden suitability have been developed in North America [16] and originate from detailed studies of the chemical, physical and mineralogical properties of strata.

Problems of compaction on opencast ground can be avoided by the careful placement of overburden on the restored ground surface. Where trafficking is unavoidable complete cultivation is required. Such ground preparation would be expected to greatly reduce the incidence of waterlogging conditions and encourage deeper rooting of tree species planted on restored areas.

Sewage sludge has been found to be of value in the treatment of plantings of Japanese larch demonstrating poor performance. Trials have indicated that treatment redresses macronutrient deficiency and imbalance, and results in improved growth. Sewage sludge has the potential to provide a low cost and effective means of improving growth on sites, where to date, tree response has been poor.

Acknowledgement

The authors gratefully acknowledge the financial support of British Coal Opencast Executive.

REFERENCES

1. Everard, J.E., Fertilisers in the Establishment of Conifers in Wales and Southern England. Forestry Commission Bulletin No.41, HMSO, London, 1974.
2. Allen, S.E., Grimshaw, H.M., Parkinson, J.A. and Quarmby, C., Chemical Analysis of Ecological Materials. Blackwell, Oxford, 1974.
3. Chapman, H.D., Cation exchange capacity. In Methods of soil analysis, ed. C.A. Black., American Society of Agronomy, Madison, Wisconsin, 1965, pp.891-900.
4. Agricultural Development and Advisory Service., The Analysis of Agricultural Materials (Second Edition), HMSO, London, 1984.
5. Bray, R.H. and Kurtz, L.T., Determination of total organic and available forms of phosphorus in soils. Soil Science, 1945, 59, pp.39-45.
6. Smith, P.D. and Thomasson, A.J., Density and water release characteristics. Soil Survey Technical Monograph, 6, ed. B.W. Avery and C.L. Bascomb, Soil Survey of England and Wales, Harpenden, 1974, pp.42-55.
7. Böhm, W., Methods of Studying Root Systems. Springer Verlag, Heidelburg, 1979, pp.48-59.
8. Binns, W.O., Mayhead, G.J. and Mackenzie, J.M., Nutrient deficiencies of conifers in British forests. Forestry Commission Leaflet No. 76, HMSO, London, 1982.
9. Hodgson, J.M., Soil Survey Field Handbook. Soil Survey Technical Monograph. 5. Soil Survey of England and Wales, Harpenden, 1976., pp.88-89.

10. Cooke, G.W., The control of soil fertility. Crosby Lockwood, London, 1967, pp.3-13.
11. Aldag, R.W. and Strzyszcz, Z., Inorganic and organic nitrogen compounds in carbonaceous phyllosilicates on spoils with regard to forest reclamation. Reclamation Review, 1980, 3, pp.69-73.
12. Brooks, R.R., Serpentine and its vegetation. A multi disciplinary approach. Crook Helm, London, 1987, pp.32-133.
13. Scott-Russell, R., Plant root systems : their function and interactions. McGraw Hill, Maidenhead, 1977.
14. Zimmerman, R.P. and Kardos, L.T., Effects of bulk density on root growth. Soil Science, 1961, 91, pp.280-288.
15. Yeatman, C.W., Tree development on upland heaths. Forestry Commission Bulletin No 21. HMSO, London, 1955.
16. Daniels, N.L., and Amos, D.F., Generating productive topsoil substitutes from hard rock overburden in the southern Appalachians. Environmental Geochemistry and Health, 1984, 7, pp.8-15.

TECHNIQUES FOR RECLAIMING METALLIFEROUS TAILINGS

JOHN P PALMER
Richards Moorehead & Laing Ltd
3e Clwyd Street, Ruthin, Clwyd, LL15 1HF, UK

ABSTRACT

Abandoned metalliferous mines in the United Kingdom are a source of water, soil and air pollution. They can be a visual scar but often also have much industrial archaeological and wildlife importance. Reclamation techniques have to balance the protection of those factors of importance against the need to address issues of pollution, dereliction and visual intrusion. Techniques such as the use of plant tolerant material, break layers and membranes in combination with well designed tailings movement and drainage installations are often necessary to achieve a successful reclamation scheme. Current research on spoil decontamination and minewater treatment may provide further techniques for the future.

INTRODUCTION

Metal mining in the United Kingdom now accounts for a tiny proportion of the World metals output and is virtually confined to fluorospar and tin mining. However Britain was a major metal producer in the eighteenth and nineteenth centuries. At that time little thought was paid to the environmental consequences of such activities, and water and land became polluted. It has been estimated that over 4000 km^2 of land in Britain has been affected by metalliferous mining activities [1]. Many individual mines were small but densely concentrated in areas where metal rich rocks are found. For example the District of Ceredigion in Wales contains over 200 known mines at an average density of 1 every 10 km^2. Other mines were large and some individual mines have resulted in many hectares of land being grossly polluted. The dependence of these mines on water as a source of power and for ore processing meant that all were close to water courses and water pollution is a principal problem caused by these mines today. Because many of the abandoned mines are old and have been relatively undisturbed some are of industrial archaeological and wildlife importance.

In recent years there have been attempts to clear up the dereliction left by these mining operations. This is particularly so in Wales where initiatives by the Welsh Development Agency have led to a number of reclamation schemes for metalliferous mines having been completed.

This paper will discuss reclamation techniques adopted concentrating principally on the work in Wales on tailings from lead, zinc and copper mines. Research into new techniques will also be briefly discussed.

THE CHARACTERISTICS OF METALLIFEROUS TAILINGS

The processes of ore separation, although more sophisticated today, have changed little in their methodology since ancient times. Ore crushing and separating fractions out on the basis of their size and density led to products of coarse jig tailings, fine tailings and waterlogged slimes.

Many mines were worked intermittently for centuries leading to sporadic deposition of spoil and the accumulation of spoil of different characteristics in the same heap. In some instances old spoil heaps were reworked when techniques made metal recovery better. At other sites spoil was removed for use as construction materials. Like other mine sites spoil heaps at the time of mining and since, have been regarded as rubbish heaps for the disposal of all nature of other wastes. One can therefore encounter gasworks waste, tannery waste, domestic refuse and a range of other contaminating materials at such sites.

Mineralogically the spoils in Wales are derived primarily from the acid rocks of Silurian/Ordovician age in mid Wales and Snowdonia and the Carboniferous limestone of North East Wales. pHs of the tailings are in the range 2 to 8 with the higher pHs being in the limestone areas. On pyritic sites however pH can vary between these values at the same site. The tailings often contain up to 2% metal and sometimes considerably more. Very high concentrations of metals can be found in some undisturbed old spoil heaps and around are dressing areas. Trace metals can also be high, cadmium in particular is often associated with zinc minerals and unacceptable levels of arsenic can be found. Fine-grained wastes are a particular problem because:

* they often contain high concentrations of metal.
* the fine-grained nature of the waste means that a high proportion of the total metal content is often 'available' to plants.
* when dry they are subject to wind blow.
* they are easily subject to water erosion.
* many are permanently wet and have thixotropic properties and are very difficult to handle.

Many rivers in Wales are polluted to a greater or lesser extent due to metal mining. Pollution from tailings is principally due to:

1. Erosion of tailings into water courses due to surface run off
2. Seepage of metal rich water through tips

Polluted drainage from mine adits can also affect water courses. This pollution can be intermittent or continuous.

GUIDANCE ON RECLAMATION TECHNIQUES

There has been much research on the tolerance of vegetation to heavy metals and on the use of covering materials to prevent the upward migration of metals [3-8]. Techniques of establishing a vegetation cover have also been usefully summarised [9]. There has also been guidance on the levels of heavy metals which are acceptable in relation to site development and the disposal of sewage sludge to land [10, 11]. More recently guidance on the restoration of metalliferous mine sites to agriculture has been published [12] and a Department of the Environment research project is underway to provide guidance on the restoration and management of metalliferous mine sites for amenity use. This latter project is expected to draw on current experience of metalliferous mine reclamation worldwide and to adopt an holistic approach addressing archaeological, wildlife and aquatic pollution issues as well as those more directly concerned with engineering aspects and establishing a vegetation cover.

RECLAMATION TECHNIQUES

Aims

In most cases the principle aim of reclamation is to remove pollution from soil, water or air. In Wales many schemes have been promoted because of gross pollution of water courses which otherwise would be very clean. In some cases the visual detraction of the site can be so great that a reclamation scheme is promoted and a major concern is not to create pollution during the course of the work. Protection of those features of industrial archaeological, wildlife or other value is also a principle aim of any scheme.

Techniques

The principle ways in which metalliferous spoil can be reclaimed are as follows:

* Covering the waste in situ with uncontaminated material. Often the most contaminated spoils are concentrated into one place prior to capping.
* Amelioration of the wastes and vegetating with metal tolerant plants.
* Excavation of spoil and removal to a disposal area or site.
* On-site decontamination of the spoil.

In practice combinations of these first 3 techniques are often used at the same site.

<u>Covering waste in situ</u>: Early reclamation schemes concentrated on the use of coarse grained materials as part or all of a covering system in order to provide a capillary break to prevent the upward migration of metals.

Problems have occurred in some of these schemes with the downward and lateral movement of metals in water which has percolated through coarse capping layers. In the most extreme case pollution of adjacent watercourses has been greater after the reclamation scheme than before.

The principle of reducing the ingress of water into metalliferous spoil after reclamation is an important one and this is particularly the case where the material is pyritic because of the likelihood of increased oxidation of the pyrites, acid water seepage and resultant toxicity problems in receiving waters. For this reason there has been a move towards covering of spoil materials with a polyethylene membrane prior to covering with a capping material. This has the following implications for design of the scheme:

* Care has to be taken that final slopes are shallow enough to prevent the capping material slipping off the membrane.
* The capping has to be deep enough to support vegetation in dry conditions. 500 mm is the most common depth used.
* If tree planting is desired deeper layers of capping material are needed.
* Particular attention has to be paid to surface drainage.
* Afteruse of the site has to be carefully considered so that the membrane does not become damaged.

Where slopes are too steep to allow a membrane to be used a bentonite seal has been incorporated into the capping material. This is more costly than using a membrane and limits the choice of capping material as the bentonite has to be compatible with the capping material in order for the seal to be effective.

Amelioration of the waste and vegetating with metal tolerant plants:
There are grasses commercially available in the UK which are tolerant of
lead, zinc and copper in both acidic and calcareous conditions. One of
the consequences of the use of metal tolerant species on spoil is that
the spoil material remains at the surface and the runoff from that site
may contain unacceptable levels of heavy metals. Additionally if
establishment is slow after remodelling of the spoil, as it can be with
some of the grasses used, then erosion of spoil can occur prior to grass
establishment. The most suitable use of such plants is on mildly
contaminated spoils and original soils which are destined for amenity
use. Recent work with sewage sludge both in North America and in the UK
has suggested that this may be a very useful ameliorant for use on these
wastes [13].

Excavation and removal: Removal of metalliferous waste off-site is
rarely used because of the quantities of material involved and the
difficulties of transport and disposal of the material. Within a large
site excavation and concentration of waste and underlying contaminated
soil into a disposal area with consequent exposure of uncontaminated or
mildly contaminated subsoil is often practised successfully. The exposed
subsoil can then be vegetated through amelioration and choice of plant
material dependent on the contamination status of the ground.

Decontamination of metalliferous spoil: On-site decontamination of
metalliferous spoil has many attractions, the principle one being that
the removal or immobilisation of metal may be to such a degree that the
site may no longer be considered contaminated. Until recently, however,
there has been virtually no work on the removal of metals from
metalliferous mine sites as a means of decontaminating them, although
repeated cropping of plants growing on mine waste to remove metals has
been suggested.

There has however been much work on the refining of techniques for
reprocessing spoil heaps in order to remove metals for commercial sale
and over the years these techniques have produced progressively greater
yields of metals. These commercially available techniques are therefore
a good starting point for investigating ways of decontaminating spoil.

The extent to which metals can be removed from spoil is dependent on the mineralogy of the spoil, the particle size of the feed material and the process used. Although much metal can be removed from a metal-rich spoil the waste from the reprocessing activity may be no less contaminating. This is because its particle size may have been much reduced resulting in a much larger surface area and consequently greater availability of metal.

An additional factor relevant to old spoil heaps is that the minerals in the heaps have often become weathered and oxidised and that techniques appropriate for unoxidised ores or spoils may not be as effective after oxidation.

The extension of commercial reprocessing techniques to spoil decontamination is being investigated by us in association with the University of Cardiff School of Engineering under Welsh Office funding. Investigations so far indicate that recoveries of up to 90% can be achieved on some oxidised wastes using non-conventional leaching techniques. Other more conventional techniques have recovered 65% of metal. An encouraging feature is that particle size reduction has not been necessary to achieve these results.

More problematical is the fact that wastes after reprocessing with less than 0.7% metal have been difficult to achieve. Although this level of metals are a significant improvement on the feed material they are still high. Work on the availability of these residual metals in the long term is yet to commence and work on further techniques of decontamination is continuing.

<u>Control of water</u>: Control of water is essential both during and after metalliferous mine reclamation schemes. The key to the reduction of water pollution is prevention of its pollution in the first place. Treatment should always be a last resort. Too often in the past acid metal rich waters have been found draining from spoil heaps or mine openings and the solution has been to introduce a bed of limestone and sedimentation tanks to treat the water rather than to see if the water could be diverted round polluting spoil heaps in uncontaminated ground or

if water could be prevented from entering the mine workings in the first place. Where limestone beds have been used to treat acid waters in the UK they have been notoriously unsuccessful in the long term, the limestone becoming coated with reaction products and rendered ineffective very quickly. Limestone beds are a satisfactory means of treating mine drainage during a reclamation scheme but even then should only be used once all attempts have been made to divert water away from pollution sources.

Recent work in North America has suggested that wetland or organic matter based systems may be effective in removing metals from metal-rich water [14]. This may be another important technique for the future.

CONCLUSIONS

Research over a number of years has provided a basis for reclamation schemes on former metalliferous mine sites to be carried out successfully. Most schemes incorporate covering of the waste after concentration into one place. Covering systems may include use of a break layer, membrane and bentonite as well as uncontaminated capping material. Revegetation of wastes contaminated with metals is also carried out using metal tolerant species.

Research work on the decontamination of metalliferous spoil may lead to an important new technique for dealing with these wastes. Control of water is essential during and after reclamation schemes on metalliferous wastes. Where water is unavoidably polluted wetland techniques may provide a useful method of decontamination.

REFERENCES

1. Thornton, I., Geochemical aspects of heavy metal pollution and agriculture in England and Wales. In Inorganic pollution and agriculture. MAFF Reference Book 326, HMSO, London, 1980, pp 105-125.

2. Cairney, T., Soil Cover Reclamations. In Reclaiming Contaminated Land ed T. Cairney, Blackie, London, 1987, pp 144-169.

3. Humphries, M.O. and Bradshaw A.D., Genetic potentials for solving problems of soil mineral stress: heavy metal toxicities. In *Proceedings of a Workshop on Plant Adaptation to Mineral Stress in Problem Soils*, Beltsville, Maryland, 1976, pp 95-105.

4. Johnson M.S., McNeilly T., and Putwain P.D., Revegetation of metalliferous mines spoil contaminated by lead and zinc. *Environmental Pollution* 1977, **12**, 261-276.

5. Jones A.K., Bell R.M., Barker L.J. and Bradshaw A.D., Covering for metal contaminated land. In *Management of uncontrolled hazardous waste sites*. Hazardous Materials Control Research Institute, Silver Spring, Maryland, 1982, pp 183-186.

6. Richards, Moorehead and Laing Ltd. *A review of the capping material trial on metalliferous mine spoil at Minera*. Report to Wrexham Maelor Borough Council, 1988.

7. Smith R.A.H. and Bradshaw A.D., Stabilization of toxic mine wastes by the use of tolerant plant populations. *Transactions of the Institution of Mining and Metallurgy Sector A*, 1972, 81, 230-237.

8. Williamson N.A., Johnson M.S. and Bradshaw A.D., *Mine Wastes Reclamation*, Mining Journal Books, London, 1982.

9. ICRCL. *Guidance on the assessment and redevelopment of contaminated land*. ICRCL 59/83. Interdepartmental Committee on the Redevelopment of Contaminated Land, Department of the Environment, London, 1987.

10. Department of the Environment, *Code of Practice for Agricultural Use of Sewage Sludge*, HMSO, London, 1989.

11. ICRCL, *Notes on the Restoration and Aftercare of Metalliferous Mining Sites for Pasture and Grazing*, ICRCL 70/90, Interdepartmental Committee on the Redevelopment of Contaminated Land, Department of the Environment, London, 1990.

12. Metcalfe, B., Establishing long-term vegetation cover on acidic mining waste tips by utilising consolidated sewage sludges, in *Acid Mine Water in Pyritic Environments*, ed. P Norton, International Mine Water Association, Lisbon, 1990 pp 255-267.

13. Wildeman, T, Machemer, S, Klusman, R, Cohen R and Lemke, P, Metal removal efficiencies from acid mine drainage in the Big Five Constructed Wetlands, in *Proceedings of the 1990 Mining and Reclamation Conference and Exhibition*, Ed J Skousen, J Sencindiver and D Samuel, West Virginia University, Morgantown, West Virginia, USA, 1990, pp 417-424.

REVEGETATION OF RECLAIMED LAND

PETER SAMUEL
National Minerals Specialist
ADAS
Castle House, Newport Road, Stafford ST16 1DL

ABSTRACT

A wide range of physical and chemical conditions are encountered on derelict sites that are detrimental to vegetation establishment and its subsequent afteruse. Successful reclamation requires accurate diagnosis of the constraints that exist and the development of appropriate strategies for ameliorating ground conditions to meet afteruse objectives. Even following treatment severe limitations may still exist and these need to be borne in mind when considering future land use options and establishing vegetation. Careful planning and implementation is therefore essential since poor choice of options or failure to deal appropriately with site conditions can result in reduced benefits to the community and increased maintenance costs. In this paper consideration is given to the requirements for successfully revegetating sites. Aspects covered include site investigation; evaluation and utilisation of soil and wildlife resources; site amelioration; vegetation establishment and associated land management.

INTRODUCTION

The costs of reclamation are normally eligible for grant aid. However long term management costs are borne by the landowner. This can place a substantial financial burden on a local authority with many sites to maintain. It is therefore essential that reclamation schemes are carefully designed to provide maximum afteruse benefits without incurring unnecessary expenditure. In relation to these goals this paper reviews the critical steps and considerations necessary to achieve revegation to 'green' afteruses. It starts by considering the importantance and conduct of site investigations.

SITE INVESTIGATION

Systematic site appraisal to compile an inventory of ecological, landscape and soil resources is an essential prerequisite to successful reclamation. Information on the nature and extent of existing landscape, wildlife and 'soil' resources enables the feasibility and cost effectiveness of alternative afteruse strategies to be properly evaluated from the outset . It enables recognition of opportunities for reducing revegetation costs by ulitising existing resources such as naturally colonizing vegetation areas of conservation/visual amenity value. Also of using substrates that can be readily reclaimed in situ with minimal treatment. Valuable intelligence can also be obtained on the nature and severity of problems likely to be encountered in establishing different afteruses and on the nature and cost of likely ameliorative treatment requirements. This way, by undertaking resource appraisals, costly mistakes can be averted and the most cost effective strategy developed in accordance with reclamation goals.

Ecological and Landscape Assessment

Mapping and evaluation of existing vegetation using the Nature Conservation Review criteria can highlight areas of wildlife value. Occasionally habitats of national or regional significance may be identified which if discovered should be safeguarded wherever possible. However, such surveys when combined with landscape assessments are more likely to highlight areas of much more local conservation or visual amenity value comprised of naturally colonizing vegetation which may be capable of enhancement by planting and management - possibly even extended by site amelioration. Such areas can also yield a valuable source of plants for recreating "nature" areas on similar substrates elsewhere on site. In this way by identifying these resources and working with nature a head start can be gained in "greening" a site. Substantial cost savings can also accrue due to reduced site preparation, establishment and maintenance requirements.

To achieve best results assessments of ecological resources and landscape often need to be carried out beyond the derelict area itself. This enables off site features such as woodlands, wet areas and hedgerows to be identified and consideration given to extending them into the area to be reclaimed. This not only has the effect of integrating the reclaimed site into the existing landscape but also of creating wildlife corridors thus assisting colonization by fauna and flora. Careful manipulation of on-site landform and substrates can then also be used to provide a diverse range of interlinked habitats with spin-off benefits for wildlife, visual amenity and informal recreation.

Evaluation of Soil and Soil Making Materials

By definition, derelict sites are those so damaged by industrial or other development that they are incapable of beneficial use without treatment. Notwithstanding this they frequently contain in situ substrates capable of supporting plant growth with only relatively minor amerlioration and cost. However utilisation of these materials is rarely straightforward given their tendency to occur sporadically across site and are frequently restricted to areas where they were either dumped following site development or as a result of being left in place as surrounding land was utilized. These soils may adjoin disused buildings, exist adjacent to waste tips, occur on site peripheries or even be buried below other materials. A primary purpose of site surveys is therefore to locate, characterise and quantify these soils so that best use can be made of them during reclamation. Almost inevitably they will be insufficient to meet afteruse requirements and extra materials will need to be found to make good deficiencies. Survey work should therefore also be directed towards locating and characterising substrate materials to make good this deficit. Investigations also need to highlight if any materials are particularly unsuitable for plant growth. Analyses will indicate whether these can be ameliorated by appropriate treatment or whether it is better to discard them totally. To be forwarned is to be forearmed and many potential mistakes can be avoided with good intelligence. Since many mistakes could be very costly if not impractical to correct at a later stage or could significantly reduce the range of afteruse options available, money spent on pre-working survey is always money well spent.

To achieve maximum benefit it is essential that soil resource evaluation strategies are geared towards the specific requirements of individual sites. In England and Wales the derelict land category includes a wide range of site types encompassing abandoned industrial and domestic waste tips, old mineral workings both wet and dry, spoil tips of various kinds, derelict industrial and housing sites and disused transport land. Although some substrate characteristics will need to be evaluated on all sites; others are only relevant in specific instances. Here specialist knowledge as to what to look for is essential as failure to make appropriate physical and chemical observations, can have disastrous and very costly long term consequences. In terms of general approach survey work is likely to be carried out in at least three stages. This comprises an evaluation of documentary records followed by reconnaissance survey of substrate physical and chemical characteristics and then the main survey. From the outset close liaison needs to be achieved between technical specialists via the project manager to achieve best results. During site evaluation liaison between the soil scientist and those responsible for geotechnical surveys is particulary important since examination of cores from boreholes can yield valuable information on the nature and availability of soil making materials at depth. This will enable decisions to be made on the feasibility of retrieving useful materials to make up soil deficiencies during reclamation.

Documentary records and interviews of key witnesses can be very helpful in developing an appropriate survey strategy. Information on the location of buildings, processing plant, tips, routeways, underground storage structures etc and the types of activity carried out are all valuable when trying to assess the likely nature and location of possible contamination. However it should be remembered that accidental spillages, run-off, and tipping may also have arisen that could have given rise to more widespread problems. Since assessment of contaminants requires expensive analyses such documentary evidence enables chemical analyses to be targeted. It also ensures that critical limitations are not overlooked. Serious omissions could if unnoticed, threaten the success and most certainly the cost of subsequent revegetation operations. Very useful guidance on the nature of contamination associated with certain types of site is given in ICRCL Guidance Notes series published by the Department of Environment. These cover landfill sites, gas works, sewage works and farms, scrap yards and metalliferous wastes.

Both reconnaissance and detailed surveys require assessment usually by hand auger followed by trial pits. The latter enables materials to be assessed below auger depth and enables auger data to be properly validated. It also enables more precise determination of characteristics such as soil structure and stone content which are of considerable importance for reclamation planning. Hire of a back-hoe excavator is recommended for trial pits since mechanical digging saves time and gives much better exposures than hand dug pits.

No uniformally applicable specification can be given for site assessment since these vary according to the nature, size and complexity of individual sites. Furthermore survey requirements become refined as survey work progresses. Sampling during the reconnaissance phase seeks to characterise the extent, location and range of materials on site and is influenced in part by documentary records (described earlier). However an experienced surveyor is also alert to subtle changes in substrate characteristics, differences in vegetation growth, bare patches, signs of discoloured or contaminated soils, made ground or waste tipping which may also indicate significant differences in reclamation potential. Findings from the reconnaissance survey then determine requirements for the detailed survey phase. This enables effort to be devoted to characterising materials that have reclamation potential and those which are borderline. On many sites best value for money is often achieved by adopting different sampling densities for different parts of the site.

Assessment of physical characteristics such as structure, texture, depth, stone content, drainage and substrate colour are routinely made using field observation techniques. However to assess chemical characteristics samples are collected and sent for laboratory analyses. As required determinations are then carried out on organic matter content, pH, nutrients, salinity, heavy metals and/or any other pollutants suspected of being present in zootoxic or phytoxic concentrations. However the exact analyses undertaken are site, even mapping unit dependant so as to ensure that critical properties are targeted.

Once survey work is completed maps and an inventory of soil resources are compiled to give guidance on the location, quantities and characteristics of on-site materials and their potential for different afteruses. Guidance is then given on discarding useless materials, safeguarding others and also on the feasibility of fulfilling afteruse objectives with the resources available. Alternative afteruse strategies may also be proposed for consideration. The soils report also gives guidance on problems likely to be encountered in using specific materials and on the need for ameliorative treatments during ground preparation, vegetation establishment and/or aftercare maintenance. Together with other survey reports this data provides a sound basis for subsequent decision making in relation to the cost effectiveness of different reclamation strategies. This is particularly important since decisions made on the utilization of different materials during ground preparation can have profound implications for subsequent vegetation establishment and management costs. Also on likely long term returns from site utilisation.

GROUND PREPARATION AND SOIL REINSTATEMENT

Landform

Since waterlogging can severely effect plant growth and make crop/grassland management difficult, careful consideration needs to be given to design gradients. For agriculture, surface waterlogging can be a serious problem where reinstated slopes are less than 2% unless substrates are free draining. Problems are particularly likely to occur in wetter climatic areas if surfaces are uneven. This will lead to ponding in hollows which will be further exacerbated if soils are compacted when reinstated. At the other extreme steeply sloping land is associated with increased erosion. Restrictions are also encountered on the range of agricultural equipment that can be used. A limit of 20% is used in the Agricultural Land Classification of England and Wales (1) to separate Grades 3b and 4 and represents the upper slope gradient for regular arable use. Above 33% severe difficulties are likely to be encountered in safely operating 2-wheel drive tractors with fully mounted equipment. Where restoration is to forestry minimum slopes of 6° are recommended to achieve adequate drainage for tree growth and various ridging techniques are recommended for achieving these (2). Between slopes of 10° and 20° it is recommended that benches are created each a bulldozer blade width apart at 20 metre intervals to control erosion. Above 20° regrading is recommended due to the difficulties of harvesting.

Use of Soil Materials

Where significant grading presents opportunities to move substrates around site attention needs to be given as to how to create the best conditions for plant growth. Since there is usually a range of materials available, all with different properties, there are frequently many options available. This requires careful matching of soil resources with aftercare requirements since once reinstated the afteruse potential and management requirements of an area of land are to a large extent fixed by the nature of the soil materials below. Furthermore many soil characteristics cannot be changed afterwards. In matching soil resources to aftercare requirements a number of factors should therefore be considered.

Topsoil resources are scarce on most sites and thus require very careful husbanding. They are characterised by their higher organic matter content. This increases the soils ability to supply water and nutrients to plants. It also improves soil structure thus making cultivations easier and increases the bearing strength of the ground. To maximise benefits it is important that topsoils are directed to areas where vigorous plant growth is most required and potential wear and tear on vegetation is likely to be greatest. Target areas for topsoil are likely to be those devoted to uses such as agriculture, sports pitches, access areas to parks and golf course fairways. For agricultural areas, except extensive grazing land, a minimum depth of 200 mm is likely to be required, for many other uses a depth of 150 mm is often adequate. However this partly depends upon the nature of underlying substrate materials.

Purchase of topsoil is an option sometimes considered. However it is frequently expensive and money can sometimes be better spent by more careful treatment of on-site materials. This is because the quality of topsoil sold varies considerably and also because topsoils are sometimes seriously damaged when transferred from one site to another. If topsoils are to be purchased, inspection of materials is essential before they are bought. Steps should also be taken to ensure that they are only stripped and delivered when dry and that due care is taken to prevent contamination with subsoil. However even without topsoil satisfactory reclamation can usually be achieved through raising fertiliser inputs during the initial years after restoration, and use of legumes and addition of amendments such as farmyard manure, mushroom compost, sewage sludge cake or shredded bark to build up organic matter levels.

Texture: Soils are classified according to the proportions of sand (0.6-2.0 mm), silt (0.02-0.06 mm) and clay (<0.002 mm) particles that they contain. Materials with very high sand content tend to be free draining and thus capable of sustaining access onto the land for long periods of the year. However on the debit side they are demanding in fertiliser and susceptible to summer drought. This adversely affects their suitableness for agriculture or intensive amenity use. At the other extreme clays are less susceptible to summer drought. However their slow permeability limits the length and intensity with which they can be used due to damage to swards and surface soil structure under wet conditions. All other factors being equal soils with the best reclamation potential are those with a loamy texture. Their behavioural properties are intermediate between those of clays and sands. Hence most benefit is derived if these materials are utilised in areas where the most demanding afteruses are to be located.

Stone content: Surface stones are a hindrance to cultivations and increase long term wear and tear costs on machinery. They also dilute the soil so reducing the amount of available water for plant growth. In addition they can pose safety hazards for some recreational afteruses. The least stony materials should therefore be kept aside for the formation of cultivation layers particularly where regular cultivation or reseeding of reclaimed land is likely to be required or where land is to be used for sporting purposes. Where installation of underdrainage and/or periodic subsoiling or moling is likely to be necessary it is also necessary to lay down sufficient depths of relatively stone free material below the cultivation layer. This is particularly important for reclamation to more intensive agriculture, golf course fairways, sports, outfields and other amenity areas that will be subjected to intensive use. Although stone removal techniques do exist they may not be practical on very stony materials. Hence prevention rather than cure is nearly always desirable.

Acidity: Some materials exhibit extremes in pH which interfere with nutrient uptake, inhibit root growth or reduce soil microbial activity.

Some chemical wastes exhibit alkalinity in excess of pH12 whilst at the other extreme some colliery spoils contain extremely acid areas even below pH3. In the latter instance liming can provide a solution. However if the spoil contains large quantities of unweathered iron pyrites the lime requirement is likely to be substantial and the risks of acidity recurring strong. In such situations total burial of the offending material can prove to be the cheapest option.

Toxicity: Some materials may contain contaminants in potentially phytotoxic or zootoxic concentrations. Guidance on "trigger concentrations" above which levels may prove hazardous are given by the Interdepartmental Committee on The Redevelopment of Contaminated Land (3). Since some levels are hazardous for certain afteruses and not others there is sometimes scope for minimizing toxicity effects by matching materials to afteruse. For example, whereas lower concentrations may pose a threat to the use of land for allotments, higher levels can be tolerated for playing fields and open space. This is because there are no risks from digestion of contaminated produce by humans. Futhermore plant cover also provides protection from exposure to contaminants in the soil beneath. Other options for dealing with contamination include burial of materials or removal in extreme cases. Sometimes treatment in-situ can also be practical, eg liming or blanketing with non-toxic substrates. Where treatment is not feasible there are frequently higher long term costs in terms of reduced afteruse options, diminished plant vigour and/or the need for increased management units.

Soil Profile Reinstatement

Soil Structure. Plant growth is related to the amount of water and nutrients available. This in turn is influenced by the volume of soil that plant roots can penetrate which is partly determined by soil structure. Soils with a good structure have a good balance of pore sizes. In particular macropores (>50 μm diameter) which allow unimpeded penetration of roots, water and air; and mesopores (5-50 μm diameter) which hold a reservoir of moisture available to roots. They also contain an interlinked network of pores and fissures that permit root penetration to depth. This structure takes decades even centuries to develop and is very easily damaged by earthmoving operations of the type associated with land reclamation. On most sites soil damage has already occurred as a result of initial stripping operations associated with the original site development, subsequent burial of soil materials and long term soil storage. However, safeguards can be adopted to minimize the extent of further structural damage during reclamation.

Soil handling. During soil replacement it is essential the soils are only handled when dry. Otherwise soils run the risk of being severely compacted by heavy earthmoving equipment. This reduces the number of macropores and mesopores with consequential damage to soil drainage and available moisture holding capacity. This has the dual effects of making the land more susceptible to waterlogging in winter and drought in summer. Smearing of fissures as each separate layer of soil is placed adds to these problems. As every single passage of equipment is potentially damaging it is therefore important that strict site discipline is enforced to ensure that layed soils and those awaiting stripping are not traversed except when being lifted or replaced.

Since the British climate is unpredictable and frequently wet it is advisable to work in small panels when replacing soils. These should then be vegetated as soon as reasonably practicable upon completion. Attempts to reclaim large expanses in one go can result in waterlogging of significant areas. This not only slows down operations but also damages the quality of finished work. By working in panels, if rain does threatens it is advisable to seal the soil overnight by backblading or grading. Work can then resume in the morning once the soil surface has been re-opened.

Soil ripping: To loosen compacted zones and remove stones and other obstacles it is important to rip subsoil (substrate materials) before the cultivation layer is reinstated. Ripping depths being determined in part by the depth and nature of the underlying materials and also the proposed afteruse. However, where possible ripping to a minimum depth of at least 450 mm is likely to be required. Where agricultural underdrainage is required in the future, two stage ripping of the subsoil is often necessary to remove stones and other obstacles. Once when half the depth of subsoil is placed and then again when the full depth is deposited. Subsoiling is normally carried out upon placement of the cultivation layer irrespective of whether underdrainage is to be installed or not. Except where subsoils are excessively stony, this subsoiling depth needs to penetrate at least 150 mm into the underlying subsoil so as to deal with compaction caused as a result of topsoil placement. Where major topsoil destoning is required mechanical stone picking can be used to good effect. However, care needs to be taken to avoid excessive stone removal as this can cause soil structural problems.

VEGETATION ESTABLISHMENT AND MAINTENANCE

Objectives

Placed soils frequently exhibit severe physical and/or chemical constraints to plant growth and afteruse. Treatments and management strategies in the initial years following placement need to be designed to ameliorate these conditions so as to achieve successful vegetation cover, reduce long term maintenance requirements and render the land reasonably fit for the choosen afteruse.

Aftercare

Placed soils tend to be structurally weak and prone to erosion. Their internal structure can also breakdown under the influence of rain and waterlogging. It is therefore crucial that the ground is cultivated and seeded as soon as soil and weather conditions permit so as to provide suitable winter cover. This way protection against erosion is provided through the binding of soil particles by plant roots. The downward growth of roots also helps to hold open the soil structure. Extraction of moisture by vegetation during the spring and summer also improves surface ground conditions thus aiding access by machinery which in turn makes management easier.

Lime and fertiliser applications are essential for successful vegetation establishment particularly on derelict sites where nutrient levels are frequently low. Prior to cultivations samples should be taken and sent away for laboratory analysis. Recommendations on applications can then be devised based upon current nutrient levels. For agricultural purposes target levels are likely to be set at index 2 for P and K with a minimum pH of 6.0. A fertiliser strategy can then be developed to build up nutrient reserves or alternatively maintain them at current levels. For impoverished materials and/or where agriculture and intensive amenity uses are planned, relatively high inputs are likely to be needed for several years. On very impoverished substrates organic matter content may be so low that vegetation is dependent upon fertiliser nitrogen for several years until soil organic matter levels are sufficient to provide enough N to sustain plant growth. In these situations in particular and for reclamation in general, establishment of legumes is thus very important due to their ability to fix N direct from the air and make it available to plants. In this way clover can supply over 100 kg/ha/annum with consequential long term reductions in fertiliser bills.

Seed mixes. A wide range of specifications exist and these need to be carefully matched to aftercare requirements. Many agricultural and amenity mixes are based upon ryegrasses which are less demanding in their seedbed requirements than many other grasses. They are also rapid to establish, high yielding and hardwearing - all valuable qualities for land reclamation. However mixes are also available for a wide range of other purposes including extensive agricultural production, low maintenance amenity grassland and conservation. For the latter a range of mixtures are available for a number of different soil types.

Seed rates need to be varied according to ground conditions and type of mix. Typical rates for normal agricultural use being 40-45 kg/ha and slightly higher around 55 kg/ha for low production mixes. However rates as high as 80 kg/ha may be needed where ground conditions are particularly poor. Time of sowing is also important. Where clover is an important component this should normally be completed before the end of August. Just one months delay after this can reduce winter survival rates fivefold. However for some wild flora mixes, early Autumn sowing is advantageous as some herbs require low temperatures for vernalization.

Grassland Management. For agricultural purposes close sward control via grazing, mowing and topping is important to create a dense productive sward. This requires careful management to ensure that utilization and growth are kept in balance. Also grazing by livestock or access by farm machinery during wet periods must be prevented to avoid compaction and poaching damage to the topsoil and sward. To achieve this it is frequently advisable to manage land via grazing licences. This also enables the landowner, who has a long term interest in the land, to ensure that any necessary reseeding, fertilising, weed control etc operations are carried out. Where land is inherently poor such arrangements may continue indefinitely to ensure that visual amenity is not lost through unsympathetic farming. However to work well adequate manpower must be set aside to ensure that sites are adequately supervised.

Management of herb rich meadows requires a different approach. Here fertililty is deliberately kept low to encourage species diversity, hence fertilisers are not normally applied, although benefits may accrue from small inital applicatons to assist sward establishment from seed. Swards are then left to grow and are not cut until mid July at the earliest. This gives herbs an opportunity to set seed. Following cutting meadows can then be grazed, however it is essential to ensure that they are shut up at the end of January before the next seasons growth commences.

With amenity grassland cutting is often the largest single item of maintenance expenditure. Considerable savings can thus be made by making maximum use of low maintenance mixes in areas where land is not subjected to significant wear and tear. By setting aside 'wilderness' areas further savings can be achieved where cutting occurs only once or twice a year but with more frequent cutting around paths, perimeters and picnic sites etc for tidiness and to create the image that the land is being well maintained. Preservation of native vegetation that has colonized naturally and its enhancement by planting with trees, shrubs and/or herbs from pots are additional steps that can be undertaken. This approach enhances conservation and visual interest whilst simultaneously reducing long term maintenance costs.

<u>Drainage:</u> Drains to intercept surface water and transmit it off site are an important component of larger reclamation schemes. They are essential to achieve adequate control of run-off, to minimize erosion and prevent flooding or waterlogging and also benefit vegetation establishment. Water control can also be used in reverse to create water areas and wetlands for conservation, sporting and amenity use. For intensively used areas such as sports pitches, parkland and better quality agricultual reclamations, installation of underdrainage is often desirable. A variety of techniques are available involving the use of pipes, permeable fill, sand slits and drainage-layer construction. However actual design is site and afteruse specific and requires the services of professionals. Although relatively expensive such systems can yield marked improvements in surface performance and reduced maintenance costs. These include increased plant growth, improved sward quality, lower weed control costs, greater resistance to wear and tear and an extended season for grazing or human use due to drier surfaces.

Since many newly restored soils are structurally unstable and suffer from soil compaction, opportunities should also be taken as they occur to improve soil structure and the drainage of surface layers via appropriate treatments such as subsoiling, moling or spiking. Such treatments also encourage deeper rooting and give additional benefits in terms of reduced susceptibility to drought. On agricultural land subsoiling should be undertaken whenever reseeding is to take place so long as soil conditions are suitable. This is because rigorous subsoiling is frequently not feasible at other times due to the risk of causing damage to the sward.

Woodland

Planting of tree areas and woodland is likely to be a feature of many reclamation schemes. Successful tree establishment depends upon the creation of adequate slopes (mentioned earlier) and then ripping to break up compaction. Ripping to depths of 600mm helps reduce waterlogging and encourages deeper rooting resulting in higher tree survival rates and faster growth. Deeper rooting also reduces susceptibility to windthrow in later years. Sturdy transplants are recommended for planting and care needs to be taken to ensure that the root is spread out and fully covered when planted. Planting should only occur during the period from late autumn to early spring and every effort needs to be taken to ensure that roots do not dry out between lifting and planting. Sometimes trees are planted directed into bare ground, on other sites the soil is grassed a year in advance to stabilise the surface and improve visual appearance. In such cases the grass needs to be killed off around the trees when planting occurs. This is very important, otherwise tree survival is reduced and growth seriously checked by competition from the grasses for nutrients and moisture. Weed control then needs to be maintained until the trees become well established. Beating up to replace any significant tree losses during the first few years is also necessary and if losses occur in concentrated locations the causes investigated since they may be indicative of soil problems that require amelioration. Choice of tree species is dictated in part by ground conditions and also management objectives. However alders are commonly included within planting mixes due to their ability to fix nitrogen.

CONCLUSIONS

A wide range of materials are found on derelict sites - some less conducive to plant growth than others. Detailed site investigation is highlighted as an essential prerequisite to success. This enables best use to be made of the limited materials available and permits soil types and land uses to be matched together. Furthermore a variety of site amelioration and husbandry requirements are identified as important. Overall the quality of reclamation depends upon careful planning, systematic evaluation and attention to detail. Close supervision is essential throughout.

REFERENCES

1. Ministry of Agriculture, Fisheries and Food, Agricultural Land Classification of England and Wales, MAFF (Publications) 1988.

2. Wilson, K. A Guide to the Reclamation of Mineral Workings and Forestry, Forestry Commision, 1985.

3. Interdepartmental Committee on the Redevelopment of Contaminated Land, Guidance on the Assessment of Contaminated Land, First Edition, CDEP/EPTS, Romney House, 43 Marsham Street, London, 1983.

RE-ESTABLISHING EARTHWORM POPULATIONS ON FORMER OPENCAST COAL MINING LAND.

JOHN SCULLION
Department of Biochemistry,
University College of Wales, Aberystwyth,
Penglais, Aberystwyth, Dyfed SY23 3DD, UK.

ABSTRACT

Investigations were carried out into various aspects of earthworm ecology on replaced land. Re-colonization from undisturbed land was an important factor in the recovery of *Aporrectodea caliginosa* and *Lumbricus rubellus/festivus*, but not of *Allolobophora chlorotica*.

Recently replaced sites were found to be capable of supporting large populations, with a species composition equivalent to that of similarly managed, undisturbed land. The absence of an effective inoculum appeared to be the major factor limiting recovery. Material from the surface of topsoil stores may, in part, provide such an inoculum, although further work is required to optimize this procedure. Subsequent management has an important influence on the establishment and development of populations. Regular and adequate supplies of food, such as occur under grazing, appeared to be important for all species. Field drainage and, particularly, fissuring associated with secondary drainage treatments greatly increased numbers of deep-burrowing species.

INTRODUCTION

Earthworms contribute to the improvement of soils replaced after opencast mining [1,2]. They are also a significant element in the food resource of a range of terrestrial vertebrates [3]. Therefore, in planning site working and subsequent rehabilitation, consideration should be given to ways in which an early recovery in populations might be achieved.

Soil stripping, storage and reinstatement practices [4,5] all contribute to a large reduction in earthworm populations. Recent studies in Britain [6,7] have indicated that a full recovery in earthworm populations following soil replacement may take longer than previously had been envisaged [8].

The rate of recovery appears to vary from one earthworm species to another. Dominant species shortly after reinstatement tend to be small and to have a surface dwelling habit. It may be 10 years or more before the larger, deep-burrowing species become numerous [6]. These larger species are more beneficial for soil rehabilitation and for site ecology. They encourage soil structural development to greater depths and, by virtue of their size, provide a more 'cost effective' food source for foraging mammals and birds.

Findings presented here arise from an ongoing investigation aimed at identifying practical means of encouraging early establishment and rapid development of earthworm populations in replaced soils. Data reported relate to natural recovery in populations, to the effect of land use on this recovery and to methods of establishing viable populations on newly reinstated land.

METHODS

Work was carried out on the north-western margins of the South Wales coalfield, in a high (>1500 mm per year) rainfall area. Soils present were predominantly clay loam in texture and all had been stored during mining. Agricultural land studied consisted of long-term grass-clover leys, either grazed by sheep, or cut for silage with grazing in Spring and in Autumn. Where manurial treatments were applied these consisted of urea (50 kg N ha^{-1}) and poultry manure (100 kg N ha^{-1}))

Selected areas were monitored on the Rockcastle East site which was replaced in 1972 (SN259214). This investigation aimed to assess the importance of colonization from undisturbed land in the recovery process and the effect of different land uses on population growth. Colonization data were obtained during 1978, the first growing season following the five year rehabilitation period, from two fields which contained areas of undisturbed land. Following an initial survey of the site in 1980, pairs of fields with similar management were selected and monitored periodically to assess population trends. Undisturbed land was included in this survey. Field drainage had not been installed in any of the fields sampled at the time of the 1978 and 1980 surveys.

The 'management packages' chosen for comparison represented a range of practical options for using replaced land and included :-
1. An extensive grazing regime with no manurial inputs or field drainage.
2. A moderately intensive grazing/cutting regime with applications of poultry manure and nitrogen fertilizer to give annual nitrogen inputs of approximately 150 kg per hectare. Field drainage treatments were not carried out because the uneven surface resulting caused problems when cutting.
3. A moderately intensive grazing regime with manurial inputs similar to those for option 2. Here underdrainage was

installed and agricultural subsoiling carried out to direct excess water to these drains and to create a framework of large cracks within the compact soil. Under a grazing regime, an uneven surface is less of a problem.
4. Groups of alder (*Alnus glutinosa*) within mixed, predominantly deciduous woodland planted in 1979 at spacings of 1.8 m. Percentage canopy cover in the areas sampled ranged from 0 in 1979 to 90 in 1990.
5. Management as for option 2 but on adjacent, undisturbed land.

On the Glyn Glas site (SN590140), replaced in 1984/5, a study was carried out to assess the potential of direct inputs of earthworms into newly replaced soils. A method of inoculating these soils using material from the surface of topsoil heaps was also evaluated. The areas studied were drained and subsoiled, received applications of poultry manure/urea and were grazed by sheep.

In the first trial, large numbers of earthworms were collected during ploughing of local, undisturbed pasture. These were introduced at intervals into recently drawn mole plough slits. Inputs consisted mainly of five species; *Lumbricus terrestris* (22%), *Aporrectodea longa* (26%), *Aporrectodea caliginosa* (23%), *Lumbricus rubellus* (11%) and *Allolobophora chlorotica* (10%), with the overall input equivalent to 69.8 m^{-2} for all species. Earthworms were introduced on three occasions up to March 1986 and sampled in October 1988. The land management programme included all practices - subsoiling, grazing and poultry manure- previously found to encourage earthworms [2,9,10]. Undisturbed pasture, receiving similar manurial inputs, was sampled the following month to provide a comparison.

In the second trial, topsoil from the surfaces of two heaps was replaced by motor-scraper in strips as part of normal site reinstatement practice (see [5] for details). Earthworms in these 'inoculation strips' were monitored, as was their spread into the intervening soil.

For all data reported here, earthworms were sampled by formaldehyde extraction [12]. In some cases population estimates using formaldehyde extraction were checked against a combined handsorting/formaldehyde sampling. For most species, the latter procedure was 5-10 times more efficient, although findings were similar regardless of sampling method. When identifying earthworms (13), *Lumbricus rubellus* and *Lumbricus festivus* were grouped because of difficulties in distinguishing between immature specimens of these two species. All abundance data include both mature and immature earthworms.

RESULTS

Natural Colonization
The extent to which earthworm populations decrease systematically with distance from undisturbed land provides some indication as to the relative importance of colonization and development *in situ*. Data presented refer to the position

six years after soil replacement. Points sampled varied from 3 to 85 m from undisturbed land.

Three species - *Allolobophora chlorotica* (57%), *Aporrectodea caliginosa* (21%) and *Lumbricus rubellus/festivus* (18%) - accounted for 96% of the population on replaced land. The deep burrowing species *Lumbricus terrestris* and *Aporrectodea longa* were not recorded. Two adjacent sampling points, 80-85 m from undisturbed land, were anomalous in having large populations containing an unexpectedly high proportion of *Aporrectodea caliginosa*.

TABLE 1. Earthworm colonization into replaced soils from undisturbed land (Rockcastle East 1978); regressions of earthworm indices ($\log_{10}$ per m^2) on distance (m) from undisturbed land.

y		Distance (m)
Live weight (g)	$y =$	$-0.005x + 1.09^*$
Numbers of:-		
Allolobophora chlorotica	$y =$	$+0.007x + 0.31^*$
Lumbricus rubellus/festivus	$y =$	$-0.003x + 0.84^*$
Aporrectodea caliginosa	$y =$	$-0.011x + 1.24^{**}$

$n = 25$ $^* P < 0.05$ $^{**} P < 0.01$

Earthworm biomass decreased exponentially with distance from undisturbed land (Table 1). For the three numerically co-dominant species recorded the position was more complex. Numbers of *Lumbricus rubellus/festivus* and of *Aporrectodea caliginosa* also declined exponentially with distance from undisturbed land, the decrease in numbers of *Aporrectodea caliginosa* being particularly marked, as indicated by the larger, negative regression coefficient. In contrast, numbers of *Allolobophora chlorotica* tended to increase at greater distances from undisturbed land. This reversal of the trend apparent in other species may be explained partly by reduced competition for food resources at greater distances from undisturbed land.

Establishing Populations by Direct Input

Inputs (1985-6) consisted almost entirely of large, mature individuals and few of these would have survived until sampling. Compared with the initial inputs, live weight, total number and abundance of all introduced species was markedly higher when sampled in October 1988. Furthermore, the age structure of each species suggested that breeding populations had become established in all cases.

The weight of earthworms recovered from input plots on replaced land was greater than that on similarly managed, undisturbed land in the area. The species composition was broadly similar on the two areas, although the replaced land had fewer *Aporrectodea caliginosa* and more of the two deep-burrowing species *Lumbricus terrestris* and *Aporrectodea longa*. Inter-species competition may explain these differences in relative abundance on the two areas sampled.

TABLE 2. Earthworm populations (m^{-2}) in replaced soils subject to direct inputs and in adjacent undisturbed pasture under similar management.

	Input plots	Undisturbed land	LSD*
Live weight (g)	58.8	42.6	5.29
Numbers of:-			
A. chlorotica	21.7	27.2	6.11
L. rubellus/festivus	25.0	20.1	5.32
Ap. caliginosa	8.8	41.0	7.44
L. terrestris+Ap. longa	21.5	10.1	3.75

LSD* = least significant difference $P < 0.05$

Establishing Populations Using Material from the Surface of Topsoil Stores

Data for 'inoculation strips' are given in Table 3 along with values for points less than or greater than 20 m from these strips. Results were grouped in this way on the basis of assumed rates of spread of 5 m per year. Previous studies of earthworm colonization have found a similar rate of spread [13].

Populations in strips showed a marked increase over the three growing seasons since the previous sampling in 1986 [5]. Between strips, earthworms were absent from all but a few of the points sampled in 1986 [5]. Here again, earthworm abundance was markedly higher compared with previously recorded values.

TABLE 3. Earthworm populations (m^{-2}) on Glyn Glas site (1984/5 reinstatement) sampled in October 1988.

	Strips	<20 m from strips	>20 m from strips
Live weight (g)	15.0_a	15.5_a	7.4_b
Numbers of:-			
A. chlorotica	24.4_a	22.9_a	18.0_a
L. rubellus/festivus	12.3_a	11.0_a	2.8_b
Ap. caliginosa	9.1_a	1.9_b	0.8_b
Ap. longa + L. terrestris	2.0_a	0.2_b	0.1_b

Within rows, means with a common subscript did not differ at a 5% level of probability.

There was clear evidence of inoculation strips acting as a source of colonization into intervening soil. Live weight, total number and abundance of *Aporrectodea caliginosa* and *Lumbricus rubellus/festivus* showed significant ($P < 0.01$) negative correlations with distance from strips. No such relationship was found for *Allolobophora chlorotica*. In the fourth year after soil replacement (Table 3), earthworm populations, similar in weight to those of the strips, were recorded on land <20 m from strips. However, numbers of *Aporrectodea caliginosa* and of *Lumbricus terrestris*+ *Aporrectodea longa* were significantly higher on strips. This result provides some indication of the rate of population spread from strips. At distances greater than 20 m from strips, population weight and numbers of *Lumbricus rubellus/ festivus* showed a further significant decrease compared with the grouping closer to strips.

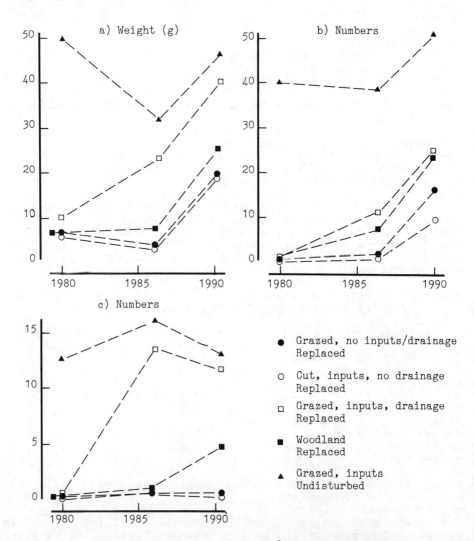

Figure 1. Changes in populations (m^{-2}) under different land uses; a) Live weight b) numbers of *Aporrectodea caliginosa* and c) numbers of *Lumbricus terrestris* + *Aporrectodea longa*

Management effects

Changes with time in earthworm populations on replaced land under different land uses are illustrated in Figure 1. Data for adjacent, undisturbed land are also given in order to place in context trends on replaced land and to indicate changes relating to, for example, weather conditions in a particular year. Data are presented for live weight, as an indicator of population size, for *Aporrectodea caliginosa* which burrows to intermediate depths and for the two deep burrowing species *Lumbricus terrestris* and *Aporrectodea longa*. In 1980, eight years after reinstatement, populations on replaced land were small with weight 10-20% of the undisturbed figure and numbers of intermediate/deep-burrowing species less than 5% of the equivalent value for undisturbed land. It should be noted that 1980 was the first growing season since reinstatement in which management varied systematically.

During the period 1980-86 population weight increased markedly only on grazed fields which had been drained and which received manurial inputs; the weight of earthworms recorded increased from 20% to 74% of the undisturbed value. Numbers of *Aporrectodea caliginosa* (46% of undisturbed) and of the two deep burrowing species (89% of undisturbed) also increased markedly on these fields only. Apart from the larger numbers of *Aporrectodea caliginosa* in woodland, there was very little change in populations under the other replaced land uses.

Between 1986 and 1990 there was a general increase in population weight and numbers of *Aporrectodea caliginosa*, but numbers of the deep-burrowing species tended to decrease. At this point, 18 years after reinstatement, population weight on grazed and drained fields receiving manurial inputs was 85% of the undisturbed value, with only relative numbers of *Aporrectodea caliginosa* (39%) markedly lower. For the other land uses, population weights were less than 50% of the undisturbed level, whilst the equivalent figure for deep-burrowing species was less than 10%.

DISCUSSION

For *Lumbricus rubellus/festivus* and *Aporrectodea caliginosa* colonization from undisturbed land appears to be an important process in population recovery. *Allolobophora chlorotica*, in contrast, appears to recover *in situ*. Numbers of *Lumbricus terrestris* and *Aporrectodea longa* were too low to allow any conclusions to be reached as to the importance of migration from undisturbed land for these species. However, given the particular vulnerability of large earthworms to mechanical damage, it is unlikely that many would have survived the mining process. These findings are consistent with previous studies [6,7] in their ranking of species as early or late colonizers of replaced land. Colonization from inoculation strips followed a similar pattern with regard to species. With the tendency towards larger areas of worked land and rates of population spread of the order of 5 m per year

[13], the absence of an inoculum is likely to become a major factor limiting population recovery. Since soil storage inevitably leads to a drastic reduction in earthworm populations [5], this problem might be alleviated, where practicable, by the adoption of progressive reinstatement practices which reduce the need for soil storage.

Recently reinstated land has the potential to support a normal earthworm population (Table 2) provided a) a suitable inoculum is provided and b) management practices are favourable to earthworms. Given differences in the timing of sampling it would be inappropriate to infer too much from the larger populations obtained on replaced land. Nevertheless, it is of interest that numbers of the deep-burrowing species *Lumbricus terrestris* and *Aporrectodea longa* were particularly high in replaced soils. It is clear, therefore, that soils on the Glyn Glas site were capable of supporting large numbers of deep-burrowing earthworms, if suitably managed.

Material from the upper surface of topsoil heaps can provide a source of colonization for the species *Lumbricus rubellus/festivus* and *Aporrectodea caliginosa*. The weight of earthworms in strips was approximately 32% of the average for input plots reported in Table 2. This difference was accounted for mainly by the much smaller numbers of *Lumbricus terrestris* and *Aporrectodea longa* found in inoculation strips. Material used for strips originated from topsoil heaps which had very low numbers of these two species [5]. It is, therefore, unclear whether or not this procedure represents an effective method of re-establishing deep-burrowing species within a site; modifications may be required to the management of topsoil heaps and to their handling at reinstatement to achieve this objective. This aspect of site working practice is the subject of current research.

Land management practices have a major influence on the development of any population after it has been established. As far as deep-burrowing species are concerned, data from experimental plots suggest that regular vertical fissuring of the soil by subsoiling is critical [10]; high bulk density has been suggested as a factor [7] limiting the spread of *Lumbricus terrestris* on opencast land. On replaced land, only those fields which were drained and subsoiled showed any marked increase in numbers of deep-burrowing species (Figure 1c). Drainage treatments at Rockcastle East were not carried out until six years after reinstatement and this would have delayed recovery in numbers of deep-burrowing earthworms. In contrast, underdrainage was installed on the Glyn Glas site soon after reinstatement and prior to seeding of the site. Food supply was probably the controlling factor determining population size under the different land uses. Regular and substantial returns are present only where a productive sward is grazed. Under the other land uses monitored, food supply was limited or inadequate for a large part of the year. Results already obtained indicate practices favourable to earthworms; these include, in addition to regular subsoiling, grazing rather than cutting [2], use of organic manure rather than mineral fertilizer [9], and avoidance of cultivation [14].

With favourable management, land newly reinstated after

opencast mining has the potential to support normal populations of earthworms. With modifications to the management and handling of topsoil heaps, material from their surface may be used to provide a source of colonization at intervals throughout a site. This procedure offers the prospect of establishing near-normal populations on reinstated land within the current aftercare period. However, it should be emphasized, that measures to establish populations are worthwhile only where the intended afteruse is favourable to earthworms.

REFERENCES

1. Vimmersted, J.P., Earthworm ecology in reclaimed opencast coal mining sites in Ohio. In Earthworm Ecology, ed. J.E. Satchell, Chapman & Hall, London, 1983, pp 229-240.
2. Stewart, V.I., Scullion, J., Salih, R.O. & Al-Bakri, K.A., Earthworms and structure rehabilitation in subsoils and in topsoils affected by opencast mining for coal. Biol. Agric. Hortic., 1988, **5**, 325-338.
3. MacDonald, D.W., Predation on earthworms by terrestrial vertebrates. In Earthworm Ecology, ed. J.E. Satchell, Chapman & Hall, London, 1983, pp 393-414.
4. O'Flanagan, W.C., Walker, G.J, Walker, W.M. & Murdoch, G., Changes taking place in topsoil stored in heaps on opencast sites. Nat. Agric. Adv. Serv. Quart. Rev., 1963, **62**, 85-92.
5. Scullion, J, Mohammed, A.R.A. & Richardson, H., The effect of storage and reinstatement procedures on earthworm populations in soils affected by opencast mining. J. Appl. Ecol., 1988, **25**, 233-240.
6. Armstrong, M.J. & Bragg, N.C., Soil physical parameters and earthworm populations associated with opencast coal working and land restoration. Agric. Ecosyst. Env., 1984, **11**, 131-143.
7. Rushton, S.P., Development of earthworm populations on pasture land reclaimed from opencast coal mining. Pedobiologia, 1986, **29**, 27-32.
8. Brook, D.S. & Bates, F., Grassland in the restoration of opencast coal sites in Yorkshire. J. Brit. Grassl. Soc., 1960, **15**, 116-123.
9. Scullion, J & Ramshaw, G.A., The effects of various manurial treatments on earthworm activity in grassland. Biol. Agric. Hortic., 1987, **4**, 271-281.
10. Scullion, J. & Mohammed, A.R.A., Effects of subsoiling and associated incorporation of fertilizer on soil rehabilitation after opencast coal mining for coal. J. Agric. Sci, Camb., in press.
11. Sims R.W. & Gerard, B.M., Synopsis of the British Fauna (New Series) No. 31. Earthworms, 1985, Linnean Society of London.
12. Raw, F., Estimating earthworm populations by using formalin. Nature (London), 1959, **184**, 1661-1662.
13. van Rhee, J.A., Development of earthworm populations in polder soils. Pedobiologia, 1969, **9**, 133-140.
14. Scullion, J. Mohammed, A.R.A. & Ramshaw, G.A., Changes in earthworm populations following cultivation of grassland

on undisturbed and former opencast coal mining land. *Agric., Ecosyst. Env.*, 1988, **20**, 289-302.

RECONSTRUCTION OF LIME QUARRIES IN THE MEDITERRANEAN REGION OF SPAIN

M. Paz ARAMBURU MAQUA, Rafael ESCRIBANO BOMBIN
Dpto. de Proyectos y Planificación Rural, E.T.S.I. de Montes
Universidad Politécnica de Madrid

ABSTRACT

A study was made of the plant species which colonize the lime quarries of the Mediterranean Region of the Spanish, via the analysis of the environmental factors and the mining characteristics of 20 quarries chosen from a total of 312. A statistical analysis was conducted of principal components with 28 species and a hierarchical grouping was made of the quarries. 17 species are of interest for land restoration and of these, 5 could be used in the whole area of study.

INTRODUCTION AND OBJECTIVES

In 1989, a study was conducted of the lime quarries located in the Spanish Mediterranean area, sponsored by the Instituto Tecnológico Geominero de España, of the Ministry of Industry and Energy of the Government. In Spain, quarry mining accounts for 80 per cent of the national mining industry and 51 per cent of the quarries are for the production of lime. 35 per cent of the lime quarries are located in the climatic area studied. The Mediterranean climate, in its strictest sense, affects over 40 per cent of the surface area of Spain. The altitude, the distance from the sea and the presence of mountain barriers afford the Mediterranean climate a large local variability. There are problems in this area with regard to the vegetative development of the plants. For this reason, the objective of the study centred upon the detection of plant species well adapted to the said environmental conditions, capable of developing and colonizing the lands degraded by mining of lime quarries. The aim was to recommend these species for the revegetation projects of these zones to facilitate the natural colonization processes and propitiate the landscape integration.

METHODOLOGY, ANALYSIS AND CONCLUSIONS

The study was conducted in four stages. The first stage included phytoclimatic zoning of the Mediterranean region and inventory of the main lime quarries. The second stage, the sampling, consisted on selection of quarries, design of sampling programme and field work, the pilot sampling and the data collection. The third stage was the data analysis and processing leading to the selection of plant species and to the establishment of groups of quarries. The stage four were the conclusions.

The phytoclimatic zoning included the climatic and phytological aspects of the Mediterranean region. In order to select the quarries, an index of frequencies was used which related the surface area of each phytoclimatic zone, the number of quarries in each one and the total number of quarries in the Mediterranean region studied. Criteria of location, accessibility of quarries and use of the lime material obtained, were also used. Finally, 20 quarries were selected; in the different zones of each quarry, an inventory was made of the plant species present. The number of plants, grass and wood species present in the inventories inside the quarries totalled 282. In the selection of species, only those present in more than four inventories were considered. The result was a total of 28 species. The significance of each species was sought through an analysis of principal components. The quarries were classified in 4 groups, using the significance of the species. Group 3 was the best defined, without regard to the others. The species which characterize it are the most evolved and are to be found in abandoned quarries. Group 2 includes species specific to very degraded areas, the majority being pioneer species. Group 1 includes grass species with a tendency to be located in sites with a certain seasonal humidity. Group 4 is related to 1 and 2, containing in addition species which denote a certain evolution. The species common to the majority of the plots are the Dittrichia viscosa, Conyza canadensis, Piptatherum miliaceum, Psoralea bituminosa and Marrubium vulgare.

TREE ESTABLISHMENT ON DERELICT LAND: GETTING IT RIGHT

ALAN S. COUPER
Lothian Regional Council, Department of Planning,
Landscape Development Unit, Peffer Place, Edinburgh EH16 4BB

ABSTRACT

Lothian Regional Council has been reclaiming derelict land to woodland in the Central Belt of Scotland since 1976. Establishing vigorously growing trees on retorted oil shale and colliery waste in a cold, wet and windy region is difficult and has led to the evolution of a particularly robust approach. No magic formula exists, rather it is careful attention to detail that creates a self sustaining woodland within 5-10 years.

PLANNING AND DESIGN

The technique behind this concept is to avoid the use of grass and sow winter barley and lupins instead. It does not work unless the regraded slope angle, drainage, soil grading, Ph and nutrient status, are considered first, along with footpath routes and the local authority's ideas for after use. After that one can design a woodland with two mixes, a 'shrubby' edge and a climax mix incorporating up to four pioneers, alder, birch, larch and pine.

GROUND PREPARATION AND PLANTING

Areas intended for planting start bare, any existing vegetation or weeds being entirely cleared prior to cultivation. Ripping at an oblique angle to the contours at 1.0m centres to a depth of 600mm with a winged subsoil tine follows. All stones greater than 150mm are removed. A compound NPK fertiliser and/or superphosphate is applied to redress the nutrient deficiency prior to sowing. A barley nurse crop incorporating four Rhizobium inocculated legumes, broom, gorse, perennial lupin and tree lupin, is sown by a double hand broadcast, following a mixing with sand to ensure a 75% scarification of the fine legume seeds. In comparison with the barley sowing rate of 100kg/ha, the legumes vary from 1.25-2.7kg/ha. After sowing they are incorporated to a depth of 5mm by raking or light harrowing.

Bare root transplants from a Scottish nursery are dipped in an Alginate solution immediately they are lifted and dipped in water prior to planting. They are planted in pits at 1.0m centres, parallel to the ripping lines at an offset of 100mm. The pits are backfilled with a 1:6 mix by volume of planting compost. Some transplants, like the willow, are then cut back to a suitable bud 150mm above ground. A spring dressing of a slow compound release fertiliser is applied.

AFTERCARE

In the winter following sowing of legumes, a soil acting residual herbicide is applied to control weeds. Certain tree species are cut back to 250mm above ground level, to encourage bushy growth, the prunings from willow inserted in the ground as cuttings. Particular attention is paid to regular firming up after severe gales and to watering during drought conditions. Although the SDA only assist with four years of maintenance the Regional Council has a policy to manage woodlands on reclaimed land for 10 years, which with Lothian's climate is essential.

CONCLUSIONS

The benefit of this 'bare earth', barley and lupin approach are;
- besides relief of ground compaction, ripping introduces a 'micro' sub surface drainage system.
- the winter barley stabilises the surface with little risk of superficial erosion.
- the fertiliser nutrients and ground moisture go to the transplant.
- the barley and lupins very quickly create a sheltered micro climate for the transplants.
- the lupins introduce maximum nitrogen input into the ground.
- the green, colourful when in flower, and 'bushy' appearance is very attractive.
- the 'woody mix' quickly gains mass yet access for weeding, if needed, is easier.
- the transplants grow more quickly.

There are some risks and limitations which emphasise why it needs careful design;
- the surface is 'at risk' until the winter barley has germinated.
- slopes greater than 1:3 present a high risk of superficial erosion with these techniques and should be avoided.
- insufficiently angled ripping accelerates the surface water run off thus exacerbating erosion.
- a lack of intermediate slope drainage introduces another risk of surface erosion.
- more site inspections are necessary to avoid fire risk when winter barley ripens.
- tree lupin is a prolific self seeder and tends to infest beyond woodland edges.
- tree lupins will die out after 5 years.

Measurement of tree growth is ongoing but from visual analysis this technique works well. A further refinement on shallower slopes has been to forest plough in addition to ripping which creates a more regimented appearance but an even better micro climate.

ACKNOWLEDGEMENTS

The author wishes to thank the Director of Planning for allowing publication of this technical note and to M. Wood, S. Verner and G. Hedger of the Unit for their helpful comments and assistance.

THE USE OF SEWAGE SLUDGE AS A FERTILISER IN THE AFFORESTATION OF OPENCAST COAL SPOILS IN SOUTH WALES

A.J. MOFFAT, N.A.D. BENDING and C.J. ROBERTS
Forest Research Station, Alice Holt Lodge, Farnham, Surrey, GU10 4LH, UK

INTRODUCTION

A shortage of soil is a common problem in the restoration of old opencast coal sites in Wales, especially those that have been worked more than once. As a consequence, trees planted directly into overburden materials often suffer from nutrient deficiencies, especially nitrogen and phosphorus. Artificial fertilisers can be used to redress infertility, but their effects are often ephemeral and they are costly to apply. However, sewage sludge has shown promise as a forest fertiliser at an opencast site in Scotland [1], and recent research has examined the potential for utilisation of sludge in the South Wales coalfield.

METHODS

Digested liquid sewage sludge was applied to the Tredeg opencast site in October 1988, following restoration in 1982 using Carboniferous coal shales as soil forming materials. Sludge was applied using a retracting reel system which comprised a diesel drive hose drum machine with a trolley mounted sprinkler [2]. Sludge was applied over the trees (Japanese larch, *Larix kaempferi*, 6 years old) in a trial consisting of three rates: 75 m^3 ha^{-1}, 150 m^3 ha^{-1} and 250 m^3 ha^{-1}. Assessment plots (total 14) were established in each of the treatments, including 6 untreated controls. Tree heights were recorded at the time of sludge application in 1988, and at the end of the 1989 and 1990 growing seasons. Analysis of foliage was performed at the end of 1989.

RESULTS

Figure 1 shows the response of larch to sludge treatments. The average tree height increment increased by a factor of two in plots receiving 75 or 150 m^3 ha^{-1}, and by a factor of three in plots receiving 250 m^3 ha^{-1} compared to untreated controls.

Figure 2 demonstrates that the addition of sludge dramatically increased foliar nitrogen and phosphorus concentrations, and brought the trees out of deficiency. Sludge additions also increased the ground cover, promoting vegetation and hence initiating soil formation.

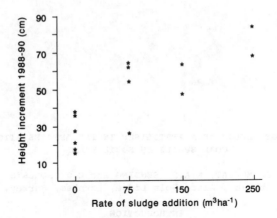

Figure 1. Response of Japanese larch to sewage sludge additions at Tredeg.

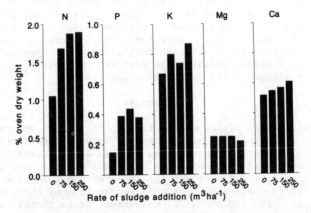

Figure 2. Foliar nutrient concentrations in Japanese larch at Tredeg 1989.

CONCLUSION

The research at Tredeg suggests that sewage sludge is a useful amendment to opencast coal spoil lacking topsoil, though the longevity of the response to sludge additions requires further study.

REFERENCES

1. Bayes, C.D. and Taylor, C.M.A., The use of sewage sludge in the afforestation of former opencast coal sites: Clydesdale Forest Trials. Water Research Centre Report PRU 1774-M/2, August 1988.
2. Daw, A.P. and Stark, J.H., The use of sewage sludge in the afforestation of former opencast sites: progress report on the Maesgwyn and Tredeg trials. Water Research Centre Report PRU 2032-M, January 1989.

MINIMAL SOIL AMELIORATION AS A TRIGGER FACTOR IN THE REVEGETATION OF A DERELICT LANDSCAPE NEAR SUDBURY, CANADA.

KEITH WINTERHALDER
Department of Biology,
Laurentian University,
SUDBURY, Ontario, CANADA P3E 2C6.

Barren soils occupying extensive areas near Sudbury, Ontario have a pH of less than 4.5 and total copper and nickel contents approaching 1,000 µg/g [1]. The barrens are a legacy of 100 years of sulphur-dioxide pollution, copper and nickel particulate deposition, fire, soil erosion and enhanced frost action. Despite improvement in atmospheric quality since 1972, the soils remain toxic to most plants, severely inhibiting root growth. Natural recolonization is limited to metal-tolerant ecotypes of *Deschampsia caespitosa* (L.) Beauv., *Agrostis scabra* Willd., *Agrostis gigantea* Roth. and *Betula pumila* L. var. *glandulifera* Regel.

Detoxification of the barren, acid, metal-contaminated soils can be achieved by a single, manual, surface-application of ground dolomitic limestone at a rate of 10 t/ha. Dolomitic limestone appears to be more effective than calcitic limestone, which can induce magnesium deficiency. The current revegetation technique is entirely manual, surface lime application being followed by 0.4 t/ha 5:20:20 fertilizer and 45 kg/ha of a grass-legume seed mixture [2]. To date, the Regional Municipality of Sudbury has treated 3,000 hectares of barren land in this way, hiring over 3,000 students and unemployed individuals on government job-creation funds [3]. The treatment triggers immediate colonization by native woody plant species of the "pioneer" type, especially *Betula papyrifera* Marsh. (White Birch), *Populus tremuloides* Michx. (Trembling Aspen) and *Salix* spp. (Willows) (Figures 1 & 2). Seeds of these species are blown in from the stunted "semi-barren" woodlands that surrounds the barrens, as well as from stunted, relict individuals on the barrens themselves.

Figure 1. Barren area and workers, 1983   Figure 2. Same area, six years after treatment

Despite the absence of further limestone or fertilizer application, growth has continued over a twelve-year period. Over time, the importance of sown grasses on seeded areas decreases, while that of sown legumes and volunteer native woody and herbaceous species increases (Figure 3).

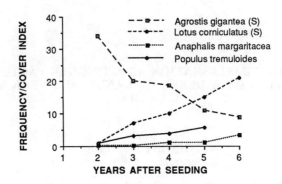

Figure 3. Change in importance of seeded and volunteer species (Cover/Frequency Index = [Relative Cover + Relative Frequency] /2). *Populus* was not measured in the sixth year. (S) denotes seeded species.

In the absence of applied fertilizer and seed, colonization by birch, aspen and willow occurs at an even greater density (Table 1). Sufficient nitrogen exists in pre-disturbance residual soil organic matter to support growth for several years. Phosphorus deficiency only occurs on sandy, low-organic soils.

TABLE 1
Cover/Frequency Index* achieved in two years by woody colonizers on a fertilized and seeded limed plot, and on an unfertilized, unseeded limed plot.

Species	Seeded Plot	Unseeded Plot
Populus tremuloides Michx.	1.5	29.1
Salix spp.	0.9	20.1
Betula papyrifera Marsh.	0.0	4.5
Populus grandidentata Michx.	0.0	3.2

* (Relative Frequency + Relative Cover)/2

It has been concluded, however, that seeding is beneficial, in that it leads to a better spacing of woody colonists, as well as providing nitrogen-fixing legumes.

REFERENCES

1. Hutchinson, T.C. and Whitby, L.M., Heavy metal pollution in the Sudbury mining and smelting region of Canada. I. Soil and vegetation contamination by nickel, copper and other metals. Environmental Conservation, 1974, **1**, pp. 123-132.

2. Winterhalder, K., The use of manual surface seeding, liming and fertilization in the reclamation of acid, metal-contaminated land in the Sudbury, Ontario mining and smelting region of Canada. Environmental Technology Letters, 1983, **4**, pp. 209-216.

3. Lautenbach, W.E., The greening of Sudbury. J. Soil and Water Conservation, 1987, **42**, pp. 228-231.

SECTION 5 : LAND USE AND MANAGEMENT

RAVENHEAD RENAISSANCE - A CASE STUDY

PROFESSOR GRAHAM ASHWORTH CBE
Chairman
Ravenhead Renaissance Ltd
PO Box 68 St Helens WA9 1LL

ABSTRACT

Ravenhead Renaissance is the public and private sector Consortium land owning interest in St Helens Lancashire which has provided the framework for a co-ordinated range of developments across some 250 acres of land to the south of the town centre. The majority of the sites are derelict. The following paper details how Ravenhead Renaissance has established itself as a role model for urban regeneration. The projects which are detailed in the paper are ongoing and provide a very useful illustration of how derelict land can be brought back into economic use through co-ordinated effort.

ST HELENS AND RAVENHEAD RENAISSANCE

St Helens

If you were to look at an old photograph or map of St Helens from the turn of the Century you would see a concentrated hotch potch of buildings, railway lines and industrial uses covering the area south of the town centre, pushing into agricultural land beyond.

Even 10 years ago there was a legacy of 150 years of concentrated development in the same area resulting in the most derelict urban landscape imaginable.

St Helens is now famous for glass, but founded its success on mining, chemicals and manufacturing. Within a very short space of time during the late 1800's the area was transformed from a small town, albeit well located for the Liverpool and Manchester conurbations, to a prosperous and innovative centre for glass manufacture But the transformation laid a foundation for an increasingly derelict landscape up until very recent times.

Nevertheless, St Helens is an area where a number of important 'firsts' in both manufacturing and related activities have been achieved, some back in history, others more recent. The historic 'firsts' include England's first industrial canal in 1754 and the first railway trials of Stephenson's Rocket at Rainhill. The more recent 'firsts' include the first enterprise agency in the country through the St Helens Trust, and now with Ravenhead Renaissance, what is probably the first true partnership between the private and public sectors for a locally based initiative to tackle the problem of derelict land.

My paper is about how St Helens is removing dereliction, not only in terms of schemes on the ground which I will illustrate, but also in the mechanisms which have been brought to bear through Ravenhead Renaissance to promote and implement new economic schemes.

Whilst there is now a pattern of new development laid on the carpet being unrolled across land in St Helens which has a recognisable shape and inter-relating pattern, the basic ingredients needed weaving together in a confused and competing set of circumstances. The town is one which has suffered badly from a lack of inward investment but has achieved replacement from predominantly one industry - glass. This has brought with it as Dickens may have perceived "the best of times and the worst of times" with a reliance on a single industry which, as its economic fortunes have see-sawed, produced substantial new plant but job losses of some 20000 people between 1979 and 1987. It has also had to cope with a view, now proved to be completely untrue, that the private and public sector could not reconcile differences of approach to turn the fortunes of the town to mutual advantage.

Ravenhead Renaissance
So how did the change of fortunes all begin and an end to dereliction become possible?

One Sunday afternoon in the Autumn of 1987 I received a phone call asking me to consider chairing a consortium of private and public interests that was to address the problem of urban regeneration in St Helens.

On further enquiry the Consortium seemed a strange mixture of unlikely collaborators. The problem it sought to address was formidable. I discovered that St Helens had inherited some 1800 acres of dereliction and had an unemployment profile more commonly associated with Northern Ireland. It was depressed. But I emphasis **was**. The Consortium and the efforts of the Borough Council have begun to change that dramatically. In fact the Consortium began in adversity. A number of planning applications went before the Council (including one from itself) all associated in one way or another with retail activities. The only way to resolve them seemed to be for the Secretary of State to call the applications in and/or hold an inquiry. Realising that would be a long drawn out procedure, the parties concerned decided to try to roll all the proposals into one plan and secure its adoption as the vehicle for the regeneration of the central area of St Helens. The core of the area was Ravenhead, the process was intended to lead to the rebirth of the town: the project naturally became Ravenhead Renaissance.

Ravenhead Renaissance is a company limited by guarantee with provision for 33 directors. The Board (made up of three representatives of each participating body) meets quarterly and sets broad policy outlines. The Steering Committee (just one representative from each group) meets about every six weeks.

My task as independent Chairman is to seek to moderate the discussions, resolve any conflicts, provide some kind of standard around which to rally and set some kind of style for the Consortium. Occasionally I have some private discussions with individual participants to secure some shift in position or attitude, but it is only occasionally. So far it has worked extremely well. I do not recall us yet have a vote on any issue and I have witnessed some most remarkable statesmanlike gestures and decisions on the part of the participants.

The Projects
So much for the principle of development, what about schemes on the ground?

So far we have helped attract well over £60m new investment into the area and secured £9.5m City Grant. The schemes that we have are independent of each other but together build upon that which has gone before in a pattern of ever increasing investment and importance to the local economy.

Development started with the British Gas Retail Park at Warrington Road where British Gas North Western took a very brave decision in 1987 to carry out reclamation of a 12 acre former gas works site at their own expense and then to develop and let 100,000 sq ft of retail space. This they were given planning permission to do on the understanding that elsewhere within the Ravenhead Renaissance area a foodstore by Pilkington, linked with a new hotel, would be promoted on industrial land within the town centre known as the Hotties, and that at some time in the future the Council's own derelict land at Kirkland Street, just to the west of the Ravenhead area, would be brought to the market for further retailing.

The British Gas site was built and opened within 12 months, so that in November 1989 companies such as Halfords, Comet, B & Q and other household names were represented in St Helens. The scheme has been a considerable success and has, I feel, brought a physical extension to what is regarded as the town centre.

At the same time the Warrington Road site was underway, Safeway won, by competitive tender, the right to build a 65000 sq ft store on the Pilkington land at the Hotties and this important project was open for business in July 1989. The second largest Safeway store in the country, it has also proved to be their third most successful trading location. It has a striking architectural frontage much vaulted and most likely to be copied elsewhere in the area as more developments take shape.

Linked with this development and important for the re-use of a derelict site is to be a new 84 bedroom 4* hotel alongside the canal. The whole area of the Hotties is underlain with shallow mine workings, extensive foundations and voids. Derelict land has pushed to the very heart of the town centre exemplified by the Hotties. Ravenhead Renaissance helped Pilkington to formulate a significant City Grant application for the hotel which in economic terms has been linked with the Safeway store investment. Our efforts resulted in a £3m City Grant in May last year so that a hotel may be constructed.

Elsewhere at the Hotties and something which we are working very hard to secure is a major new Visitor Centre as a celebration of glass which will bring back into use the remaining derelict land at this part of the town as well as providing a significant visitor attraction in the North West. Rather than consider just heritage and looking back at what has been we are concentrating on a science and arts centre with interactive exhibits and overall a £10m project - in manageable phases - to bring back into recognition the features which have made St Helens synonymous with manufacturing technologies in glass.

Ravenhead Renaissance is not a developer, it is a facilitator of schemes and nowhere is this better demonstrated than at Greenbank where one of the greatest recognitions of the partnership approach during the last twelve months has seen the award of £6.3m City Grant for the reclamation and development of a major town centre site, bringing £27m of new private sector investment to the area and setting in motion the provision of over 300 new homes for sale.

The award of grant, which was made in October 1989 was the largest made by the Government up to then and reflects the importance of both the scheme, and the ways in which Ravenhead Renaissance is helping co-ordinate the programme of developments in St Helens.

Greenbank

Greenbank is one of the most significant land reclamation projects in Europe. The proposal was to reclaim 48 acres of land and former factory works and to replace them with 309 houses and 12 acres of new parkland.

The project is in two stages. The reclamation stage started in December 1989 and will require the handling of up to four million cubic metres of material. This includes the treatment of up to 100000 cubic metres of waste material and the extraction of at least 50000 tonnes of coal from old workings on the site. Once reclaimed, 309 houses will be built for sale and a new park created as an essential element of the environmental upgrading now taking place in St Helens.

The organisation of this very significant scheme was assisted by Ravenhead Renaissance, who helped three of its Consortium members, Pilkington, Milverny Properties and the St Helens Metropolitan Borough Council, to form a joint venture company called Greenbank (St Helens) Ltd. Greenbank has placed the reclamation stage of the project with AMEC Regeneration and has entered into agreements for the building of new houses with Fairclough Homes, also part of the AMEC Group.

Already well underway, the project is bringing dramatic changes to the landscape in this part of the town. The scheme is making a major contribution to the problems of dealing with derelict and difficult inner city sites in a self-contained and economic way. The linking of private and public sector interests has worked well at Greenbank and, it is believed by Ravenhead Renaissance, can be repeated elsewhere.

Linked with what is happening at Greenbank is the need to store overburden at an adjacent site. The area known as Ravenhead Park has been partially reclaimed but to all intents and purposes is derelict. When we started our projects this 45 acre site could have been developed for industrial sheds. It would have seen a very low return and would have required a great deal of grant. We have moved from industrial development to business/leisure prospects, and now leisure proposals backed by high quality low density business uses in a new landscaped environment.

What is also very important about this location is that it fronts the proposed M62 link road into town and we have been playing an important part in helping Consortium members Milverny Properties, AMEC Regeneration and the Local Authority to secure a land reclamation project, equally important as Greenbank, that will reclaim the line of the road in advance of its construction and prepare and restore land at Ravenhead Park for new development. The project will also provide a greenway link into the town centre and by reclamation processes similar to those at Greenbank, will, for once and for all, reclaim an area known as Lyons Yard for much needed open space close by the town centre. This area in itself is partially contaminated, has some old coal workings to contribute to the cost of reclamation, but is incapable of being properly enjoyed for open space as it currently exists.

We are not the technological experts on the scheme but our role has been to nurse and assist methods of overcoming problems to ensure that the scheme can go ahead.

The Town Centre
I have talked about derelict land but I think it is also important to realise that in St Helens much of the basic infrastructure is qualified as derelict. An ambitious programme of new schemes through the Unitary Development Plan is proposed by the Local Authority including new shopping.

Our Visitor Centre proposals will link closely with what has been achieved by another Consortium member, MEPC Investments Ltd, to revitalise town centre shopping in St Helens. MEPC inherited the shopping centres from Oldham Estates, which when first opened in the early 70's must have been old fashioned even then. MEPC have invested £5-6m in bringing the shopping centres into the 90's and are justifiably proud of what has been achieved. If I tell you that on a recent headcount done on the same day of the week as two years ago a 75% increase in shoppers was recorded, you perhaps begin to see what a dramatic change is taking place in people's perceptions of what the town has to offer.

New Areas for Consideration
We are now being asked to consider other areas and are being seen rather like Heineken, which purports to reach the parts that others cannot.

Already we have added a 50 acre site to our area for consideration and it is most likely will provide new industrial development north of the town. Most recently we have been in discussion with British Coal and other land owners of a former colliery site to see how we may bring about the restoration and subsequent development of an importantly located site close to the link road. Generally, we are being seen as an ever increasingly important role model for what can be achieved to bring about an end to dereliction as well as urban regeneration in St Helens.

What we are achieving can be achieved elsewhere and we host an increasing number of visits from both UK and overseas "urbanists" for them to see our collective success at first hand. Of the 1800 acres of derelict land I mentioned at the beginning of my speech which covers nearly 20 sites, 15 locations are now the subject of discussion for new development.

The M62 link road which will be completed by 1993, is being seen as a good road in rather than being a good road out and is itself being constructed through derelict land, and along which Ravenhead Renaissance projects are assisting to bring about economic prosperity for the area.

Links with others in St Helens
It would be wrong to assume that Ravenhead Renaissance is the only activity underway in St Helens to bring about an end to dereliction. The Metropolitan Borough of St Helens, as an Authority, fully recognises the negative impact for investment that derelict land can present, as well as the psychological effects it has for the local population.

Environmental policies have been pursued on a variety of fronts. The major component remains, nevertheless, the partnership with the private sector. The proposals for wasteland to woodland will, during the next few years, see the planting of well over two million new trees along the southern corridor into St Helens and the Ravenhead Renaissance area. We are actively promoting a first class environmental treatment of all our sites and already the planting of new trees in other initiatives is bringing a much needed improvement.

The Council's own policy for nature which was produced in 1986 set out a range of policies to protect and enhance the considerable wild life resources of the Borough. The Clean Up Campaign was launched in 1989 but before both of these initiatives, in 1982 Operation Groundwork was established in St Helens and Knowsley.

The success of this pioneering groundwork area - another first for St Helens - has subsequently resulted in the establishment of the Groundwork Foundation and a number of Groundwork Trusts throughout the country. Operation Groundwork has been a learning experience for the Local Authority as well as the private sector and has assisted in changing perceptions on nature as well as the problem of derelict land once schemes have been undertaken.

As Ravenhead Renaissance is a locally based initiative it must and does link with other agencies which are actively promoting the well being of the area. Co-operation is all.

CONCLUSION

Many of Britain's urban dereliction problems are being addressed by Urban Development Corporations of differing sizes. In many situations where land assembly is critical or political forces are irreconcilable they are undoubtedly the only way to achieve regeneration., But in other places, like St Helens, very substantial areas (comparable with those designated for the U.D.C.'s) can be regenerated by the imaginative and active partnership of public and private sectors. We know because we've done it. Not without pain and sometimes only by the skin of our teeth but nevertheless done it. The excitement of the management of it has been matched by the thrill of seeing the improvement on the ground.

LEISURE BASED AFTER-USE OF RESTORED LAND

D K HEMSTOCK, B.Sc., I.Eng.
National Consultant, Amenity and Recreational Drainage
ADAS, Block 7, Chalfont Drive, Nottingham NG8 3SN

ABSTRACT

With the present decline in emphasis on agricultural production and increased interest in leisure based land use, new restoration techniques are required by those working in this field to cope with the particular demands and problems of each end use.

Amenity and recreational facilities have certain common problems; compaction, poor drainage, poor ground cover, etc and have often been placed on marginal or low grade land, eg washland, reclaimed areas, poor soils. They may have been stripped and recontoured giving rise to similar conditions faced after minerals extraction or landfill operations.

All of these can give rise to difficult, long term use limitation problems.

The specialised techniques that have been developed to tackle heavily used and disadvantaged leisure based areas are discussed, with particular reference to soil manipulation, drainage and aftercare aspects.

The range of water-controlling techniques from simple underdrainage and soil loosening to the 'ideal soil' concept are detailed, using golf course construction and other leisure facilities such as parkland and sports fields as examples. Existing machinery and new developments are also highlighted.

The need to restore a site from <u>soil stripping</u> stages to overburden or fill replacement onwards, carefully planning for the particular amenity or recreational use is stressed.

INTRODUCTION

The briefest analysis of the distribution of disturbed and derelict land indicates a close link between geographical position and population density; industrialisation placed production adjacent to raw materials, and the workforce next to the factory. Waste material has very rarely been transported any significant distance to a point of storage, and remains a feature of many town and city landscapes.

As a result, we are commonly working close to high population densities when dealing with disturbed or derelict land. And it is generally true that in these areas lies the greatest demand for amenity or leisure based facilities, not surprisingly. There is now an increasing trend towards considering an amenity afteruse for a wider range of 'worked' land; on one hand it is unjustifiable in many cases to restore land to agricultural use when agricultural production is no longer a priority for many, and on the other the demand, public relations and promotional factors give extra weight to amenity restoration projects.

Agricultural restoration techniques are now well developed, but quite demanding from the point of soil depths, soil handling, drainage and aftercare treatment required.

An amenity afteruse of worked land may be less critically dependent on soil type and depth, but it is often much more demanding from a management approach. Once seeded to grass, an area may rarely if ever be completely reseeded, remaining undisturbed from a cultivation point of view. But, set against this is the much heavier trafficking by feet and wheels in all weather conditions, in fact the desirability is for a resilient, attractive surface which is subjected to a very adverse set of growing conditions.

The end use may be as a general purpose open space, parkland, showground, multi-purpose area; or more specifically for, say, golf or winter sports pitches, but all have potentially heavy trafficking as a common feature of their use. This results in essentially a long term struggle against compaction of the surface and all that follows; surface water, poor grass cover, disease and weed problems.

However, this is not a new struggle, and an awareness of the techniques available to improve and maintain amenity and leisure areas is essential in the proper planning of a worked-land restoration project. Techniques have been developed to enable intensive use of turf grass areas situated on the heaviest of soils, with all the disadvantages one could ever wish to avoid; including subsurface obstructions, high water tables, lack of fertility and so on. They may be relatively expensive to implement when compared to perhaps an agricultural system, but the higher returns possible measured in money or social benefit terms often make the expense worthwhile.

DISCUSSION

As in any land management programme, an amenity based restoration programme requires the careful integration of various aspects; soil drainage involves more than simply installing a pipe system for instance, a secondary drainage soil de-compaction system is also required together with development of soil structure through vigorous root growth. Without a rounded approach a great deal of money and effort can very easily be wasted for the want of a little bit of extra attention, particularly to the aftercare programme.

Soil

Taking this rounded approach from the initiation of a project involves initially the problem of soil handling and/or identification of off site sources of soil or soil-making material.

If soil is to be stripped either from on or off-site areas, stored and re-distributed, it should be in the best possible condition prior to the stripping process. This may involve a de-watering operation to keep the soil as dry as possible, or improving structure and condition from the earliest possible stage by soil loosening, green manuring etc. If the soil is in good condition when initially handled, it is more likely to be in reasonable condition when finally replaced.

If soil is in short supply, then it makes obvious sense to place the better soil in good depth at places of highest potential wear. On a park for instance this would be at access points, or where the sports pitch or regular events are to be situated. On a golf course concentration would be on the fairways, particularly around the putting green where wear is greatest and appearance is most important.

The opportunity may exist to vary conditions on site, leaving areas with no topsoil in order to encourage wild flora and tree establishment, ie areas of low maintenance; 'rough' on the golf course and unmown margins on parkland. It is worth considering placement of particular types of soil in selected areas to help establish a variety of potential habitats for locally occurring flora.

Particular advantage can be taken of the presence of earthmoving equipment when a golf course is the intended after-use for a site. Well-shaped mounds and hollows are important on a course, particularly around the putting green for play and appearance reasons. Flowing contours blend into the surrounding land but inevitably mean large areas of topsoil strip and re-distribution of subsoil when constructing a course from virgin land, which in turn is a major expense within a project. If material is to be moved around on a reclamation or restoration project, it should take relatively little extra work to 'shape-up' fairways and greens surrounds with the available equipment. Final finishing work is a skilled job for a proficient machine operator in conjunction with the close supervision of a consultant, in order to ensure detailed contour plans are accurately transferred onto the ground.

A certain amount of extra expense is inevitable when forming a good golf course on worked land rather than an informal park or grassed open space. However, taking advantage of recontouring work which may be necessary anyway provides a possibly very valuable facility at a reduced cost. Most local authorities appear at present to be struggling to provide facilities for golf for the public, and are actively seeking methods to finance and develop facilities which may increasingly involve previously worked or proposed working of land.

Once the ultimate use of an area has been determined, preferably fixed at an early planning stage, experience of the requirements of that type of use can be incorporated into a cost-benefit analysis of recontouring, soil and drainage options.

A typical question to be addressed might be whether to use an existing heavy soil and to intensively drain and manage it, or look to a more easily amended soil-making material to take its place.

Drainage

As with agricultural restoration, a good drainage system is essential as a basis for successful aftercare. A feature of amenity area drainage methods is the extensive use of permeable materials such as sand and gravel up to the surface and on the surface of intensively trafficked areas, removing water directly from the surface and buffering the soil from the compacting influence.

A traditional piped underdrainage system with a secondary treatment of regular moling or subsoil loosening is still useful for outfield areas on sports pitches, golf course fairways and parkland if the soil is suitable. Any subsurface obstructions, stone for instance, will prevent such operations in many cases because of the inevitable disruption to the turf surface. Predictable, even surfaces are needed for ball-games in particular. Even if a 'pipe plus soil-loosening' system can be installed, the surface zone may rapidly be re-compacted, sealing off what may be a perfectly operational drainage system underneath very rapidly.

Sand or sand gravel slits can combat this pattern, effectively taking the place of moling or subsoiling. The general aim is to preserve a porous medium at the surface which allows rapid removal of surface water [Diagram 1].

Compaction can still occur between slits and has to be treated regularly. The surface of the slits may also be 'capped' by soil being smeared over them by trafficking or surface flow of soil in suspension. This type of system is quite expensive to install and subsequently to maintain, but is very commonly used on sports pitches and golf courses. Design and specification for a system for a particular area and use is critical if problems are to be avoided, such as settlement and migration of particles, poor surface traction, erosion of the slit surface material etc.

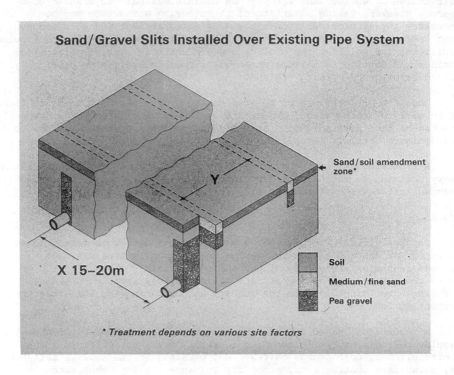

Diagram 1. Cross section of a typical sand/gravel slit system with a top-dressing of sand or sand/soil mix.

The equipment commonly used to install sand slit systems uses either a chain-trencher or cutting wheel, with soil being removed via an elevator from the trench, directly into a trailer or hopper. Sand or gravel or a combination is then used to fill in the trench to the surface. However, if conditions are very hard or there are subsurface obstructions, it may not be possible to install such a system without serious wear problems on the cutting edge or rest of the machine, or severe disruption to the surface.

'Trenchless' pipe installation has for a long time been favoured by drainage contractors on British Coal opencast sites, the machinery being less prone to damage due to the size and power of the equipment and minimal moving parts used. Similar equipment is becoming available for narrow trench installation, including sand slitting and the narrower grooving systems.

For instance, using weight and vibration to cut a slot, the type of equipment indicated in Diagram 2 can counter the problems presented by subsurface obstructions by breaking through, or riding over them. The

essentially downward force used avoids the hooking or lifting action of
other types of equipment, which can lead to a severely disrupted surface
and high machinery wear. The vibration of the surrounding area enables
lateral rather than upward displacement of soil from the slot forming
process. At high frequencies of vibration, the soil can be made to
accommodate quite large lateral soil displacements due to the development
of a 'fluidised bed' effect, according to initial trial work results.

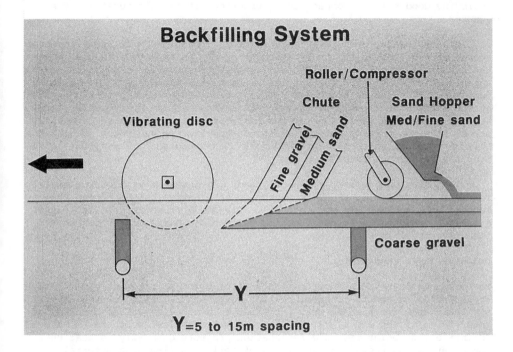

Diagram 2. Newly developed equipment for 'trenchless' sand slitting uses an
essentially downward-acting force to form a slot, which helps installation
work on stony and hard or contaminated soil.

De-compaction

The principle behind this type of equipment can also be applied to a soil
loosening operation, with a series of such cutting discs acting in a
similar way to cultivation discs, but deeper below the surface. Soil
loosening is a very important maintenance requirement but it can be very
difficult to carry out if the soil is stony or very hard, or as is often
found, 'laminated' to form a very strong, compact layer (formed during the
spreading operation).

At present, soil loosening equipment designed for amenity areas, ie loosening without excessive <u>surface</u> disruption, is often carried out by spiking equipment. This equipment generally operates by either pushing in a spike or knife type of tine, using either a roller or rotating crank principle to insert the tines into the surface of the soil. The latter is often slow and expensive, and prone to damage, the former often ineffective, as the slots formed are narrow and easily sealed and re-compacted. Moling and subsoiling can be very effective, but it may be very difficult to set up the equipment to give an acceptable even surface finish, especially if stones are present. As a result of the difficulty of achieving good soil de-compaction in practice, it is unfortunately often ignored.

Stones, waste, and other remnants of previous land uses are a constant source of problems to the reclamation process. An amenity afteruse, where the surface is stone-picked, seeded and possibly never touched again may be appealing in its simplicity, but the surface will at some point require some type of loosening treatment. Areas where stones must not be present at the surface, eg sports pitches and close-mown areas in particular, are treated in various ways. One of the simplest methods to remove even small stones from the surface quickly is to use specialised equipment such as the Lely Burrovator, which can sift and sort stones and soil, laying the latter over the former in one pass.

If contamination is particularly bad, for instance with broken glass remnants or similar, then there may be no alternative but to screen the surface with another cleaner material. Diagram 3 shows a cross-section of a typical drainage layer construction which has a buffer layer finish, possibly to act in such a way.

Drainage-layer Construction

The drainage layer construction method is used for golf greens and other areas including sports pitches, bowling greens etc. Design of the layers and their specification is very important to avoid the various problems of inter-layer mixing and migration, perched water tables, root breaks, etc. An interesting point for land reclamation projects generally is that once the after-use is known, and the types of soil or soil-making material determined, the opportunity can be taken to set aside particular material for a particular use.

Some projects have made use of site-originated stone seivings, gravel layers, or imported clinker or ash to form the drainage layer of sports pitches, for instance, which then requires a rootzone and blinding layer which may again be available on site.

Sandy material or lighter soils if present on site are always worth retaining for particular use, as a surface layer or for mixing with imported material to form a rootzone perhaps.

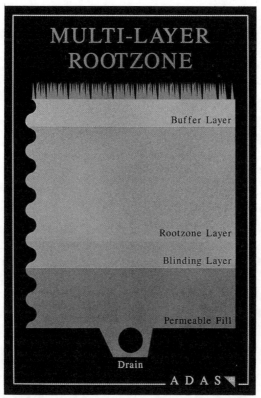

Diagram 3. Cross-section of a typical 'drainage layer construction' for greens and sports pitches. A buffer layer may be included to keep the surface free of larger particles harmful to maintenance equipment, and possibly users of the area.

The solution to many problems arising from intensive use of poor soil is to bypass the existing soil, using intensive slit-drainage and surface-buffering layers, or to amend the soil by changing its physical characteristics by incorporating sand or other material.

Ultimately, the surface is amended to suit the purpose, which may also include switching from natural turf to artificial surfacing if grass alone is unable to support the wear imposed upon it.

CONCLUSION

The two processes of sports and leisure facility construction and land restoration often have a profound effect on the physical and mechanical capabilities of the soil surfaces involved. The two processes have many aspects in common (1).

Conditions on heavily worn amenity areas are particularly arduous for plant growth, doubly so when low-grade or disadvantaged land is pressed into use.

Techniques developed mainly for the sports pitch, aimed at achieving more concentrated use and better performance from a range of grassed surfaces, can be effectively used as basic tools in amenity land restoration.

The myriad problems encountered in the reclamation world require a flexible approach in solving them, and a wider range of techniques and understanding can only increase this flexibility, particularly if existing techniques can be adapted and proven.

More research is needed into the particular problems applying to amenity after-use of worked and restored land to help make projects more successful, long term.

The manipulation and amelioration of problem soils and soil making material requires more study. The concept used in golf green rootzone design, of manipulation of soils to produce an 'ideal rootzone' with particular physical characteristics, could often be extended to the reclamation world. The work of various researchers on the concept can be brought together to give broad ranges for the soil physical parameters to be aimed for (2).

The further development of equipment which can cope with hard and stoney ground, for drainage and soil de-compaction especially, is necessary

to enable the improvement of standards of finish on reclaimed and restored surfaces put to amenity after-use. Stage one may be the major reclamation effort, but stage two, aftercare, makes the most of the original effort in providing good conditions for plant growth. Without the confidence to use specialised techniques, or the equipment to successfully execute them, the potential of an area may never be fully realised.

REFERENCES

1. Scullion, J. and Steward, V.I., Drainage of Sportsfield and Restored Opencast Coal Mining Land. Welsh Soils Discussion Group, 1982, Report No 23, pp. 1-14.

2. Ward, C.J., Sports Turf Drainage: A Review, J. Sports Turf Res. Inst., 1983, 59, pp. 9-28.

TINSLEY PARK LIVES AGAIN

by David Hunter and Martin Stott of British Coal Opencast

1. Tinsley Park is situated at the centre of a large populated area between Sheffield and Rotherham. The M1 motorway, arterial trunk roads, rail facilities and all the essential services make this particular site of strategic importance to South Yorkshire. For many years it had been recognised that the region lacked an airport to complete the broad infrastructure needs. Over 70 years ago, Sheffield City Council considered buying the First World War airfield at Coal Aston and since this land was eventually turned over to housing, various schemes have been considered and rejected. Because the area is generally hilly the locations available to meet modern aviation standards for the siting of an airport were limited. Level sites had been developed and some had subsequently become derelict 'problem sites'. One such site was Tinsley Park which, in 1987, was targeted for a closer look.

2. *History of the Site*

2.1 *Coal Mining* - From the beginning of the industrial revolution the whole area has been extensively mined for coal, first from shallow workings then later from the deep mines of the Tinsley Park Collieries. Each phase of mining brought different problems to the site.

The excavation of bell pits and room and pillar workings left near-surface voids or collapsed ground. The diagrams figs 1 and 2 illustrates the nature of the problems of ground instability, this creates problems clearly requiring special treatment if that land is to be used for construction purposes. Photographs 1 & 2 give a unique insight into the nature and scale of coal extraction once the surface has been stripped off at Tinsley.

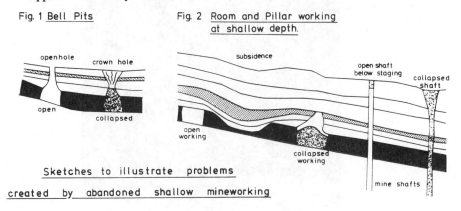

Fig. 1 Bell Pits

Fig. 2 Room and Pillar working at shallow depth.

Sketches to illustrate problems created by abandoned shallow mineworking

Colliery spoil heaps amounting to about 2.8 million cubic metres, and about 90 disused mine shafts and adits, would have presented difficult obstacles for the redevelopment of Tinsley Park. Subsidence of the workings themselves will have substantially occurred by now, so this is not a material consideration.

2.2 *Steelworks* - The first reclamation took place in 1963 when the English Steel Corporation built a large new steelworks on the north western third of the site.

The areas not reclaimed for construction were used for tipping and slag reduction processes related to the operation of the steelworks. The steelworks were constructed with very deep foundations including vast basements. After the closure of the steelworks the buildings were demolished but the foundations were left in, leaving large basements which would present problems for the redevelopment of the site.

The waste from the works tipped on site over the years comprised flume dust and slag as well as other hazardous materials which were identified by consultants to contain alkali and heavy metals. Disturbance and the safe handling and disposal of these materials required a Licence from the Waste Disposal Authority. The large melting shop and billett mill at Tinsley Park closed in the mid-1980's bringing the loss of jobs to over 25,000 in the steel industry in Sheffield. Photograph 3 shows the site as it was 1988.

2.3 *Natural Environment* - Despite the heavy industrial nature of the site, it was found to contain a number of bogland areas of special interest including dragon flies and plants. In addition, the impoverished soils have lead to the development of suitable conditions for orchids. The identification of the areas of ecological interest at an early stage enabled a conservation and rescue strategy to be agreed with the Local Authorities and the Wildlife Trust. This resulted in the protection of Bee Orchids and the rescue of Sphagnum Bog and Woodland flora together with the transplanting of over 1500 trees comprising mainly Oak and Birch. This was achieved by utilising staff from the City Ecology Unit and Countryside Management Unit.

3. *The Role of British Coal Opencast*

3.1 The initiative for the redevelopment of Tinsley Park came from Sheffield City Council in 1987 in their search for an airport site.

The site was in the correct location but suffered from several physical problems together with difficulties associated with land ownership. The potential existence of opencastable coal reserves beneath the site led to a series of meetings between the Local Authorities and British Coal Opencast to explore the possibilities of a mutually beneficial project. The willingness to co-operate came as a welcome change to Opencast staff who had previously been unsuccessful in attempts to discuss opencast schemes with the former Metropolitan County Council (abolished in 1986).

3.2 Sheffield's objective was to have an operational airport in time for the World Student Games in 1991, however to achieve the very tight timescale, Planning Permission for the opencasting would have been required by the 1st of January 1988. Owing to the complexity of the site it was not possible to submit the Planning Application to the Local Authorities until November 1988 because detailed site investigations, access road design, and working method assessment had to be carried out. This application ran into difficulties by which time the Sheffield Development Corporation had taken over planning responsibilities.

3.3 During this consultation period British Coal had identified the key role which opencasting could play in the redevelopment of this site. The physical tasks were principally to harness the technology to reshape the topography to one suitable for an airport by moving a vast volume of material, to stabilise pitfallen and degraded land by using heavy compaction techniques, and to remove deep concrete basement foundations and contaminated ground. Apart from this the role of 'honest broker' developed, whereby complex land negotiations were undertaken by British Coal to pull the many interests in the site together. Infrastructure needs for the opencast site, such as the access road, were designed and built to the specification required for the subsequent redevelopment, at no cost to the Local Authority or the developer, and planning gain associated with the Section 52 (now S.106 under the 1990 Act) attached to the opencast planning permission would solve some of the environmental issues.

3.4 A major advantage arising from the integration of Opencast Mining and Derelict Land Reclamation is that all material in the surface layers has to be removed in order to reach the coal. This gives the opportunity to pre-plan and phase the removal of wastes and other materials such as foundations in such a way that they can be incorporated at the appropriate position and depth during the placement of excavated rock behind the coal extraction operations. It also allows the final contours to be planned to be suitable for the intended afteruses. Stabilisation of old mine workings and shafts would need to have been carried out by drilling and grouting at an estimated cost of £1 million. The removal and stabilisation of the billet mill foundations alone were estimated to cost up to £500,000 and would have been a major cost restraint on redevelopment. The removal and encapsulation of hazardous wastes could not have been achieved economically either within the site, without jeopardising its development potential, or by transporting from site due to cost.

3.5　All of these issues have been addressed and incorporated into the Opencast Mining scheme in such a way as to cause negligible impact on either the mining or airport development. The deep burial of the hazardous wastes are subject to rigorous control under a waste disposal licence granted to British Coal and monitored by the South Yorkshire Hazardous Waste Unit. This arrangement has secured a permanent and economic solution to the final disposition of this material which would have been extremely difficult to achieve otherwise.

3.6　On Tinsley Site the earthmoving is geared to a target figure of 250,000 m^3/week of rock and other materials immediately above coal however, rehandle of overburden dumps and additional excavation to create safe excavation profiles could raise this figure at peak times to between 400,000m^3 and 500,000m^3/week.

The scale of such an operation carried out by modern plant, see photograph 4, can absorb costs for ground preparation which would deter or inhibit many developers when comparing the comparative costs of development on alternative sites.

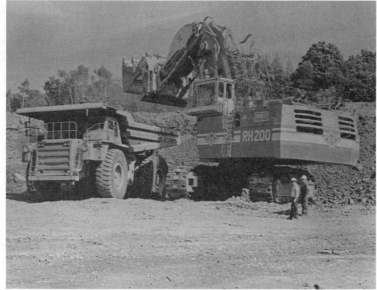

The existing land form and ground conditions at Tinsley were such that preparatory work for the construction of the airport might have been as expensive as the civil construction.

4.　*Planning, Site Investigation and Assessment*

British Coal made a great effort to submit its planning application speedily in view of the pressure from Sheffield City Council and subsequently the Development Corporation to mine the coal and restore the site before set dates. These dates were later incorporated into the Section 52 Agreement legally binding the signatories. Ironically, it was the negotiation of this Agreement which delayed a start on site. As a result of the imposition of these dates the winnable quantity of coal was reduced, however because of successful negotiations between the mining Contractor, Budge Mining Ltd, and the Development Corporation to construct and operate the airport some of this coal will be recovered.

4.1 *Geological structure* - An assessment of the quantity and quality of the coal including estimates of the amount of old workings, details of the extent and type of contaminated ground, all had to be carried out before the working method, and the basic restoration landform could be worked out. This was carried out by British Coal staff in conjunction with specialist Consultants and Contractors. The multi-discipline project team comprised Geologists, Engineers, Planners, and Surveyors. Desk studies were carried out to collect known information and a total of 1600 boreholes were drilled to a maximum depth of about 140 metres using both open-hole and coring techniques, each hole was then geologged. In order to assist with the interpretation of the information British Coal's Geomodel computer programme was utilised and extensive use made of aerial photogrammetry and total station surveying.

4.2 *Tendering and Contract Letting* - A Tender process is carried out on all British Coal Opencast work with Tenderers selected on the basis of experience in the mining possession and availability of the necessary plant equipment to carry out the works, the calibre of management, and most importantly the financial muscle to see through a contract worth anything between £50 - £100 million pounds.

4.2.1 The essence of the contract is the production of a set tonnage of coal per week over a fixed life of the site and other than for a number of lump sum items usually occurring at the commencement of the works the only payment made is per tonne of coal delivered to British Coal's coal treatment plant. In the case of Tinsley however, the payment for compaction is remeasurable and paid for at a tendered rate. These payments have to cover strict compliance with all health and safety, and environmental conditions.

4.2.2. The Tinsley Opencast Mining Contract was awarded to Budge Mining Ltd and work started on 17th July 1989 and involved an extremely tight working schedule as the runway area had to be mined out, compacted and brought back to formation level by May 1992. This requirement will allow the handing over of the airport area so that work can commence on the airport construction contract.

5. *Technical Issues*

5.1 *Compaction* - Compaction is required within the overall scheme on replaced overburden under the runway, terminal buildings, access roads, and new factory units. All supervision of compaction is carried out using a separate contract with independent firms of Consulting Engineers to certify that compaction has taken place to the specification. British Coal however have retained their position of Engineer to the Contract.

To emphasise the complex nature of contractual arrangements on Tinsley on certain areas both British Coal and British Steel will employ their own consultants to supervise the works. With one of these consultants also acting on behalf of the Sheffield Development Corporation under a separate contract.

The runway crosses the main mined area (A) with depths of backfill varying from 32m in the west to 94m in the east. Area (B) of the mining scheme will be redeveloped for factory space in agreement with British Steel with compacted backfill depths ranging from 5m to 74m in depth. Other areas of built development and roadways will require compaction of surface layers to take account of raised levels of a few metres.

Compaction is to a method specification within the requirements of the Department of Transport "Specification for Road and Bridgeworks" however this can only give guidance, as British Coal recognises that this specification is not entirely appropriate in an Opencast context and some relaxation might be achieved when a larger data base of monitoring results over time is available from previous compaction projects. For the time being however a conservative policy has been adopted which will ensure maximum stability of compacted land by adopting the full recommended specification.

Problems have arisen on site where 100% compaction of excavated material is required due to the variable nature of banded coal measures strata varying from easily degradeable mudstones and siltstones to the inevitable larger blocky sandstones. Maximum acceptable lump size being determined by layer thickness which means that if an uncompacted tipping area is not available then expensive crushing of larger lumps is required to achieve appropriate lump sizing.

Layer thicknesses of between 300mm - 500mm are the norm but trials have been carried out up to 750mm with varying degrees of success using the Case Vibromax 1802 self propelled vibratory rollers on other sites. On Tinsley to cope with the compaction volumes of about 4000m^3/hr a team of 5 no Bomag 217 rollers will be required.

Compaction achieved by the required number of passes per roller weight is confirmed by the use of Nuclear Density gauges and water displacement tests. Sand replacement tests have been abandoned due to the irregular nature of test holes caused by the make up of the compacted material.

After initial compaction, significant stages in the further settlement of the material are observed to occur due to the recharge of the natural ground water table. Steep draw down curves have been observed at short distances in from the exposed face of the compacted material and total immersion reduces the strength and initiates particle boundary determination of the affected material.

To establish further settlement, and settlement rates, a comprehensive monitoring regime will be carried out using piezometers, extensometers and settlement points comprising concrete levelling control blocks set at surface level. Ongoing settlement is a function of depth and also differential settlement at the junctions with undisturbed strata. This is carefully recorded by Surveying during the mining operation and the lay out of development amended accordingly to avoid potential problem areas.

5.2 *Environmental Control* - In common with all planning consents for Opencast Coal, Tinsley Opencast Site operates under a tight schedule of 20 environmental planning conditions.

5.2.1 *Noise* - Under the original planning permission work is permitted between 7am - 10pm weekdays although this by amendment has been extended to 6am having proved that this would not increase noise from the site received at occupied properties also 6am to 1pm on Saturdays with no work on Sundays or Bank Holidays. Only work of an emergency nature is allowed outside these hours and the Local Authority must be informed.

5.2.2   *Dust* - The presence of large quantities of flume dust tipped on the surface over many years from the adjacent steelworks was a source of concern to both local residents and Sheffield City Environmental Health Department.

Funded to an extent by British Coal the Local Authority embarked on a comprehensive dust monitoring and analysis scheme to establish if during the working of the site dust either from the mining operations or the movement of these wastes was likely to create a problem. The results published to date in their interim reports confirm that dust from neither of these sources is a problem. From the end of 1990 the only potential source of dust is from the mining operation itself as all exposed flume dust has been successfully incorporated into the backfill.

5.2.3   *Blasting* - A small amount of blasting is taking place on site on specific horizons. It is not considered that this will cause any problems as results to date have shown vibration levels at the nearest structures of less than 1mm/ppv.

To summarise, in spite of the considerable concerns expressed by local residents at the outset British Coal and Budge Mining have demonstrated that it is possible to conduct an operation of this scale in a way which has minimal impact on those around.

5.2.5.   *Waste Management* - Steelworks waste loosley tipped over the original site surface has been mixed and incorporated with uncompacted areas of the opencast backfill. The opportunity has been taken also to remove and incorporate the contents of a large slurry lagoon of similar material within the existing steelworks site.

Freshly produced flume dust from the steelworks is deposited in a licensed transfer area by British Steel where it is mixed with wet material from the opencast site, transported and placed into the backfill. On completion of the site a void will be left of about 1 million m^3 to take all foreseeable future waste from the steel plant. All of these operations are licensed with appropriate conditions by the South Yorkshire Hazardous Waste Unit.

5.3   <u>Shafts, mineworkings and concrete foundations</u> - In common with standard British Coal practice all shafts continuing below basal seam level are capped in reinforced concrete and the edges of all coal seams on the excavation perimeter sealed with selected clays.

6.   *The Final Development*

Development of the Tinsley Park site is an essential part of the new vision for Sheffield. Sheffield's strategic regional location is shown on fig 3 and the Sheffield Development Corporation's declared aim is to create a vital new balance of land uses, activities and jobs, within a special quality of infrastructure. Fig 4 indicates the basic elements of the final development for Tinsley Park.

6.1   *The Natural Environment* - An essential part of the landscape structure of the area will be the 70 acres of new open space to the south of the airport runway. The greater part of the derelict land was devoid of topsoil so suitable root medium, recovered during the opencasting, will be spread on the backfill. Following the opencast restoration, a five year programme of rehabilitation to plant trees, shrubs and hedgerows together with associated fencing and drainage works will be undertaken by British Coal.

Fig. 4   TINSLEY PARK OPENCAST SITE
The Basic Elements of the Final Development

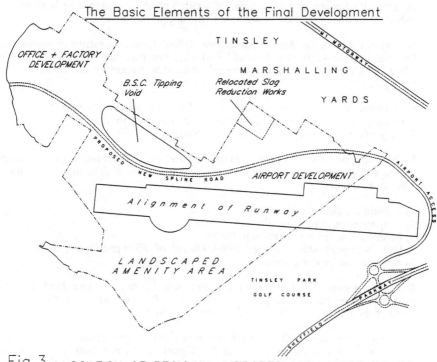

Fig. 3   LOCATION OF REGIONAL AIRPORTS AND MAJOR ROADS
Showing Sheffield's Strategic Regional Importance

6.2 *The Sheffield/Rotherham Airport* - Apart from attracting investment and large-scale development to Tinsley it is anticipated that the beneficial impact will be felt in Rotherham and other areas of South Yorkshire. A range of business and industrial uses are proposed in addition to the airport terminal.

As part of the Planning Agreement required British Coal to construct an access road for coal haulage onto The Sheffield Parkway, this road has been constructed to adoptable standards to a greater specification than necessary for access for the mining.

By retaining this road a ready made access has been provided for the Airport. An extension of this road will eventually be constructed to link right through the Tinsley Site to the heart of the Lower Don Valley.

In the spring of 1990 the opencast mining contractor, Budge Mining Ltd of Retford signed a deal with the Sheffield Development Corporation. The main elements of this agreement are

1) to build the airport and runway
2) to operate the airport for 10 years
3) building new spine road through the site
4) build between 800,000 sqft and 1,000,000 sqft of office and factory space
5) carry out landscaping works.

6.3 *Built Development* - The strategic importance of development land in close proximity to an airport cannot be denied. Some 100 acres will be available to the north of the runway in addition to the airport terminal.

The Tinsley Park area will remain the dominant site for specialised large-scale steel production and complimentary processes within the Lower Don Valley. As an alternative to the previous practice of tipping waste above ground, the opencast operations will leave a properly shaped 1 million cubic metre void for the nearby steelworks

The views expressed in this paper are those of the authors and not necessarily those of British Coal.

INDEX OF CONTRIBUTORS

Anders, I.J., 193
Aramburu Maqua, M.P., 387
Ashworth, G., 397
Atkinson, S.L., 329

Bailey, D., 96
Barratt, P., 336
Barry, D.L., 319
Bell, R.M., 215
Bending, N.A.D., 347, 391
Bentley, S.P., 146
Bridges, E.M., 40
Bromhead J., 321
Buckley, G.P., 329

Cairney, T., 196
Card G.B., 125
Couper, A.S., 50, 389

de Silva, M.S., 299
Doubleday, G., 82
Dutton, C., 86

Edwards, S.J., 256
Escribano Bombin, R., 387

Failey, R.A., 215
Farberov, J.F., 204

Gagen, P., 84
Gahir, J.S., 86
Gordon, T., 201
Gray, C.D., 106
Greenshaw, L.M., 146
Griffiths, D.G., 93
Gunn, J., 96

Harold, P., 336
Hartley, D., 65
Hemstock, D.K., 402

Hobson, D., 323
Hobson, D.M., 75
Hodgkinson, R.A., 202
Hunter, D., 411

Jefferis, S., 310
Jones, D.L., 228
Jones, H.L.M., 86

Leonard, M., 235
Lopez-Real, J.M., 329
Loxham, M., 241

Mabey, R., 3
Maddison, J.D., 117
Moffat, A.J., 347, 391
Musgrove, S., 248

Otlacan, L., 88

Palmer, J.P., 357
Pearson, C.F.C., 256
Pereversev, R.A., 204
Privett, K., 235
Pulford, I.D., 269

Roberts, C.J., 347, 391
Roche, D., 321
Roche, D.P., 125

Samuel, P., 366
Scott, M.J., 206
Scullion, J., 377
Slingsby, A., 289
Smith, M.A., 279
Snaith, B., 84
Sofronie, R., 88
Statham, I., 206
Stenning, A.S., 146
Stott, M., 411

Sury, M., 289

Taylor, N.J., 135
Thomas, B.R., 146, 299
Tillotson, S., 158
Treadgold, P., 158

Uren, J., 174

Vierhout, M.M., 208

Williams, L.M.S., 210
Wilson, E.J., 325
Winterhalder, K., 393
Witherington, P.J., 183

Yong, C.T., 174